"Phillip Ackerman-Leist has been in the trenches of food systems change for many years, from farm to school. Now he has elegantly laid out the principles of how to redesign foodsheds for greater food security, justice, and energy efficiency, while engaging communities in making tangible innovations on the ground. He is undoubtedly in the best place to address these issues, since Vermont communities have accomplished more food relocalization than those in any other state."

—**Gary Paul Nabhan**, *pioneer in the food relocalization movement, author of* Coming Home to Eat *and* Renewing America's Food Traditions

"*Rebuilding the Foodshed* introduces readers to local food systems in all their complexities. In moving from industrial to regional food systems, communities must consider an enormous range of factors, from geographic to socioeconomic. Difficult as doing this may be, this book makes it clear that the results are well worth the effort in their benefits to farmers and farm workers as well as eaters."

—**Marion Nestle**, *professor of Nutrition, Food Studies, and Public Health at New York University and author of* What to Eat

"The future of food is local. But how do we transition from our current globalized, supermarket-centered food world to one that's human-scaled and ecosystem-friendly? This book shows how communities across America are reclaiming the ability to feed themselves. It's inspiring as well as informative. If you eat, you really should read it."

—**Richard Heinberg**, *author of* The End of Growth *and* Peak Everything

"Now that it's not just acceptable but fashionable to write about local food systems, lots of people do it. Few pay close attention, however, as Ackerman-Leist does in this volume, to the variously shaped components successful local systems will require and the multiple efforts around the country working to create them. A wise, informed, and thoroughly useful book."

—**Joan Gussow**, *author of* Growing, Older *and* This Organic Life

"By now we have all learned that local food is about much more than food miles. Philip Ackerman-Leist has eloquently helped us to understand just how comprehensive the concept is: how our food system must be redesigned if it is to be reliable and resilient, how that design must be guided by principles of ecology, justice, health, and humility, and how to put such theories into practice for farmers, chefs, consumers, and communities. A practical guide for anyone interested in imagining our food systems of the future."

—**Frederick Kirschenmann**, *author of* Cultivating an Ecological Conscience: Essays from a Farmer Philosopher

"*Rebuilding the Foodshed* provides a critical compass as we tackle the next set of challenges to build secure food systems and resilient communities. Ackerman-Leist is a skilled writer, thinker, farmer, and educator who raises and deals with the toughest questions about how to truly make the way we grow, harvest, and distribute our food fair, sustainable, and humane."

—**Judy Wicks**, *founder of the White Dog Café, cofounder of BALLE, the Business Alliance for Local Living Economies*

THE COMMUNITY RESILIENCE GUIDE SERIES

LOCAL DOLLARS, LOCAL SENSE

HOW TO SHIFT YOUR MONEY FROM WALL STREET TO MAIN STREET AND ACHIEVE REAL PROSPERITY

By Michael H. Shuman

Foreword by Peter Buffett

POWER FROM THE PEOPLE

HOW TO ORGANIZE, FINANCE, AND LAUNCH LOCAL ENERGY PROJECTS

By Greg Pahl

Foreword by Van Jones

REBUILDING THE FOODSHED

HOW TO CREATE LOCAL, SUSTAINABLE, AND SECURE FOOD SYSTEMS

By Philip Ackerman-Leist

Foreword by Deborah Madison

REBUILDING THE FOODSHED

HOW TO CREATE LOCAL, SUSTAINABLE, AND SECURE FOOD SYSTEMS

PHILIP ACKERMAN-LEIST

Foreword by DEBORAH MADISON

A COMMUNITY RESILIENCE GUIDE

POST CARBON INSTITUTE
SANTA ROSA, CALIFORNIA

CHELSEA GREEN PUBLISHING
WHITE RIVER JUNCTION, VERMONT

Editor: Joni Praded
Developmental Editor: Daniel Lerch
Project Manager: Patricia Stone
Copy Editor: Nancy Ringer
Proofreader: Alice Colwell
Indexer: Shana Milkie
Designer: Melissa Jacobson

Printed in the United States of America.
First printing January, 2013.

10 9 8 7 6 5 4 3 2 1 13 14 15 16 17

Our Commitment to Green Publishing

Chelsea Green sees publishing as a tool for cultural change and ecological stewardship. We strive to align our book manufacturing practices with our editorial mission and to reduce the impact of our business enterprise in the environment. We print our books and catalogs on chlorine-free recycled paper, using vegetable-based inks whenever possible. This book may cost slightly more because it was printed on paper that contains recycled fiber, and we hope you'll agree that it's worth it. Chelsea Green is a member of the Green Press Initiative (www.greenpressinitiative.org), a nonprofit coalition of publishers, manufacturers, and authors working to protect the world's endangered forests and conserve natural resources. *Rebuilding the Foodshed* was printed on FSC®-certified paper supplied by Thomson-Shore that contains at least 30% postconsumer recycled fiber.

Library of Congress Cataloging-in-Publication Data
Ackerman-Leist, Philip, 1963-
Rebuilding the foodshed: how to create local, sustainable, and secure food systems / Philip Ackerman-Leist.
p. cm.—(A community resilience guide)
Includes bibliographical references and index.
ISBN 978-1-60358-423-4 (pbk.)—ISBN 978-1-60358-424-1 (ebook)
1. Food supply. 2. Local foods. 3. Food security. I. Title. II. Title: How to create local, sustainable, and secure food systems. III. Series: Community resilience guide.

HD9000.5.A314 2013
338.1'9—dc23

2012043955

Chelsea Green Publishing
85 North Main Street, Suite 120
White River Junction, VT 05001
(802) 295-6300
www.chelseagreen.com

To Erin
When I say "local," I think of home,
and I always come back to you.

A long habit of not thinking a thing wrong,
gives it a superficial appearance of being right,
and raises at first a formidable outcry
in defense of custom.

—Thomas Paine, *Common Sense*

Introduction to the Community Resilience Guides

In the twenty-first century, we face a set of interconnected economic, energy, and environmental crises that require all the courage, creativity, and cooperation we can muster. These crises are forcing us to fundamentally rethink some of our most basic assumptions, like where our food and energy come from, and where we invest our savings.

While national and international leadership are key to navigating the bumpy road ahead, that leadership thus far is sadly wanting. And, in any case, many of the best responses to these challenges are inherently local.

Thankfully, a small but growing movement of engaged citizens, community groups, businesses, and local elected officials is leading the way. These early actors have worked to reduce consumption, produce local food and energy, invest in local economies, and preserve local ecosystems. While diverse, the essence of these efforts is the same: a recognition that the world is changing and the old way of doing things no longer works.

Post Carbon Institute has partnered with Chelsea Green Publishing to publish this series of Community Resilience Guides to detail some of the most inspiring and replicable of these efforts. Why "community resilience"? *Community*, because we believe that the most effective ways to work for the future we want are grounded in *local* relationships—with our families and neighbors, with the ecological resources that sustain us, and with the public institutions through which we govern ourselves. *Resilience*, because the complex economic, energy, and environmental challenges we face require not solutions that make problems go away but *responses* that recognize our vulnerabilities, build our capacities, and adapt to unpredictable changes.

These are frightening, challenging times. But they are also full of opportunity. We hope these guides inspire you and help you build resilience in your own community.

Asher Miller
Executive Director
Post Carbon Institute

Contents

Foreword

My first editor, years ago, used to warn me, "Whatever you do, don't look at the big picture! You'll just get overwhelmed."

The big picture for me at that time started with a national book tour—a terrifying experience for a former monastic—then went onto rebuilding my life, starting with where to live and what to do. I did get overwhelmed when I tried to wrap my mind around the entirety of my future, but later I was able, bit by bit, to walk into the big picture.

Today I'm not sure we have time for an amble when it comes to reshaping food systems (nor am I sure my editor's advice was entirely right, though perhaps it was appropriate at the time). I've come to see that while we all work in our own areas, we accomplish more when we work in concert with others, which brings me to *Rebuilding the Foodshed*.

This is definitely a big-picture book, one that requires us to dig in and look deeply at all aspects of the food system and to examine assumptions we might not even realize we held. Thinking and writing about what brings which foods to whose plates at what costs is a tall order. As I read through, I found myself asking many questions, realizing how important systems thinking is to our food future, and wondering just how one finds their place in the food web. It can be overwhelming, but also oh so necessary.

Imagine a piece of cloth and go to a point on its edge, a corner maybe, if it has one. Let's call that Local Food. Now, as we continue to cast our eyes along the edge, imagine finding several different definitions of local food printed there: definitions that might or might not be sound; definitions related to distance, for example, or definitions that expose flaws not noticed before but inherent in various models. But that's just the beginning.

Pick up the cloth by that local food corner, and now with the whole cloth in view we can begin to make out some more words—among them carbon sequestration, rotational grazing, land reclamation, waste streams, landscape, synthetic nitrogen, peak phosphorous, food insecurity, heat-recovery systems, community gardens, prisons and food, dead zones, soil fertility, poverty and health, food sovereignty, farmers. And that's just for starters.

All of these words link to complex topics that have some relationship to "local." It's essential to understand those relationships in order to begin to build an understanding of our food system and how it might be changed for the betterment of people, health, animals, and environment.

As our cloth becomes increasingly covered with words and arrows linking one topic to another, with graphs and stories both good and appalling, it becomes obvious that rebuilding our food system is going to be far more complex than we may have imagined. It's not really surprising that Philip Ackerman-Leist is an academic, an activist, a farmer, and, I strongly suspect, a good cook, a good talker, a fine listener, and a thinker who can delve into all the aspects of this ungainly challenge and take them on one at a time in order to describe what rebuilding our food system actually entails.

Thinking like a system requires one to be courageous and unflinching, because we have to look at each segment and partial piece and consider its role thoroughly and carefully, even those parts we don't like. Like Walmart. Or pesticides. Reading *Rebuilding the Foodshed* required diligence and a willingness to find the flaws in my own thinking. I know what I've valued in food: local *and* organic, farming that builds soil, landscape and sense of place, and the farm as habitat are all incredibly important to me, as are school gardens, biodiversity, community gardens, and my own backyard efforts. These are all areas I've worked in, cared about, written about, and funneled into my life as a cook, chef, and food writer. But how do they connect to other parts in the larger picture? Could they connect better and more deeply? Is connection necessary? (It is.)

In the rural community where I live in Northern New Mexico, a rather typical American sentiment is sometimes expressed. People like to feel that they are rugged individuals—independent people. I've even heard one of my neighbors proudly describe herself with those very words. But of course we aren't, not really. We might be artists, writers, and independent thinkers and creative types, and we might compost our waste and grow some food, but we all get in our cars and drive to the farmers' market or Whole Foods with the greatest of ease, unless the road happens to be icy. And even if we were rugged individuals, where does that get us? I recall how, in the novel *Independent People* by Icelandic writer Halldór Laxness, the burning desire for independence ultimately enslaved the protagonist, the rigidly autonomous

Bjartu. So fierce was his dream of self-sufficiency and freedom that he failed to cultivate connections to others, and that became his downfall.

What especially impressed me in *Rebuilding the Foodshed* (though I could easily have tagged each page with a sticky note or more) is that Ackerman-Leist stresses the importance of being in a conversation with others, including those who are not necessarily like-minded, if change is to take place. Communities that manage to survive and prevail display a resilience that is ultimately based on the ability to have those conversations, to listen and speak and reason.

Indeed, the last part of the book shows that what's common to highly effective movements seeking to change the food system is not necessarily a bright new idea but rather *the conversation*—the ability to meet with and work with other groups most likely doing different things but together aspiring to rebuilding the foodshed. It makes sense that the most impressive efforts are often not national in scope, but regional. They work in the context of their place and on behalf of the people who live there, and who have a stake in what happens. That's what gives them the power to be effective.

This encouraged me to take a closer look at my own community. I saw more clearly how different parts of our complex culture are in dialogue, from the farmers' market to foodbanks; from policy committees to childrens' programs; from the community visioning group "Dreaming New Mexico" to the water-management organization Acequia Association; from some very good agricultural extension agents to a group of young gardeners who meet once a month for a potluck and information sharing. It's essential to take part in a larger conversation, whether you are a farmers' market manager, a member of Slow Food, the head of an agricultural think tank, a native farmer, or even a cookbook writer. It takes all of us to rebuild the foodshed, that's for sure, and this remarkable book is a most timely guide.

Deborah Madison

December 2012

Acknowledgments

This book would not have happened were it not for the steadfast endurance of my wife, Erin, who heard "It's almost done" more times than I care to admit. I always did think it was almost done. And then I (or my steadfast editors) would realize once again what a vast topic local really is. It was also possible because our children, Asa, Ethan, and Addy, did their best to share their home with someone too often deep in thought. My deepest love and thanks to all of you.

Meanwhile, the farm and homestead depended on the deft hands of apprentices Ollie, Jose, Tomer, and Simon, who kept things going outside in all kinds of weather while I huddled up inside—sometimes envying them, sometimes wondering why they stuck by us, all with nearly constant smiles. Others, like (but none quite like!) Maggie Ackerman, provided me with untold hours of quiet writing time by taking the kids for a hike or a swim—or hours of play at her cabin in the woods.

One could not ask for a more careful or insightful editor than Daniel Lerch, publications director of the Post Carbon Institute, to whom I am enormously grateful. He even found a kind way to let me know when my wordplay just went over the top. In sum, I'll miss his "=)". Sometimes, it meant I nailed it; other times, it was the quick and easy letdown.

Thanks also to Asher Miller, executive director of the Post Carbon Institute, whose astute eye and crisp thinking always came to the fore in his manuscript reviews. The same goes for Joni Praded of Chelsea Green Publishing, who also seems to have an uncanny instinct for chiming in with precisely what is needed at a given point, even in a crunch. Copy editor Nancy Ringer provided more careful corrections than I hope my students or colleagues will ever realize, so I thank her wholeheartedly for sparing my reputation. Dan Sullivan was also kind enough to use his editorial skills in reviewing part of the manuscript early on.

Geography clearly has played a central role in this book, but it would be much more abstract were it not for the brilliant mapping efforts of friend and colleague Dr. John Van Hoesen, the MacGyver of maps (and just about everything else technical in nature!).

I would be remiss not to express gratitude to my colleagues and students at Green Mountain College for their patience and support in this project, and my journey from sustainable agriculture to food systems began in the college dining hall with Chef Dave—thanks, Dave! Shep Ogden has long been a compatriot in thinking about the viability of new sustainable food system models, and much of what is written here reflects our longwinded dialogues about how to design a graduate program that can help move these ideas forward. Credit should also go to the first cohort in our new Masters in Sustainable Food Systems, students who have risen to the challenge while also challenging me to think harder about the possibilities for change in our food systems.

And, although they are too far away (but it was me who moved north), it is my parents who are behind the heart in this book. They have never been willing to believe that the status quo is acceptable, simply because someone out there always deserves far more than they have been served. It was from them that I learned the taste of justice.

A final thanks to all who shared their ideas and stories with me. Hearing those perspectives has been a true privilege. I only hope that I heard and cast them in ways that are not only accurate but also affirming for the important work you and others are doing.

Preface

If you're interested in local food, then you're probably as interested in geography and history as you are in taste. As a result, you probably wonder just how it is that we got here. How in the world did we move from a time when almost all of our food was local to now, when accessing local food can be such a challenge in just about every region of the United States? Sure, there are local food hot spots, and anyone with enough determination can almost always find a certain amount of local dairy, meats, and produce available for purchase. But why are local and regional foods generally the "alternative" instead of the mainstay of food systems in this country? As is the case with most agricultural endeavors, it takes a little digging to get started.

Whether you are inclined more toward geography or history, the right map can be an important reference point for helping you get your bearings. That's where Armour's Food Source Map from 1922 comes into play (see map 2 in the color insert following page 180). Not only does the map provide a beautifully depicted overview of food production in different regions of the United States during the early twentieth century, but it also provides an intriguing set of clues about what was happening to our food system at that historical juncture.

Beneath a banner proclaiming, "The Greatness of the United States Is Founded on Agriculture," the text that accompanied the map in its original version began as follows:

> Because of the many kinds of farm products raised in the United States, and because of the large amount of those products, the United States is the greatest self-sustaining nation in the world. Within its borders it can raise, and manufacture from its natural products, everything that is needed to keep every industry busy, to give employment to millions of people, to feed all of its own people, and still have much food left over to help feed other nations.
>
> Agriculture is the most valuable of all industries. Of the agricultural products for which the United States is famous, livestock are the most important and the most valuable.

The map's text goes on to describe the ways in which particular regions of the country are best suited for raising certain types of livestock (hogs, cattle, sheep, and poultry), but it also points out that certain regions such as New England are not well suited to agriculture yet are ideal sites for the manufacturing industry. The producer of the map, the Armour meatpacking company, makes no bones about its intent here. As a business, it has a service to provide, not just to the individual consumer but to the country at large:

> The service of the packing house brings a great part of the farmers' products to the door of the person who wants to buy them. It has made the best of food available to any person in any part of the United States, thereby enabling many of those persons to spend their time in making useful articles, or in doing other work that is of value to the progress and the culture of the nation.

That probably sounded like a great idea to the citizens of the Roaring Twenties. The horrors of industrialized warfare displayed on the fields of Europe in the previous decade had given way to the wonders of an industrialized economy that brought foods from all over the country to bustling East Coast cities, and quality manufactured Northeast goods to the bursting frontier. The nation was transforming from a patchwork of somewhat isolated local economies into a national growth powerhouse.

The natural inclination is to ask, "But where did these ideas of centralized production and national distribution all start?" Given the enormous complexity of our food system, there is no single answer. But there is one particularly good story that begins to shed some light on how increasingly industrialized forms of agriculture began to displace local and regional food systems, and it stems from the Armour legacy.

As you might expect, the modern assembly line plays a critical role in this story, although in this case we might more accurately call it a "disassembly line." Meatpacking plants—highly efficient facilities with assembly lines designed to ensure the maximum amount of speed and the minimal amount of skill on the part of the workers—were key drivers in the development of modern industrialized agriculture. Perhaps you're thinking that these assembly lines were based on Henry Ford's pioneering approach to manufacturing. In reality, Armour's assembly lines in the packinghouses of the late nineteenth century were the basis for Henry Ford's oft-touted "invention" of the assembly line. Granted, the meatpackers' assembly lines were designed to take apart pigs, cattle, and sheep instead of putting together automobiles, but they preceded Ford's advancements by several decades. If you, like I, didn't get the straight story in your grade-school history classes, then it's high time that you get to know a little about Philip Danforth Armour, one of the premier architects of our industrialized food system.

Born in 1832 on a farm in upper New York State, "P. D." Armour set out to make his fortune as part of the California Gold Rush at the young age of nineteen. He purportedly set off to walk from New York to California with four other men and was the only one among them who made it—that gives you a sense of his determination. Although he had intended to be a miner, he ended up making his first financial windfall by digging sluices to bring water to other miners. As he built his resources and reputation from this venture, he was able to hire miners who had given up on their own prospects to help him dig more sluices for more miners. By the time he departed California at the age of twenty-four, not only had he learned how to manage a large number of workers, but he left with his first fortune—somewhere between $6,000 and $8,000, an enormous sum at the time by any measure.[1]

P. D. then went to Wisconsin to sell hogs and dry goods with a business partner, which set the stage for a business venture with his brother, Herman Ossian "H. O." Armour, a successful grain trader in New York. P. D. was convinced that the Union forces were about to win the Civil War and that the price of pork barrels would then drop precipitously, so he and H. O. sold futures for $40 per pork barrel. When the floor ultimately did fall out of the market and pork barrels collapsed to $5 per barrel, they amassed a stunning windfall of $2.5 million. P. D. then bought his way into his brother's grain trading business and they set their sights on Chicago, which was quickly emerging as the primary nexus point for goods moving between the East and West Coasts. Given the American heartland's ample supply of grains and forage, the stage was set for Chicago to become the nation's leading center of commerce for grains, as well as the slaughtering and butchering of hogs.

FIGURE P-1. Cattle Pens, Union Stock Yards. Image courtesy of Lake County (IL) Discovery Museum, Curt Tech Postcard Archive.

At the time, the emerging meatpacking industry was focused primarily on hogs because the pork could be "packed" and preserved with salt, whereas beef and mutton were more difficult to preserve and quick to spoil, particularly in the warmer months. (Some salted "jerked beef" comprised a much smaller portion of the work in these packinghouses,[2] but it was the development of refrigeration technologies that would later bring cattle and sheep into this rapidly developing and increasingly centralized commodity stream.[3])

"Armour and Company," as it came to be known, was located on fourteen acres just west of the Union Stock Yards of Chicago, the giant hub of the fast-growing livestock industry in the United States (see fig. P-1).[4] There was no year from 1893 to 1933 during which fewer than fifteen million head of livestock were delivered to the Union Stock Yards.[5]

The enormous scale of these stockyard and meatpacking operations demanded efficiency both in the slaughtering process and in the subsequent use of every possible by-product involved. It was Philip Armour who developed the assembly line approach to slaughtering and butchering hogs in order to expedite the rendering of a live animal into its ensuing components—virtually all of which were eventually put to use in some form or another. "Everything but the squeal" was the moniker of the day in these

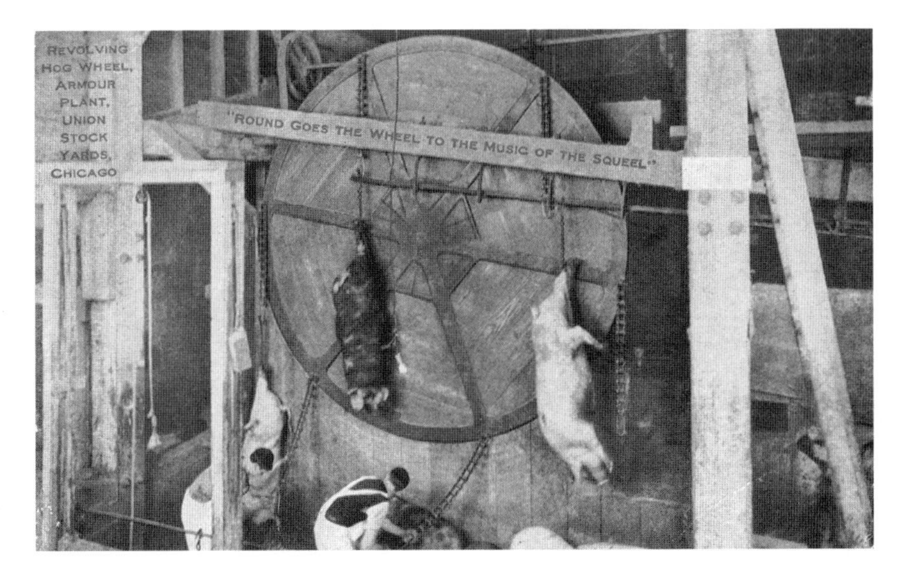

FIGURE P-2. Revolving Hog Wheel, Armour Plant, Union Stock Yards. Image courtesy of Lake County (IL) Discovery Museum, Curt Tech Postcard Archive.

plants. Armour had taken note of some very rudimentary organizational approaches to efficient slaughtering and butchering in Cincinnati packinghouses during their heyday in the first half of the nineteenth century, a time when the city dubbed itself "Porkopolis." By the time of the Civil War, however, Chicago had surpassed Cincinnati in pork processing.[6]

Armour was on a continual quest to create an efficient production line for the processing of hogs—an approach that would quickly be adopted by a variety of other modern industries well beyond the scope of meatpacking. He was pioneering key concepts such as the division of labor, mass production, continuous flow, and standardized units of production.[7] Despite the grim nature of it all (or perhaps in part because of its gruesomeness), Armour's packinghouse floor became a Chicago tourist attraction, as depicted in the illustration of the "revolving hog wheel" (fig. P-2) on a popular postcard of the era. Attached to the giant wooden wheel by a chain around its hind leg, the hog was lifted upside-down, squealing and thrashing, to a hanging rail above. The live pig began a gradual descent on the rail to its ultimate demise and dismemberment, beginning with a "sticker" who slit the animal's throat and let it bleed out before the army of workers down the line reduced it to packaged parts and separated by-products.

The tasks were all carefully divided up among workers along the length of the rail, maximizing speed as well as consistency of the final product. As you might expect, the diminution of skill involved meant that labor could be had for cheap, and language skills were not even all that critical. In fact, the less capacity the workers had to communicate with others, the more influence and control the company could have other them. Therefore, immigrant workers consistently comprised a large component of the labor pool for Armour and its meatpacking competitors of the time. Armour also hired a large number of black workers from Southern states, including many who were brought in as strikebreakers in 1904.[8]

Armour, Swift, and other meatpacking plants of this era had an obvious interest in maintaining a labor pool with little capacity for union organizing, and they utilized the frictions between different cultural groups to their own advantage.[9] There were ample reasons for these meatpacking workers to rebel, strike, and unionize: They were working and living in dangerous and deplorable conditions, often with terrible wages. Virtually all of the laborers for the Union Stock Yards lived in or near the New City section of Chicago's South Side, the area surrounding the Union Stock Yards and other industries. It wasn't until a public transportation system was established that the stockyard laborers were able to live outside the squalor of the area's garbage-filled streets and offal-polluted waterways.

It is worth pointing out, however, that not all of the workers were on the packinghouse floor or relegated to the living conditions of Chicago's South Side. Given the immense volume of business generated by the meatpacking industry, thousands of clerical workers and managers were also needed to track the multiple and far-flung business arrangements. The office space for these workers was influential in defining the Chicago skyline. Highly pragmatic and relatively uninterested in aesthetic embellishments, the Armours and other early meatpacking industrialists demanded form that followed function, such that they ultimately influenced the early skyline with some of its first skyscrapers and drove the development of the "First Chicago School" of architecture.[10]

One clear benefit of the implementation of the assembly line approach to livestock slaughtering and butchering was that it discouraged waste of any parts of the animal simply by rewarding entrepreneurial efforts that could convert a by-product into a useful consumer good. Gelatin, fertilizers, tallow, cooking oils, leather products, margarine, medicines, glue, pet food, knife handles, brushes—the spin-offs seemed limitless, and growing scientific knowledge (especially chemistry) and rapid developments in engineering dramatically expanded the market opportunities.

Sometimes, the meatpackers sold the by-products to other smaller manufacturers, many of which surrounded the stockyards, and sometimes they developed product lines of their own. The Armour name is but a peripheral shadow of its original stature in the industry today, but Dial soap—originally an Armour product—remains on store shelves as a reminder of just how pervasive the meatpacking industry was in the lives of ordinary Americans, regardless of whether they actually ate meats processed by these companies.

The industry spawned other technological developments that would utterly transform the American food system. Armour pioneered the canning of meats, for example, an enormous development for consumers in an era when home refrigeration was not widespread. But it was the use of refrigeration in the meatpacking industry that was one of the most dramatic game-changers in transforming a mosaic of relatively local food systems into a carefully coordinated supply-and-demand food system that was national in scale.

Refrigeration by means of ice had certainly played a role in various aspects of the food system, including the meatpacking industry. In the 1840s, ice was being used to transport milk and other dairy products from rural areas into cities, but the rapid evolution of specially designed refrigerated railroad cars in the 1850s and 1860s helped make the transport of perishable products more economically viable. Unlike dairy products, meats suffered in appearance and taste if they came into direct contact with ice. The early versions of refrigerated railroad cars designed for the transport of meats relied upon the movement of air over ice placed at either end of the railroad car, with the cold air settling into the lower midsection of the car. Armour was quick to adopt these new transport technologies. In fact, when the railroads balked at investing in refrigerated cars, Armour proceeded to manufacture its own, eventually building twelve thousand of them. But it was the development of mechanical refrigeration—that is, controlled refrigeration—that really made the difference.

FIGURE P-3. Armour Refrigerated Car. Photograph courtesy of City of Vancouver Archives, reference code: AM1506-S2-1-: CVA 447-1676. Photograph by Walter E. Frost.

Mechanical refrigeration was first adopted and used extensively by breweries, starting in the 1870s. By 1891, virtually all breweries in the northern reaches of the United States were using refrigeration in order to maintain the consistency and longevity of their products. And by the cusp of the twentieth century, the largest meatpackers—Armour, Swift, Morris, Wilson, and Cudahy—were utilizing mechanical refrigeration in all relevant aspects of their operations.[11] They could now keep their processing facilities cold year-round, meaning that slaughtering and meat preservation were no longer just cold-weather activities, as they had been in the past. Branch cold-storage facilities made distribution simpler and more reliable, and temperatures could be taken down as far as −30 or −40 degrees Fahrenheit. In sum, meats—and especially more perishable items such as organ meats—could be transported farther distances with less threat of spoilage, and the meatpackers penetrated more distant markets with increasing effectiveness.

Of course, the increasing spread of household iceboxes and then refrigerators throughout the early part of the twentieth century galvanized the public's new buying and eating habits. The new distribution possibilities were changing consumers' expectations. One example of this consumer shift stemmed from the fact that it made no sense to have railroad cars empty for their return trips to the Midwest. So Armour worked to fill the refrigerated cars, bringing in goods such as prized California produce. As a result, seasons, distance, and perishability became increasingly inconsequential for consumers. Armour and his peers had found a way to bridge the gap between producers and consumers on a national scale—and to amass enormous fortunes as the middlemen in this great exchange of commodities and currency.

Ninety years later, it is instructional to ponder the implications of the 1922 Armour Food Source Map, both the image itself and the banner headline, "The Greatness of the United States Is Founded on Agriculture." Expert in distribution, Armour and Company didn't hold back in getting its map out there. The company distributed it to schools "to aid in teaching Geography and to make clear the importance of Agriculture."

The instructional value of this map is perhaps more salient now than it was in 1922. It depicts a moment in time when localized food systems were still intact and vibrant, and a centralized, overindustrialized global food system was not necessarily inevitable. Modernization certainly *was* inevitable, but the technological innovations were increasingly geared to national and international wholesale markets, and consequently the wholesale dissolution of far too many local markets. Our society chose one path and not the other. I'm not sure precisely what led us, or who led us, or to what degree greed guided us more than plain old shortsightedness. I am certain, however, that we left a lot of good options behind. Of course, the consequences are much more obvious now than they were ninety years ago.

These consequences can be summed up in one word: erosion. That is, the erosion of topsoil, values, communities, and even democracy. Relationships became contracts, communities transformed into centers of production, and animal husbandry devolved into livestock management. The ultimate results of this erosion are ghost towns in the Midwest, food deserts in our urban areas, more than one-third of our food wasted, and the key workers in our food system living in poverty—even, ironically, malnourished.

We opted for largeness instead of largesse, and the expense of our expanse is now writ clear in the dead zones off our shores and the crash of biodiversity on and around our farmlands. As a farmer, an educator, and a parent, I want to believe in the phrase "The Greatness of America Is Founded on Agriculture." But I also don't want to forget—or let my children and students forget—that the land upon which we farm was ultimately wrested from the Native Americans. And I want us all to remember that any such greatness relies upon not only the careful stewardship of fertile expanses (as well as tiny garden plots) but also the careful tending of community.

If agriculture is to be "Great" in any capital G kind of way, then we simply have to view ourselves more as food citizens than as food consumers. Citizens will studiously *delegate* decisions about food and agriculture to people who will be stewards of the common good; consumers will merely *relegate* these decisions to someone else. Better yet, citizens will take it upon themselves to create more just and resilient food systems. And not just *local* food systems, but *community-based* food systems. In that spirit, be forewarned: this book is ultimately as much about "our food" as it is about "local food." It's also about what the term "local," with all its nuances, really means and how we can create foodsheds that make sense in the face of enormous energy challenges and increasingly severe weather events.

Take a hard look at that 1922 map one more time. Then look at the maps that accompany it in the color insert. What happened over the course of the past ninety years?

It's simple. In our first attempt to get this modern food system right, we ceded the future. This time, let's seed the future—with clear visions of how we can define our food systems, instead of having them define us.

Introduction

Going local. It all seemed so easy. But how do you define "local food"? Well, you can just start with an imaginary string. Select a point—the center of the family table—and stretch the string from there to the point at which "local" ends and something else begins, using the string as a radius to circumscribe the "local" circle.

No, no, that doesn't quite work. It's just a different kind of circular reasoning. Hmmm. Maybe "local" should be a given distance, a town boundary, a county boundary, a state boundary, a culturally distinctive area, a watershed, or even a funky "foodshed"? Tidy, perhaps, but probably too simple.

Okay, so let's try "food miles"—that makes it less arbitrary. Stretch the string from point A, the center of your table, to point B, the farm. Hmmm . . . but the food product went from the farm to a processing facility to a storage warehouse to a distribution center and then to the grocery store, to where you had to drive to pick it up. Or at least you chose to drive, even though you could have easily ridden a bicycle. Oh, heck, forget it. Let's just all start using the same figure of "the average food item in the United States travels approximately 1,500 miles to get to your table." Problem solved. Temporarily, at least.

Meanwhile, there's a split screen displayed on the nearby computer, showing Webster's online dictionary on the left so you can look for definitions of "local" and Google Maps on the right so you can see what a 1,500-mile-radius from your home address looks like. Suddenly, a headline flashes across your computer screen as a news alert: "Local Trumps Organic." As you stare into the screen, pondering the complexities of it all, a tweet from Oprah abruptly appears, informing you that she is now at her favorite farmers' market buying Chioggia beets ("Oh, the splash of color they'll make on a salad with those concentric circles of red and white!"). No sooner has your attention been diverted by Oprah's digitized epiphany than a beep from your computer indicates that a new word has just been added to the English lexicon, providing a much-welcomed (and somewhat self-congratulatory) label: "I'm a *locavore*!" At last, self-actualization with a community flair! But wait, is that new word spelled with or without a second *l*?

Thinking about our local food radius isn't an exercise in circular reasoning. It is, in fact, an important starting point for thinking about the role of local foods in our daily lives and our communities. But we can't stop there. The ultimate goal is for us as individuals and as communities to think more complexly about community-based food *systems*. Part of that thinking involves cultivating our imaginations and seeding our aspirations with relevant examples—some of them from nearby, others imported from distant lands and eras. The stories of these examples serve as touchstones and springboards; they are tales of hope and, on occasion, of caution.

The good news in the renaissance of more localized food systems is that hope and appropriate scale tend to be close allies. Individuals and communities discover empowerment through the promise of even the smallest of intentions, and small successes pave the way to even bigger dreams. Yet there is a curious irony in the fact that the drivers of this hopefulness frequent the downside of so many different bell curves. We face shortages of oil, water, fertilizers, productive land, agricultural biodiversity, and even farmers. Then, as if agriculture isn't already challenging enough, we find the weather and the climate becoming increasingly volatile and unpredictable. Despite these challenges, a pragmatic optimism is rising among advocates for more sustainable and localized food systems.

Naive? I don't think so. The rapid rise of environmental constraints that challenge a safe and reliable food supply requires that we intensify the quest for sustainable food production, particularly in our home regions. The social inequities and health problems so evident in the United States force us to reexamine the links between our national food system and the problematic aspects of our individual diets. And the economy is like the weather, volatile and unpredictable, requiring us to seek and create shelter in the security of the familiar—our local communities.

Probability and possibility intersect here. The probability that all of these challenges—environmental, social, and economic—will increase in volume and velocity brings us to the brink of possibilities, both positive and negative. The default response—a response but by no means a solution—is to maintain the status quo. In contrast, one critical and creative response (albeit not a panacea) is the rebuilding of community-based food systems. The work involved in developing these local food systems requires that we not just passively *accept* these inevitable changes, but that we find ways to *adapt* to them. This adaptive approach, in the vocabulary of some forward-looking thinkers with their shirt sleeves rolled up, embodies the concept of resilience. Resilience theory dissuades us from dichotomizing humans and ecological systems and encourages us to adapt to changes, even when they come in the form of disturbances and shocks, in constructive ways.

While the challenges to the global food system are daunting, I find the opportunities and the momentum for reweaving the strands of locally based food systems into the fabric of our communities to be tremendously exciting. From my vantage point as a farmer, a professor, and a local food systems advocate, I believe the prospects for positive change are remarkably encouraging. And as someone sitting astride the half-century mark, I see more reason for optimism in the next half century than what I have seen and experienced in food and agriculture this last fifty years.

Growing up in North Carolina, I saw national fast-food chains begin to replace local cafés and restaurants during my childhood, while the neighborhood Piggly Wiggly grocery store ("Hoggly Woggly," we kids used to call it) began to replace its regionally sourced fresh foods with expanding aisles of processed foods. In the public schools, those of us bound for college but interested in farming and vocational skills were, in essence, shown a fork in the road and told that our career decision was a choice between two divergent paths, with no possibility for integrating intellectual challenge with a love for soil and craft. The idea of organic agriculture was anathema to the cultural paradigm—in fact, it was simply deemed illusory and impossible in most circles. Local foods, although much loved in the South, were giving way to a flurry of food industry developments. Not only were we enticed by the conveniences (items like Campbell's soup, Steak-umms, and Pillsbury biscuits) that relieved women of some of the burdens in those hot kitchens, but I also distinctly remember the allure of "ethnic foods" that tempted us to step beyond our parochial boundaries. As absurd as it seems now, I can clearly remember the enticement of "Italian food" when pizza finally came to town. Mexican food came much later—no small irony considering the fact that nearly one in ten residents in North Carolina is now of Hispanic or Latino origin, with many of them working in the state's dynamic agricultural sector.

As those transformations took hold, my generation and those following were fortunate to expand our culinary horizons (often an early critical step in embracing cultural diversity), but the links between food, place, and tradition began to dissolve. Behind the scenes, the foundational components of local and regional food systems were being dismantled at breakneck speed. Giant distribution centers and airports replaced street corners and local warehouses as hubs of commerce, while the local food businesses succumbed to the same pressures as local farms. The middlemen became the titans. Deal makers and deal breakers, these brokers relegated farmers and others to the role of price takers. By the end of the twentieth century, many of us hardly knew what a local food system looked like, much less how to begin to rebuild one.

I was lucky in that regard, however. Entranced by the possibilities of a life of farming but dissuaded by a lack of examples that fit my idealistic visions, I was fortunate enough to join an international exchange program in 1983 at Brunnenburg Castle in Italy during my junior year of college. Not only did Brunnenburg house a museum dedicated to the disappearing agriculture and foodways of the Alpine farmers in South Tirol (an autonomous German-speaking province in the Italian Alps), but South Tirol was an astounding, beautiful collection of villages with bakeries, butchers, cheesemakers, orchardists, home gardeners, beekeepers, wineries, distilleries, fresh markets, and creameries.

I had stumbled into a region of interconnected small-scale food systems built upon topography and tradition, with tight ties to agritourism. Foods from other parts of Italy and Europe could be found, too, of course, but the regional specialties dominated. And it went deeper than just the broader regional specialties. The steepness of the terrain and the relative isolation of many of the locales meant that unique food traditions could be found in single villages or throughout the length of upper-elevation valleys. A slow walk through a village was, in fact, a culinary tour in which the residents picturesquely boasted of their unique foods in their shop windows. Cheeses and charcuterie products would vary, but the telltale symbol of a valley's pride would be its traditional breads, molded into different shapes and created with varying proportions of traditional grains. Each loaf had a story to tell, and each baker's storefront was a window into village pride and sense of place.

At other times, however, my discoveries of those intensely "place-based foods" would come by way of a hushed invitation from the innkeeper or the mountain farmer to come down to the cellar to taste his own *eigenbau* ("self-made") wine, spirits, cheese, and aged meats, all raised or cultivated in most cases within a few hundred meters of the house. Although I didn't realize it at the time—and I certainly didn't have a name for it—I was getting a firsthand look at the most intact community-based food systems that I would probably ever encounter in my life.

I also couldn't foresee that Brunnenburg would become a second home for me, a place where I would send students and return repeatedly throughout my adult life. In those returns throughout the past three decades, I've witnessed a slow erosion of some of those food and agricultural traditions due to the fast-paced infiltration of regulation and homogenization into these high-elevation valleys. The European Economic Union's efforts to level the playing field among its members in terms of regulations and trade often shoved aside the traditions and specialties of centuries. Fortunately, the residents of South Tirol and many other regions across Europe sensed the gravity of the losses and began to lay claim to protecting their foodways and associated infrastructure, with at least some degree of success. Europeans clearly saw what we had lost in the United States in decades prior, and many of them also resented the fact that we had unleashed our hounds of homogenization on them with the export of our fast-food chains and supermarket economics. For the Europeans, the threat was more than a loss of foods—it was a loss of culture rooted in place.

In the early 1990s, I ended up going back to South Tirol to farm and teach at Brunnenburg for several years. During my second year there, I vowed not to leave the region—an area about half the size of Connecticut—for one year, other than my required trips to pick up students at the airport in Munich. I wanted to learn as much as I could about the farming and food traditions of that one area, and I opted to travel as much as possible by foot. The wines, the meats, the cheeses, the fruits—everything was nuanced by precise location and well-honed tradition, and walking enhanced the possibilities of unexpected observations, conversations, and culinary surprises. A distance of but ten or twenty miles would yield different tastes, so the tight geography of the region seemed enormous in terms of culinary nuance.

I was incredibly fortunate to stumble upon a part of the world that still had a rich variety of intact local and regional food systems, and it is in part those memories of traveling through South Tirol and other parts of Europe that get me so excited about the potential for the future. But I have also been privileged to witness successful examples of resurgent local food systems closer to home, successes that speak to the burgeoning potential of this kind of hard work throughout the United States. When I return to North Carolina, I am always astounded by the increasing visibility of sustainable agriculture activity and local food entrepreneurship. What a difference a few decades can make—North Carolina is now a powerhouse in promoting not only its own farm-fresh products but also sustainable agriculture initiatives. The state's early efforts in developing a "buy local" campaign and its pioneering investments in large, well-equipped regional farmers' market facilities are now complemented by a range of private entrepreneurial efforts that make eating local anything but a deprivation. "Drinking local" is also a possibility, thanks to the fast-paced growth of wineries, microbreweries, and coffee roasters throughout the state.[1]

In my home region of Vermont, I have been privileged to have worked with a diversity of talented colleagues who helped transform the Rutland area, one of the most beleaguered regions in the state, into a vibrant agricultural economy. Despite having some of the highest poverty and obesity rates in Vermont, the city of Rutland created the first farmers' market in the state to run for fifty-two weeks of the year, including a winter farmers' market that has more vendor demand than spaces to accommodate them all.

Meanwhile, during my tenure on the Vermont Sustainable Agriculture Council, I've watched conversations about Vermont's local food potential quickly transform into a legislatively supported initiative to create a statewide strategic plan for re-envisioning and reconstructing the state's food system, an effort known formally as the Vermont Farm to Plate Initiative. Finally, in my role as a professor at Green Mountain College, I've watched alums put down roots in the region and build farming and food-related enterprises, while the enrollment numbers in our related undergraduate and graduate programs rise—in parallel with the tremendous growth of such programs all across the country. The sense of a renaissance in community food systems is directly tied to the invigorating energy and enthusiasm brought forward by our youth.

It is not just the successes of these ventures in creating more resilient and localized food systems that give me hope, but also the velocity of the changes. The momentum is nothing short of extraordinary, and it should serve as inspiration to any efforts in relocalization of resources, whether the target is food, energy, or any other commodity. In all of these initiatives, the small seeds of local solutions harbor promises that national governments can scarcely dream of and seldom deliver.

However, these promises depend upon our willingness to think more complexly and to work harder than we might initially expect when stepping into the world of community-based food systems. Therein lies my biggest concern for the ultimate success of these ventures. The sustainability of these efforts is dependent upon moving beyond the hype about just the foods and into the real complexities of the systems that produce them. Otherwise, the focus never moves past marketing and into a significant transformation of the marketplace.

This situation could be described as the difference between Local Food 1.0 and Local Food 2.0. My favorite example of such a difference comes from communications strategist Duane Hallock, who describes 1.0 as a dazzling fireworks display for an adoring audience and 2.0 as a campfire conversation among those who gather to share ideas. To parallel Hallock's insightful distinction and put it into the local food systems (LFS) context, LFS 1.0 is directed to a public audience, whereas LFS 2.0 is an interactive and decentralized community conversation—not a marketing pitch. And lest we forget the significance of the era in which we live, the 2.0 version also employs a full suite of social media resources in order to expand the dialogue and the innovation.[2]

In this new era, we have the opportunity—indeed, the privilege and responsibility—to completely reimagine our community food systems in such a way that they connect people not just to their food but also to one another. Communities of all scales, scopes, and colors are beginning to recognize that food is not a commodity to be simply entrusted to large corporations and government entities. To do otherwise, however, requires creativity and collaboration—and a willingness to confront the complexities head-on.

It is this approach to rebuilding local food systems that sets this book apart from many of the other recent books related to local foods. The solutions we create cannot be simpler than the dilemmas that we face; systems thinking will take us farther than ideology. Hence, the structure of this book:

- The first part, **Dilemmas**, lays out some of the key challenges and questions inherent in understanding and describing local food systems.
- The second part, **Drivers**, takes a hard look at the justifications that are commonly put forward as reasons for rebuilding community-based food systems, as well as some important justifications that are too often missing in these discussions.
- Finally, **New Directions** offers a number of ways that the reader can support the development of sustainable food systems. This final part also offers a number of models—farms, businesses, organizations, and initiatives—that can serve as inspiration for new locally rooted efforts in one's home community.

Although there is a building-block approach to the order in which this book is structured, most of the chapters are designed to stand on their own so that any one of them can serve as the starter for those important campfire conversations happening all across the country. The reader will quickly discover that I firmly believe it is not enough simply to describe the incredible array of food system innovations out there. In order to ensure both the proper fit and the longevity of any new businesses or initiative, we have to understand how they fit into the broader systems—hence the importance of the "Drivers" part of this book, which examines how local and regional food systems relate to issues of energy, the environment, food justice, cultural and biological diversity, and the marketplace. Bring the burning questions posed in those chapters to your next local food systems campfire, and there will be plenty of fuel for a conversation that will burn long into the night. After all, anyone who appreciates systems has to embrace complexity and a good debate.

It's time to light the first match.

A Final Introductory Note: On the Use of the Word "We"

The reader will note that I use the pronoun "we" fairly liberally throughout this book. I firmly believe that the food and agricultural dilemmas faced by any segment of our population are ultimately *collective* concerns that none of us should ignore. On the other hand, I also recognize that not everyone wants to be held to the assumptions of someone else's perceived sense of "we," whether considered in a particular local context or a broader geographical discussion. More importantly, it's never ideal to feel as if one is subject to someone else's solutions, all under the guise of some undiscussed assumption of unity.

While "we" may be a pronoun born of pragmatic compromise for people like myself, it's important to state clearly that it's not enough for people like myself—with privilege and power by virtue of race, socioeconomic status, and gender—to simply be cognizant of when and how to respond to food system dilemmas. Those of us who are in such a position must also be prepared to step aside and quietly listen to more marginalized voices, the voices of those most severely impacted by nutrition and food justice issues. Single mothers, mothers with young children, blacks, Hispanics—individuals within these and other traditionally marginalized groups are among the most likely within our society to face serious struggles related to food, with too few opportunities to express their concerns and advocate for change in the food system, not to mention their overall economic situation.

In the end, I hope it is clear that my use of the collective pronoun "we" is neither casual nor careless, but rather quite intentional. Food and agricultural issues are everyone's concern, and they should constantly be examined under the bright light of any shining democracy. As such, employing the word "we" is the first step in taking responsibility for our own actions and for the well-being of the broader community. In doing so, we transform what is all too often a discussion of economics into one of democracy, based upon what is the most inalienable right: nourishment.

Part One

DILEMMAS

Chapter 1

Location, Location, Values . . .

Find your place on the planet.
Dig in, and take responsibility from there.
—Gary Snyder

With its turret, rounded arches, and heavy stone motifs, the Romanesque Revival–style building at 116 State Street in Montpelier, Vermont, appears at first glance to be a refuge for monolithic thinking, not a bastion of creativity where policymakers, regulators, and marketing specialists might work together to "relocalize" a state's agriculture and food sectors. But, as it turns out, some of the discussions in the high-ceilinged chambers and labyrinthine stairways of the Vermont Agency of Agriculture are as cutting-edge as can be found within any such state agency in the country. If nothing else, those conversations, formal and informal, tend to bring out some of the conundrums we face in the United States as we try to rebuild and even create from scratch local food systems that make sense.

Several years ago, I was sitting in at one of those discussions as a representative to the Vermont Sustainable Agriculture Council among a delightfully diverse array of farmers, policymakers, nonprofit leaders, and educators when the secretary of agriculture—someone for whom I have great respect—put forward a statement that completely silenced the typically gregarious group: "If a farm is in Vermont, then it's sustainable. We don't have any farms in Vermont that aren't sustainable."

No one knew quite what to say. The awkward silence seemed to seep its way into the cracks between our varied perspectives. We were all tackling the complexities of re-envisioning and rebuilding the Vermont food system from different angles, and it was the richness of those varied viewpoints that made the meetings so interesting. The unanimity of the silence was unusual among this group of creative, independent thinkers.

On one hand, the secretary was wisely putting forward a critical reminder that warrants constant reflection: drawing lines between farmers and labeling farms and food with terms that oscillate between ambiguous and pejorative do not help create common cause in reconstructing a state's food system. In that regard, I knew that at least several of us agreed with the secretary. Many of us in the room had seen the fallout from farmers entrenching into highly polarized camps on "sustainability" issues. It disrupted the possibility of constructive dialogue on agricultural issues, much less the potential for collaboration in tackling the strength and resilience of Vermont's food systems.

But there was something in the statement that I found unsettling. I was uncomfortable with the assumed link between an artificial boundary (i.e., the state line) and sustainability, and I suspected several other quiet members of the group were having similar thoughts. Despite our devotion to Vermont agriculture, at least some portion of us weren't ready to immediately equate location with certain values or best practices. It was just too simplistic. Many of us had visited farms in our state that were not necessarily models of good stewardship of finances, natural resources, or even animal well-being.

I couldn't quite muster the finesse to articulate my concerns that day, nor did I want to diminish the importance of the secretary's reminder not to polarize our agricultural community. However, the idea that farming practices or food products should be assigned a certain status simply due to their point of origin and without careful examination didn't sit well with me then, and if anything, I'm even less comfortable with it now.

As we work hard across this country to rebuild our local food systems, we are drawing a line of sorts, and it's not just a geographical one. It's a reclamation of spirit and intent more than a delineation of terrain. But if we're going to start claiming things—whether territories or truths—we need to think hard about our underlying assumptions and assertions. Creating community-based food systems is one of the most intellectually challenging tasks of our age. Doing it well requires a diversity of community players, perhaps a few experts from different fields, and some inevitable but carefully considered compromises. And it all has to fit the community in question and its surrounding environment. Any local food system is only as good as the intent and the practices behind it.

The current energy behind the push for local foods in communities across the United States and abroad is tremendously exciting and long in coming. Frankly, the prominence of "local food" in our vocabulary and in the marketplace is astounding. "Local food" is fresh, direct, hip, in vogue, totally simple, and unbelievably complex. The wave of interest in local foods has swept every state, people of all socioeconomic stripes, and even the full spectrum of generations, from kindergarteners to college students to seniors. The younger generation is looking forward with an eye toward food security and nutrition concerns, while the older generation is reclaiming memories, meaning, and tastes from previous decades. With the unsettling realities of the global economy, climate change, peak oil, biodiversity loss, and continued social inequities, the table is the one place where most of us—but not all (and that is a critical point in this book)—can find room and reason to celebrate. Despite the difficulty we have in defining the radius of "local," we are clear on one thing: the nucleus for local foods is ultimately the table.

"Local food" is one of the top environmental and social issues of our time . . . heck, it's just about the only bipartisan issue out there right now! It could be that we've come to the realization that food, human health, landscape, and local economies are the common interests that we all share and recognize on a daily basis. Perhaps it is because the costs of our current food systems are manifesting themselves in our physical well-being, the patterns of our daily lives, and the ecological integrity of our planet, all in ways that we can no longer ignore when we look in the mirror, glance at the daily calendar, or gaze out the window of an airplane skimming the surface of the continent. And then there is that unusually diverse array of individuals—from strategic military planners to backwoods libertarians to peak oil activists to climate change scientists—who see the need to ensure local food security in case of various worst-case scenarios. Regardless of what actually brings each of us to the local foods table, we need to savor the time we spend there together. But it is imperative that we also work hard and fast to transform this current energy and these ideas into viable community-based food systems.

While we can laud the force and vitality of the pursuit of local foods, we must also ensure that we are doing more than applying Band-aids and window dressing. Transforming our relatively nascent American values and concepts about food into enduring initiatives and infrastructure will take years to develop fully. It will also take time for these new concepts to become integral parts of how communities and local economies operate.

When we look at food systems of different scales—local, regional, national, and international—two observations are particularly important:

- All of the scales are interlinked, clearly nested within and to some degree dependent upon each of the other food systems.
- The most beleaguered of those food systems is currently the local scale, primarily as a result of decades of building out national and international food systems and markets, with too little attention to (and sometimes even outright disregard and disdain for) the ultimate negative impacts on the local scale.

These two observations are essential in creating resilient food systems because, in our exuberance to embrace local foods, we risk oversimplifying the task before us and, consequently, endangering the likelihood of success. They can also help guide us as we navigate the inevitable dilemmas we face in rebuilding food systems that fit our times, our tools, and our communities.

CHAPTER 2

The Geography of Local

Everywhere is walking distance if you have the time.

—STEVEN WRIGHT

Before we even delve into the dilemmas associated with defining "local," we first need to consider whether "local" is simply about location and distance—in other words, the tensions between the point of origin and the point of consumption. Is "local" just about physical geography?

You've probably already guessed that the answer to this question is a firm and fast *no*. The potential for a whole variety of answers is likely to keep a slew of geographers employed with tenure for the next decade or so. In fact, it is these geographers, multidisciplinary thinkers working in an underappreciated modern academic discipline, who have had some of the most helpful insights into the current local food frenzy. After all, "place" matters most to people who derive not only their income but also their intellectual fulfillment from studying what "where" means.

We humans have a particular penchant for layering words with meaning, which provides for rich expression but potentially diverse interpretations. And the marketplace requires us to create distinctive niches for our products, especially when we are the underdogs (as tends to be the case with the homegrown spirit of so many local foods). So we use "local" not just to describe, but also to justify. Later in this book, we will examine the "drivers" of local foods, different factors that can inform and shape what a community-based food system might look like. But it's important to begin here by considering the different ways in which we relate "local food" to place.

In essence, when the local food discussion is at its best, the conversation is an expression of caring. We care about our local landscape, the ecological impacts of distance, our sense of place, the people in our communities, and the manifestations of various economic relationships. It's worth thinking in more detail about the place-based presumptions that shadow the local food debate, as well as some of the dilemmas we face in justifying and building community-focused food systems.

Landscape

If the first association that comes to mind with the phrase "local food" is a bountiful stand at a farmers' market, the second association may well have something to do with the regional surroundings, whether it's the bucolic scenery, the conversion of a suburban lawn into an edible landscape, or the modern-day phoenix vision of a community garden arising from urban rubble. Most of us see buying local food as first and foremost an investment in the gardens and farms that, in many ways, frame our local communities—and also in the farmers who feed us.

The local agricultural landscape—when not dominated by agricultural monocultures and monopolies—offers us numerous values, ranging from wildlife habitat to links to the region's cultural past. If we're wise, we realize that the conservation of prime agricultural soils, as well as farm infrastructure and layout, is key to ensuring future local food security. And these soils are certainly central to what some particularly allegiant local food and food justice advocates call "food sovereignty."

In our disregard for the gradual dismantling of our local food systems, we have created virtually irreversible changes to the landscape. It's a fact that the lands of highest agricultural value—fertile fields of prime agricultural soils in central locations—often tend to be prime sites for development. We have allowed the development of those areas thanks to our shortsighted notion that food can still be grown somewhere else, perhaps even more efficiently than it can be grown close to home. It is this passive shrug of our collective shoulders that has devastated the agricultural base of many communities.

Many people not active in farming—and that's more than 98 percent of the U.S. population at this point—do not realize that when the domino effect of farm sell-offs and subsequent residential and commercial development begins in a community, farmers lose more than just arable land. Suddenly, there's not enough demand for people and companies providing agricultural supplies and services. As an example, a John Deere equipment dealer less than ten miles from me recently was cut off by the John Deere company because its sales did not meet the company's ever-increasing sales benchmark. Perhaps the decision was directly linked to the declining number of farmers in the area, but it was probably also tied to the region's economic downturn. Regardless, farmers now have to travel much farther for equipment, service, and parts. Quite a lot of infrastructure is needed to maintain a thriving local agricultural sector: feed stores, seed and equipment suppliers, slaughterhouses, large animal veterinarians, farm-savvy accountants, cold storage facilities, cooperative warehouses and distribution centers, milk haulers, and artificial inseminator technicians, among others. All of these critical services quickly evaporate as the farms themselves disappear from the landscape.

While some small-farm advocates may celebrate the absence of medium- and large-scale farm operations—and the reduced competition that it implies—the reality is that small-scale farmers often have access to top-notch tractor sales and service operations, slaughterhouses, and agricultural supply stores only because larger operations are keeping them in business. (Although it should also be noted that when extremely large agricultural operations move in, they sometimes bring their own infrastructure with them, displacing locally based agricultural service providers or, alternatively, shifting the way that these service providers do business so that the big operations receive preferential or even exclusive service.) There's strength in diversity even at the local level—and when one sector of the agricultural economy gets hit, virtually everyone involved in agriculture takes some punches. When one farm goes on the block for sale, it's often part of a cascade, with property values and taxes on other farms going up because of their "development potential."

There are linkages and layers in our local landscapes that we do not always consider. Rebuilding locally focused food systems is seldom about saving one farm or one reclaimed vacant lot, although such places can be keystones in galvanizing support for community-based initiatives. Not only does the full vista of an agricultural landscape help bring us closer to our traditional views of the local economy, but these "undeveloped" farmlands can provide a diversity of habitats that serve a variety of important ecological functions. The ecological integrity of these landscapes is important, and so are the aesthetics of it all. An agricultural landscape conveys meaning and purpose, as well as shared heritage. These aesthetics often inspire a spirit of regional pride and local affinity. At its best, an appreciation of beauty can inspire a sense of belonging. "Rootedness," that elusive American icon, seems possible in places where we have to wait a season for vegetables, years for fruits, and decades for nuts and timber. We can become local in our long and patient waiting.

Sense of Place

Philosophers, geographers, and psychologists have all pursued the haunting but somewhat ethereal "sense of place" that we humans seem to share in the deeper reaches of our minds. The culmination of sensory experience, memory, and terrain, a sense of place is perhaps most easily conveyed through and contained in the combination of taste and smell, our broadest avenue to a storehouse of memories. Local food marketers and advocates constantly appeal to our sense of place in order to generate both buying and buy-in. They work hard to convince us that the place in which we live is special and meaningful, and therefore worth investing in. They tantalize and taunt us with the possibility that we can never truly experience a place until we taste a particular fruit, a time-honored dish, a unique spirit, any of which may be replicated elsewhere but diminishes in authenticity with every mile it travels beyond a given boundary. These local specialties convince us that local foods are therapy for our culture's chronic transience.

When we become disillusioned with that transience and try to assess its negative impacts on our psyches, our local communities, and our cherished landscapes, we start to gather around the power of place. It appeals to those of us with long-term associations with a particular place, as well as those of us longing for belonging. Rootedness again takes hold—and supporting our local agricultural economy and caring for the well-being of our community seem to be the perfect manifestation of anchoring ourselves and building a home around the table.

Not only does a sense of place give us a certain comfort in the past and a confidence in the future, it can also serve as a foil against the economic and cultural homogenization so many of us feel in an increasingly globalized world—a world in which our most common links to food simply strengthen highly homogenized chain stores and food-chain clusters.[1] Philosopher E. S. Casey calls the erosion of our sense of place the "thinning of the life world," meaning that our connections and commitment to our immediate surroundings are diminished due to our transience and geographical obliviousness.[2] Geographer Robert Feagan goes so far as to call it the "annihilation of place."[3] Despite the difficulty in fully assessing or definitively naming this modern-day malady, one self-prescribed response seems to be a healthy dose of local foods, an antidote to the anonymity of mass consumerism. Some empiricists undoubtedly shudder at the thought of basing any actions on a concept so amorphous as "sense of place," but there is clearly a cultural malady in need of a prescription here, and local foods might be just what any good doctor would order.

This indulgence can be taken a step further, from a bioregional, ecological, or even hedonistic perspective. The oldest new kid on the block in promoting local foods and their link to our sense of place is *terroir*, a concept most articulately espoused—albeit with an epicurean accent—by the French foodophile. He raises his wine glass to the air, swirls with centrifugal confidence, watches the legs of the wine meander down the sweep of glass, sniffs not once but twice, utters an exclamation, sniffs again, sips, swishes, sips more confidently with an erudite grunt, and smiles slightly before twisting off a morsel of rustic bread and casually beginning to chew it with a small cube of artisan cheese that cuts the palate like the peasant's scythe—"*Aahhh . . . ça, ça c'est la Provence.*" Not only does he make it clear that the three merged tastes of wine, cheese, and bread are born of his region's sun, soil, and sweat, but he goes on to name the qualities of the hillside on which the grapes were grown, the molds inherent in the cheese, and the qualities of the local airborne yeasts that invaded the simple mixture of flour, water, sugar, and local sea salt. Then, *naturellement*, he tells a quick story about the artisans who crafted the foods. That's terroir—the linking of taste, terra, and tradition. It is, as Amy Trubek calls it in the title of her seminal book on the subject, "the taste of place."[4]

Just as a chef can transform a peasant's plate into a signature dish, France has upped the ante with terroir and even taken it up to the next level: patrimonialization, a concept that merges regional authenticity and heritage with the food products themselves, thereby creating a culinary result that cannot be replicated in its truest form elsewhere. It's a strategy of the underdogs in a globalized world. Simply put, patrimonialization creates certain boundaries of authenticity for specific foods and beverages that would otherwise be lost in the bigger mix. A wine, therefore, is not just the product of a specific variety of grape; it is also the melding of tradition, craft, and location. The local food niche just got a little tighter through the precision linking of people and place.

Relationships

Place is first and foremost the intersection between latitude and longitude, but place as it relates to local foods is also the intersection of people and their environs. And it is not just about existing relationships. In fact, one of the most compelling arguments for rebuilding community-based food systems is that it requires us to broker new relationships—relationships that help build local economies, conserve local landscapes, create entrepreneurial collaborations, enhance food security, enlighten and educate, and generate new friendships. The terms and concepts associated with the relational nature of local food systems are as prolific as summer zucchini: shortened food chains, Know Your Farmer (from the USDA, no less!), civic agriculture, community food security, community-supported agriculture, conviviality, and the list goes on.

In some ways, the emphasis on relationships inherent in the local food push is the most exciting aspect of it all: we are consciously making the choice to build new economic relationships, rekindle traditional ways of doing business, support those in need, and even invent new techology-based social networks that can, rather ironically, link neighbors. The narratives of this good work start to drive the numbers. As author and ethnobotanist Gary Paul Nabhan so eloquently puts it, we are "restorying the landscape."[5] It is through these stories that we move the local food agenda forward (although, as we'll see later, there is risk in relying solely upon the narratives and not paying close attention to the numbers).

Geographers seem to keep bandying about the rather cumbersome word "embeddedness" to describe how the surge of interest in local foods is allowing for the possibility of enhanced relationships between consumers, farmers, and others. Two phrases I find more useful are "civic agriculture" and "community food security," both of which highlight community transformation as well as a sense of belonging. Tom Lyson, a much-beloved and respected professor at Cornell University, described civic agriculture as being "associated with a relocalizing of production":

> From the civic perspective, agriculture and food endeavors are seen as engines of local economic development and are integrally related to the social and cultural fabric of the community. Fundamentally, civic agriculture represents a broad-based movement to democratize the agriculture and food system.[6]

Food systems expert Laura Delind, describing the dynamism of Lyson's concept of civic agriculture, said, "For him, local food systems are civic in nature and, as such, are instruments of place-based negotiation, collective responsibility, and a participatory democracy."[7] Suddenly, we're far beyond any simple notions of food miles or even much more complex concerns regarding food's carbon footprint. We've not only entered an important area of social discourse, but we're actually talking about changing the way we do business and how business involves strategic community decision making and planning.

While all of this heightened complexity may seem rather theoretical and abstract, it quickly becomes concrete when thrown into the hard realities of our very real "food deserts"—poor urban and suburban neighborhoods where convenience marts and liquor stores have replaced grocery stores—and poor rural neighborhoods where the racially diverse workers so critical to our food production system live on "subsistence wages" (hardly a fair term) that don't allow them to adequately feed their own children.

The term "community food security," in its earliest vestiges during the 1960s and '70s, was developed around concerns related to world hunger.

In particular, it evoked the need for impoverished communities around the world to develop and maintain their own steady and nutritious food supplies.[8] The term began a slow evolution, eventually taking on clear local implications.

For many years, the phrase was most commonly associated with the activist nature of the Community Food Security Coalition (CFSC), one of the more diverse and decentralized food security advocacy groups in the United States. Through its emphasis on community-tailored solutions, the organization has been one of the most important points of exchange for sharing and disseminating ideas on food security. (However, as I write, the CFSC has decided to cease its operations as a nonprofit coalition and pass its torch on to other nonprofits.) These other organizations have supported community-based food security programs by providing resources, networking opportunities, advocacy expertise, funding support, and even legislative initiatives, and some have been formed to keep an even keener eye on racial justice and cultural diversity (as will be discussed in later chapters). Many of these organizations share the definition put forward by community food systems specialists Michael Hamm and Anne Bellows: "Community food security is a condition in which all community residents obtain a safe, culturally appropriate, nutritionally sound diet through an economically and environmentally sustainable food system that promotes community self-reliance and social justice."[9]

But just as "local" eludes clear delineations, "community" can create confusion and dissent, too. Is it geographical or cultural or economic? In some cases a neighborhood is the building block for efforts, while at other times a shared culture may be the point of commonality. Regardless, most efforts to build community food security are geared toward reducing the distance and anonymity built into our highly industrialized food system. Shortened food chains, community-supported agriculture, civic agriculture, community food security, and a host of other approaches all seem to imply or require increased participation, and not just at the consumer level. By rebuilding relationships between neighbors, consumers, suppliers, farmers, and businesses, these movements chip away at uncertainty and distrust, not only in our food systems but in our communities.

In the end, it seems that we're searching for much more than just local food.

Distance

When we're talking local, the issue of distance is never that far away, but by now it should be obvious that this local food thing is about much more than "food miles." The concept of food miles has been helpful as a starting point, and the energy dedicated to transporting food from one place to another is not to be readily dismissed. All energy inputs into our food system need to be examined and minimized to the greatest extent possible. The problem with food miles is not that the figures are wildly inaccurate—getting any averages out of the immensely complex food system is extraordinarily challenging and fraught with necessary but problematic assumptions. Rather, the problem is that an exclusive or even dominant focus on food miles masks other larger energy sinks in our food system such as production, storage, and processing (more on these later). Furthermore, although it may be counterintuitive, the big trucks, ships, and trains that we envision when we hear "food miles" are almost always more efficient transporters of goods than a farmer's pickup or a consumer's car.

"Distance" has additional connotations and other means of measurement. Consider, for example, the correlation between proximity and familiarity. Taken a step further, the closer we are to someone geographically—a farmer in this case—the more confident we are in our own abilities to independently judge the quality and value of the products or services offered by that farmer. We can also better assess his or her contributions to things we care about—our local communities, the quality and safety of our food, the humane treatment of livestock, the treatment of farm workers, ecosystem stewardship, and so on. As the distance increases, however, we start to depend upon other means of assessment, including official certifications. Despite their shortcomings, certifications are, in essence, sanctioned brokers of our trust. They are particularly helpful when the local choice doesn't seem to be working for one reason or another.

That's the point at which we face the inevitable need to rank our purchasing priorities. What comes first? Local? Organic? Convenient? Grass-fed? It's not that we have to be militant in this prioritization. Simply involving and educating ourselves is part of rebuilding community-scale food systems, so it's probably wise for each of us to allow for some evolution in our own prioritizations. We're not dealing just with moving targets but also with constantly morphing targets: the debeaked hen becomes a cage-free hen becomes a free-range hen becomes an organic pasture-raised heritage-breed hen.

We each have to prioritize according to our own values and realities, and we have to recognize the inevitability of compromises. For my family, we find the most satisfaction with the foods that we harvest ourselves. When that's not an option, we prefer what I laughingly dub "neighborganic"—producing, purchasing, or bartering for foods produced locally and in the spirit of "organic." In this case, I'm more interested in the spirit of the intent regarding organic than I am in the letter of the law. If I know the producer, the legalities are a moot point. If we are both producers, the exchange is all the richer, as we inevitably compare notes as we make the swap, and there is a mutual respect and appreciation that changes hands along with any currency involved. But as good as it is from a philosophical perspective, "neighborganic" isn't always viable. Convenience, cost, and availability often enter into the equation. You'll find me at the local food co-op and at the supermarket, too. And when I'm in my role overseeing the Sustainable Food Purchasing Initiative at Green Mountain College, I've learned that I have to use different lenses due to the difference in scale.

Local vs. Regional

For those with middle- and upper-class incomes, it's generally not all that difficult to support "buy local" campaigns and personal ideals at the home level. But when we start trying to initiate changes within institutions—schools, colleges, hospitals, senior centers, the charitable food system—certain complexities appear quickly and with some frequency. The quandary boils down to scale: scale of consumption and scale of production. Here's an example.

In September 2009, our college's dining hall staff was preparing for a celebratory banquet following our convocation ceremony. The staff and I generally work together to ensure that the meals represent our college's environmental mission and our commitment to supporting our local economy. The meal was being planned for six hundred people—not an enormous crowd, but fairly large by our small college's standards. Chefs Dave and Tom—two extraordinary white-tablecloth chefs who (fortunately for us) made their way into college food service thanks to the regular hours and reasonable vacation periods—began planning for the event and ran some ideas by me for sourcing local products. We settled on a main entrée of chicken breast and targeted the one source in Vermont that we thought could provide the quantity of organic chicken breasts we needed. Chef Tom called in the order a week or so before the event, only to discover that this largest and most reliable source of organic poultry products couldn't come close to filling the order, at least not without several weeks' advance notice. We went to Plan B (local grass-fed beef) and Plan Sea (certified sustainably harvested seafood).

We unpacked multiple local food lessons from that one occurrence: plan and order a month out; find a way to use whole chickens or diced chicken; use the more substantial supply of naturally raised turkey we have in our southern Vermont, instead of the meager supply of organic chicken; create a database of producers of local and organic products and check on the typical quantities available at different times of the year; or just produce those products on the college farm for special events.

Those were the easy lessons, and they were all incumbent on us, the institutional consumer. But they also highlight the importance of how we employ the word "scale" as it relates to local food purchasing. We may think of "scale" in terms of "local" or "regional"—in other words, how large is our defined scope when we use those words? But "scale" can also refer to the scale of production on specific farms. Does it matter how big the farm or food enterprise is when we wrap it into our embrace of local?

And then there is the scale of consumption. Local consumption is typically much easier at the household scale than it is at the institutional level. With home gardens, community-supported agriculture (CSA), food co-ops, farmers' markets, and emerging home delivery options, we can access a variety of local foods for home use relatively easily, albeit within seasonal constraints. But in many parts of the United States, there is a lack of significant diversified agriculture, and there isn't always sufficient local infrastructure in place for processing, storing, distributing, and retailing agricultural goods once the farmers have supplied them. In part, we are facing the dilemmas resulting from the losses in what a group of advocates across the country have termed the "agriculture of the middle,"[10] or midscale agriculture that can meet the needs of a country of citizens who eat away from home almost as much as at their kitchen tables. Many such advocates, myself included, maintain that we have to build out the number of ecologically minded midscale producers in the United States in order to meet our food production needs while not sacrificing (hopefully) the ecological integrity of their local landscapes.

An element of realism confronts us, then, as we look at current local food capacity and actual levels of demand. The current marketplace and paradigm have us constrained, and some compromises are necessary. One compromise that often makes sense is to look beyond the fuzzy border of "local" and assess the possibilities available "regionally," yet another concept with soft edges but a more expansive and potentially resilient core.

A common justification for putting so much effort into re-envisioning local food systems is to build "resilient" communities—communities that can withstand unexpected blows from both Mother Nature's and Adam Smith's invisible hands in the short term, and cope with the uncertainties of climate change and peak oil over the longer term. Ensuring that basic local needs can be met despite the threats of the unforeseen means reducing vulnerability, diversifying resources, and even encouraging some redundancy. Therefore, it probably makes more sense to develop *regional* food security strategies—complemented by and linked to community-based local food systems—than it does to assume that a lonesome local strategy will get us where we want to go.

I realize that such a suggestion does not always sit well among all locavores with strong purist streaks. But when we start discussing accessibility, affordability, and resilience, our perspectives must broaden. We have to begin thinking about complex questions like "How far is too far?" with regard to the particular challenges of different geographical areas. Across the United States, ecological, demographic, and economic constraints arise in different combinations. For example, the sparse population densities and lack of agricultural diversity in areas such as the prairie states create market-based challenges for tightly defined local food markets. The southwestern states face serious obstacles to diverse agricultural production due to heat, soils, and water. Increasing the year-round production of a variety of products in close proximity to the dense urban and suburban populations in the cold and damp northern-tier states raises a whole different array of issues. In all cases, we also have to think about what people can pay for items for which there is a limited supply—with the frequent and frustrating result often being higher prices for local products than for the same products sourced from elsewhere.

And then we have to think again about resilience. What happens when disaster strikes and a local community or a whole food-producing region is devastated by floods (Hurricane Katrina in 2005), drought (southern United States in 2011 and two-thirds of the United States in 2012), long-term droughts (Horn of Africa in 2011), an oil spill (Gulf Coast states in 2010), or even radioactive contamination (Fukushima in 2011)? Strong community food systems are the foundation of food security, but it is wise to remember that well-managed regional, national, and international food systems currently contribute to a diverse food security portfolio for all global citizens.

Chapter 3

How Far Should Local Go?

"Local" is shorthand for community-based food systems.

—Ken Meter

I was on an Austrian train headed into Italy in the spring of 1984. The European borders at that time were still strictly regulated, and the narrow Brenner Pass just south of Innsbruck was well known not only as a historically important gateway between southern and northern Europe but also as a modern bottleneck for travel and trade. My father and I were traveling together, and I had forewarned him of the tedious delays at the border as the Italian and Austrian customs officials stowed away any charitable sensibilities, inflated their chests, and began their suspicious sweeps of the railcars.

The train screeched to a slow halt, such that our car was just beside a building that seemed to bulge with blue-suited Italian rail employees and customs officials. We lowered our windows to get some fresh air during the stop, only to regret it as we saw the office door burst open with an outpouring of uniforms and a cloud of cigarette smoke tinged with the essence of espresso. Two shrill whistles pierced the air, and the customs officials rattled off a barrage of inflected instructions. For some reason, they went straight to the baggage car and offloaded a large box almost big enough for a small refrigerator. We watched as the officials inspected the box warily but with caffeine-inspired animation. One officer gave the signal and another extracted a razor blade from his coat pocket and sliced open the box at its seam. A whole slew of small triangular boxes cascaded out onto the train platform. To this day, I still wonder what was in those boxes, but whatever it was apparently satisfied the lead customs inspector after he opened one. In true form, he signaled the others to repack the box, and he left to fill out the paperwork or just have another coffee.

The remaining officers looked at one another in mild disbelief and, given their gesticulations in his departing direction, were clearly denouncing the absurdity of the whole scene and their superior's stupidity in creating such a mess. More and more heads began to peer out the train windows as the comedy unfolded. The customs officials struggled for more than twenty minutes to get the triangles repacked just perfectly so that they could seal the box, load it back up, and send the train on its way.

No matter what they did, the officials couldn't get the correct configuration to pack those little triangles back into the big box, despite the suggestions, catcalls, and curses from the passengers hanging out of the train windows. Finally, facing increasing pressure from the agitated train crew and passengers, one customs official ran into an office and grabbed an even larger box, an enormous container that allowed the officials to cram all of the little triangular boxes inside without any problem.

In a sense, we face a similar scenario as we constantly struggle to create and repackage food systems that make sense. This task is nothing new. Ever since we stepped into the precarious shelter of domestication, we humans have struggled to create food systems that could sustain us—all of us—without unraveling the delicate ecological fabric that ensures future harvests. In the process, we have tried to pack our food-related values into tightly packaged words, words meant to capture and contain our intent. Inevitably, at some point, someone decides to open up the tidy box that holds our reigning assumptions and values, and the contents spill out all over the floor. We're left to find a way to repackage the driving assumptions and values in a different container and then send them on to their original destination, configured in a manner we didn't necessarily expect but still on the right track.

You could say that it's all about local motion.

Moving Words

We're obligated to think hard about food systems. Probably the most complex conundrum of the twenty-first century is how we are going to feed our human population—and it's a question that pulls us in multiple directions: the efficiencies of small scale versus large scale; a fair price for farmers and affordable prices for consumers; the optimism of scientific advances contrasted with gross miscalculations of our ability to manipulate not only massive ecosystems but also tiny pieces of DNA; the need to relocalize while maintaining a global perspective.

We're obligated to think hard about words, too. Anyone who has lived a few decades has seen how quickly movements in the guise of words come and go. It's one thing to see a word wane if you think the objectives of the "movement" have been accomplished in large part. But if an important cause disappears because we get bored, impatient, or disillusioned with a word ("organic" comes immediately to mind), it feels disappointing, perhaps even tragic. The word is attacked and minimized while the bulk of the work has yet to be completed.

When it comes to contemporary food and agriculture, we're highly dependent upon words. Words are our tools for dissecting food systems and for strategizing how best to move forward. We also charge them with the task of carrying and conveying values. A common language about food and agriculture assures us that common ground is also possible. And we tend to change our vocabulary fairly frequently. With those changes in vocabulary come changes in presuppositions and values, often accompanied by shifts in food production methods.

Think of how many words we use to describe the various parts and processes associated with our food system: local, organic, sustainable, resilient, regenerative, community-based, conventional, industrial, humane, grass-fed, cage-free, genetically modified. There is a progression to our vocabulary—a progression related to scientific understanding, ecological realizations, community needs, workers' rights, animal welfare, marketplace opportunities, economic shifts, technological advances. Sometimes our values are in tow behind these developments; other times, our values drive the change.

There are two primary ways of viewing our continually evolving understanding and articulation of what comprises "good" or "appropriate" food and agriculture. A simpler, knee-jerk and somewhat cynical reaction is to see it all as relative and fickle and therefore inconsequential. Another way is to acknowledge that our understanding of the positive and negative impacts of our food system on ecosystems, human health, justice, and the economy is constantly expanding and changing. We are therefore required to be intellectually engaged, open-minded, and committed to the difficult and unending work of creating better food systems—systems that we "know" (given our evolving understandings of an ever-shifting landscape) offer better ways of producing, processing, storing, distributing, preparing, consuming, and recapturing food resources.

Grassroots initiatives and the media have cast the alternative paradigm to a global, industrialized food system as "local." But if we believe in a changing world and a dire need for resilient solutions, then part of our preparations must include a readiness to move beyond our current understandings and vocabulary at some point in the future. While many of the core ideals may be retained, it is virtually certain that as we move forward, our current concepts of successful local food systems will be repackaged as part of our adaptation to an increasingly dynamic and globalized food system.

The Local Trap

If you're like me and you find your eyesight slowly getting worse from too much time at the computer, your optometrist will tell you to look up frequently from the computer screen and focus on objects at varying distances in order to exercise your eyes. The same is true for local food system advocates: we need to remember to look up and into the distance to maintain a healthy focus.

It's easy to forget to pause and look up when it's so clear that smaller-scale, locally based food systems are the most beleaguered of all the scales and they need our constant attention. That makes it too easy to forget that working to reconstruct local food systems is but one part of creating just, ecologically appropriate, and resilient food systems for all citizens of the world, not just for our own communities.

And there is also the danger of what some scholars have dubbed "the local trap." According to urban planners Chris Brown at the University of Kansas and Branden Born and Mark Purcell at the University of Washington, we cannot make the claim that local food systems are inherently of higher value than the other food systems simply due to scale or location. In their view, the actual driving values of a food system of any scale can be assessed only by means of a thorough examination of "the actors and their agendas"—that is, the people making decisions about food and the things they care (and don't care) about.[1] Unfortunately, the prevalent temptation among many local food systems advocates is to automatically equate scale with laudable values.

In some ways, what we are trying to do with this wordsmithing is understand the distribution of power and resources within the various scales of the food system, from local to global. An example helps demonstrate just how complicated the realities are. In several distressed U.S. cities, serious questions and allegations are being raised about "land grabs" by wealthy individuals who are purportedly acquiring cheap land for urban agricultural ventures under the banner of local food security. Community members who have been promoting local urban agriculture in these areas feel that these large, capital-intensive ventures are beginning to commandeer land, markets, and publicity—and therefore the potential for equitable food access—for profit and not necessarily for the betterment of the entire local community, particularly the poor and minorities.

And then there is the question of our commitment to essential humanitarian values on a global scale. Can we so readily dismiss the importance of the global food system in providing quick and massive disaster relief to populations devastated by natural disasters or war? Undoubtedly, we want all communities to be able to feed themselves to the greatest degree possible, but self-reliance can be a fragile if not utterly impossible proposition in many parts of the world, including areas of the United States.

The beauty of pushing for stronger localized food systems across the globe is that we are not just talking about *changing* "actors and agendas," but we are actually working to *increase* the number and the diversity of actors and agendas. At the same time we're also increasing the transparency of what the actors are actually doing, helped in large part by the fast-paced democratizing effects of the Internet. The good news is that not only are more actors appearing on the world food and agriculture stage, but there seems to be an increased interest in the health and well-being of individuals, communities, and landscapes.

Can we build and support smaller-scale, locally oriented food systems that are more likely to be just, ecologically appropriate, accessible, and resilient than food systems of larger scales? Although it puts us on the brink of the "local trap," I would like to think so. It's easier to be both informed and vigilant on the local level than it is at broader and less familiar scales. We can get our brains and our arms around local food systems—and probably also regional food systems. It's common sense that we can best steward the places, the communities, and the economic relationships that we best understand, access, and explain. We also become empowered in the process. Not only are we potentially reshifting the balance of power, but we are helping to heal the ills of distance, anonymity, and transience that have alienated us from our life source—food.

All of the presumed benefits of localized food systems depend, however, upon the composition of "we." A gathering of locals is not inherently either good for everyone or representative of the community. We should always take into account the full slate of local actors and their agendas, even when we ourselves are in the mix. Who is at the table? Who is not? Active representation from a broad cross-section of the community is critical to ensuring a positive outcome for the entire community, not just an elite few. Discriminating tastes can become discriminating viewpoints.

In the end, it's not just about where the food was produced. We must also bear in mind the impacts of its production, processing, storage, distribution, marketing, preparation, and even reclamation. *Where* matters immensely in the food system world, but so do *how, why, by whom*, and *for whom*. How is the food being raised? Why are those the chosen methods of production? Who is doing the real work in getting the food from farm gate to dinner plate? And is it ultimately food for all or just a select few? The "Drivers" part of this book tackles these questions in detail.

Sound Bites and Unsound Bytes

As we've made the case for reclaiming our food system, the necessary "sound bites" about innovative urban gardens and delicious heirloom vegetables have at times been "unsound bytes." You can tweet about a local food product in 140 characters, but there's not much complexity that can be conveyed about local food *systems* with such constraints.

Fortunately, there's no shortage of characters in our current push for relocalizing food systems. We have farmers, chefs, nutritionists, and entrepreneurs on board—even First Lady Michelle Obama broke new ground with her influential White House vegetable garden. Facts, figures, and local food logic come at us almost regularly these days through feature-length films, TED talks, news blips, pithy posters with the top ten reasons to eat locally, and questionnaires to determine whether you are a locavore or a globavore. In many ways, the recent intensity of interest in local foods can be attributed to "buy local" lingo that (thankfully) seems to have pervaded most of the country, channeled through a variety of media and personalities. However, the likelihood that we can transform the present fascination with local foods into enduring community-based food systems depends upon our persistence in calling for deeper and more complex conversations, not only in our media but also in our neighborhoods, economic development agencies, and state and federal governments.

If the media sound bites move us to dig deeper, all the better. But if we don't go beyond the blips and blurbs of fast facts and simplified justifications, we may never get to the necessary work of constructing infrastructure and systems that can enduringly feed our bodies, our minds, and our communities.

We live in a messaging environment that waxes between numbers and narratives: data that drive opinions, and decisions and stories that become the beacons of our local food dreams. Both the numbers and the narratives are important—and we must constantly assess the validity of the information and examine the assumptions upon which various studies and figures are based. Building a solid foundation for local food infrastructures and systems will ultimately require the best information possible for our thinking and planning.

Local Food Moment, Momentum, and Movement

We seem to be living in a local food *moment*. The key challenge for us is to capture this local food *momentum* and transform it into ideas, investments, and infrastructure. Despite all of that, there may be reason to be skeptical of a local food *movement*.

I've lived through a period in which any proposals for an alternative food system seemed wacky if not downright unpatriotic, and food and agriculture were deemed "anti-intellectual" enterprises (such was the phrase used to portray my efforts at introducing food and agriculture into a liberal arts curriculum years ago, and similar prejudices still persist at some universities with more hubris than humus). Given that, I am ecstatic to be living in a time when we are not only openly questioning the trajectory of our food and agriculture systems but also seeing these questions pervade virtually every region and socioeconomic strata in our country. Knowing how dark and frustrating it seemed just a few years ago, when farming was supposedly what you did because you didn't know any better, and food was what unenlightened and enslaved women did, it's a joy to see food and agriculture once again become points of serious conversation and intellectual engagement here in the United States.

But as the interest level surges, is a local food movement really what the world needs? I'm not so sure. With such radically different bioregions spread across the United States, the enormous variety of cultural influences on food systems, the distinctively varied agricultural sectors driving regional economies, the almost unfathomable variances in population densities, and the radically different dilemmas and resources facing urban and rural communities, we should question any "one size fits all" movement. Any attempt to put a bunch of individually shaped pegs into a big square hole almost certainly means that specific community approaches and priorities will get lost in the mix. If the rationales and strategies for building local food systems become too formulaic, they risk losing their relevance to specific communities' needs, interests, and resources—and the efforts are unlikely to endure. Furthermore, due to the dynamic nature of the food and agriculture sectors, the longevity of any given locally oriented system will depend not on its hardened facade but on its ability to transform and adapt.

On the other hand, there is power in numbers and there is reason to band together not with a singular voice but with the power of a chorus in order to achieve certain goals that a bunch of small potatoes in a French-fry nation can't impact. There are many policy issues for which strong voices in favor of rebuilding locally based food systems are badly needed, some of which will be discussed later in this book. Primary among these issues is the recurring national Farm Bill, which sends funding and associated mandates to agencies, states, communities, and institutions for programs that are either integrally linked to local food efforts or in diametric opposition to such efforts. Another example is health and sanitation regulations, which, even when well intentioned, often create unnecessary market-entry barriers for new concepts and products and threaten existing local food businesses and practices. These regulations tend to be developed in response to needs and concerns related to large-scale production, ultimately posing significant challenges to smaller-scale producers and processors.

At the international level, an alliance of local food underdogs can bring important voices to the table (or to the streets) when trading practices and regulatory policies are discussed and created at the perennial global summits. Furthermore, trading stories of the local dilemmas faced in different parts of the world can yield some creative insights into tackling economic, regulatory, cultural, and environmental challenges. Slow Food International has probably set the bar for the most creative and celebratory integration of "grassroots" and "global" through its biannual gathering of food communities and producers. Dubbed "Terra Madre," this enormous gathering highlights, celebrates, and fortifies the cultural centerpieces of community-based food systems around the globe. Slow Food's dual emphasis on local autonomy and global exchange is exemplary in a polarized era of local versus global.

Ultimately, the greatest concern in defining this moment and all of its momentum as a local food movement is that movements can be both exclusive and short-lived. Resilient community-based food systems can be neither, and reestablishing such systems is not simple or whimsical work. *Rebuilding a local food system is the work of a lifetime, and the vigilance required to sustain it is the task of generations.*

Part Two

DRIVERS FOR REBUILDING LOCAL FOOD SYSTEMS

Chapter 4

Energy

Food is energy. Food provides energy. Food requires energy. Food and energy are virtually synonymous. They even share a common unit of measure. But that doesn't mean that they are in balance. To the contrary. And nowhere is that imbalance more evident than in the United States.

As soon as one opens wide and espouses the need for a food system that's balanced in terms of health, equity, and ecology, it becomes apparent that much of the discussion is about how to extract one's ecological footprint from one's mouth. The problem is that, in terms of energy, our ecological footprints are estimated to be somewhere between seven and ten times the size of our mouths. In other words, it takes seven to ten calories to produce and deliver the equivalent of a single calorie of food in the United States.[1] These food system calories eventually add up to an estimated 19 percent of America's total energy consumption.[2] (It is important to note here that we typically measure calories in our diet as a "small calorie," the amount of energy needed to raise one gram of water one degree Celsius. When we measure energy on a larger scale, we call it a "kilocalorie" or a "large calorie" and denote it with a capital *C*, as in "Calorie," since it is defined as the amount of energy needed to raise one *kilogram* of water one degree Celsius.)

Do we simply go retro? Techno? Heck, no. A total historical reversal to preindustrial conditions is just as unlikely as a technological absolution for our modern-day petroleum-based gluttony.

The energy behind human civilizations was once a product of the food supply. But we are at a point in human history in which food is predominately a result of nonhuman energy inputs. The prospect of bringing food and energy closer to a one-to-one ratio of *calories invested* to *calories derived* is extraordinarily complex, and it has direct links to the call for creating more sustainable and resilient food systems. Today in the United States, these food and energy questions comprise a quandary that most of us can ponder in relative comfort, without the imminent threat of being unable to feed ourselves due to costs, energy constraints, or shortages. And yet, even as we relish the extraordinarily low cost of food in the United States, certain threats do lurk in the background. The energy supply that feeds our food system is at short-term risk of disruption by natural disasters, international conflict, and economic turmoil. The long-term impacts of worsening climate change, dwindling petroleum supplies, and increasing global population pressures are looming realities that we may try to ignore but ultimately cannot avoid. We have already seen how spikes in food prices can create social unrest with the seeming velocity of the flick of a match.

Such inquiries into food security should not be viewed as mere intellectual exercises or myopic self-preservation interests. Perhaps the most compelling reasons to grapple with our precarious food/energy imbalance are sheer justice and altruism.[3] People who are "food insecure"[4] are generally far too busy trying to convert their own personal energy into food dollars to spend much time researching and thinking about the national food and energy dilemma. The onus is upon those who are concerned enough to care and are able to do something about it. As actor Alan Alda once said during a graduation speech to a group of medical students, "The head bone is connected to the heart bone—don't let them come apart."[5]

Energy Fields

I am an optimist and a good-natured (I hope) skeptic. But from my vantage point as a farmer and an academic, few things worry me more about the human condition than the intertwined fragilities of our food and energy supplies—and our habits that exacerbate the amount of energy consumed between farm and fork.

I struggle to make sense of the food/energy dilemma most every day, although I would by no means characterize those days as gloomy or my attitude as morose. Rather, my days tend to be filled with sunshine, pastoral landscapes, solar panels, healthy livestock, laughing children, and inquisitive students. But the energy-to-food ratio is a constant theme, starting with the morning milking on our off-the-grid farm. Our grass-fed herd of American Milking Devon cattle get either fresh pasture or good-quality hay every morning—no grain, but plenty of gain. The milk pails are washed with solar-heated hot water while the early morning lights in the house are powered by yesterday's sunshine. (We are almost entirely solar-powered, with fossil-fuel backups providing about 20 percent of the additional energy we need.) We'll use one of our two Kubota tractors to do the morning's heavy lifting or towing, but the goal is to use them as little as possible and, when feasible, not at all.

When the chores are completed, by me or often by one of our apprentices, I admittedly leave home in a gas-guzzling four-wheel-drive vehicle and head out sixteen miles to my job at Green Mountain College, where I oversee the college's Farm & Food Project. As I pull up, students are usually walking to and from the farmhouse and the various outbuildings that comprise the college farm complex, often toting milk pails or vegetable bins as they wrap up morning chores there. Their farm—and it *is* theirs in many ways—is much like mine at home, an experiment in trying to minimize energy inputs and maximize food output. However, their work is more rigorous in its analytical aspects, thanks to the research oversight headed by my colleague Kenneth Mulder, one of the few PhDs in the United States who is also an expert at using oxen in agriculture.

The farm's focus is to probe ways toward a food system that eschews fossil fuels as much as possible—and indeed, all of the activities on the farm seem to orbit the question of our overblown American diet. Draft animal equipment, photovoltaic panels, a solar hot water system, greenhouses, ergonomic hand tools, and bike tractors dot the farm. Students' experiences with these techniques and technologies contrast sharply with the predominant realities of our current food system, which has us guzzling kilocalories of diesel energy in our tractors and gorging on excessive calories of food energy from our kitchens.

My favorite view from my office window in the second floor of a restored farmhouse is the summer scene of the oxen cutting and bringing in the hay for their winter ruminations. Other days, I gaze out the window and watch Kenneth and the students work in the vegetable fields that are his research plots. He has divided the vegetable production into three plots, each powered

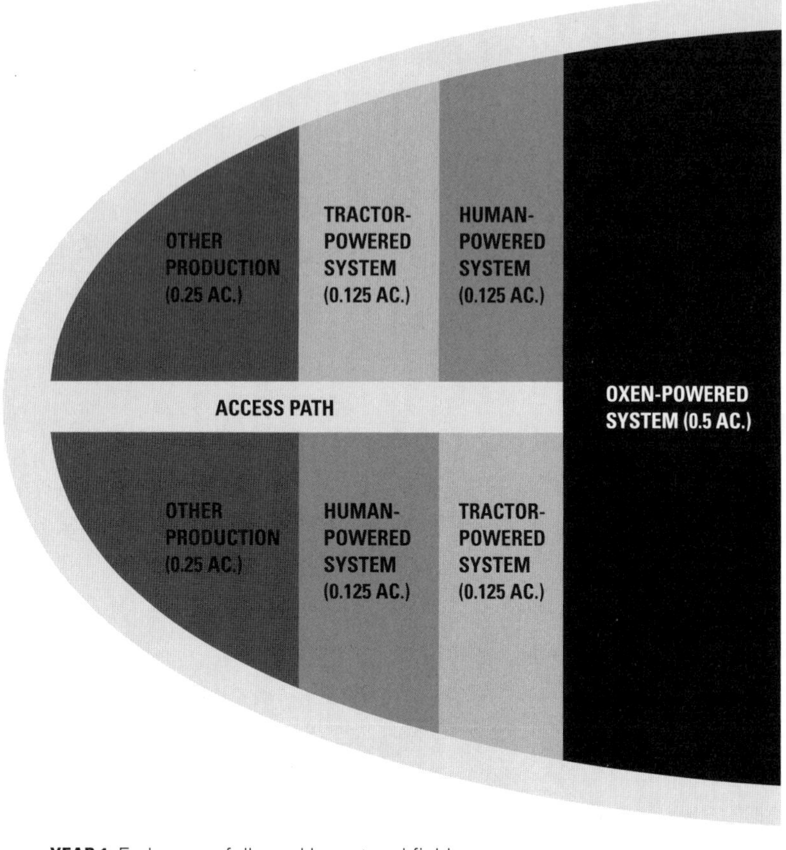

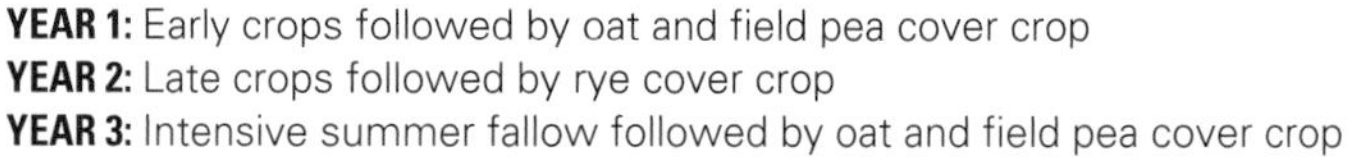

FIGURE 4-1. Long-Term Ecological Assessment of Farming Systems (LEAFS) Research Plots at Green Mountain College. Image courtesy of Kenneth Mulder, Green Mountain College.

by a different system (see fig. 4-1). The easternmost section is cultivated, planted, maintained, and harvested exclusively by human power and the use of highly efficient hand tools. The middle section relies upon a combination of human power and a BCS walking tractor, essentially a highly versatile tiller with a variety of implements ranging from a sickle bar mower to a potato harvester. The western plot catches the most attention, as it is the market garden section powered primarily by the oxen and their accoutrement of fancy new (yes, generally new, and also quite efficient) tillage equipment.

FIGURE 4-2. Oxen Math. Used by permission of Bethanne Elion.

This research project, dubbed LEAFS (Long-Term Ecological Assessment of Farming Systems), is Kenneth's brainchild, a means of evaluating all of the energy inputs and outputs within each system. The goal is to develop a database of ten years of experimentation in order to discover the energy requirements of each system and to assess its efficiencies and challenges.

One of the more amusing aspects of it all is watching students work with stopwatches and scales in order to monitor their own energy inputs and each plot's productivity. Even the energy expended by the oxen in pulling different pieces of equipment is measured by means of a dynamometer, a device placed between draft animals and any load that they pull as part of a task on the farm. The dynamometer sends a signal to a computer in the oxen-driver's backpack, indicating precisely how much energy the oxen are exerting every second. This information is then transferred to a Google Earth map so that the oxen energy can be recorded both in joules (a unit of energy) and on a map that details the different levels of energy expended on certain tasks and in specific locations.

Efficiencies can also be measured in a variety of ways. For this long-term ecological study, Kenneth has opted to analyze efficiency in terms of labor, land, and energy, and his figures are based on wholesale organic vegetable

TABLE 4-1. Land, Labor, and Energy Efficiency of Three Methods of Organic Vegetable Production in the LEAFS Research (Year 1)

	LABOR EFFICIENCY		LAND EFFICIENCY		ENERGY EFFICIENCY	
	kg/hr	$/hr	kg/ha	$/ha	kg/MJ	EROEI*
Human	6.60	$22.94	16,253	$56,497	4.30	5.1
Walking Tractor	8.84	$30.79	15,365	$53,503	1.61	2.3
Oxen-CS**	6.63	$23.86	3,098	$11,144	3.49	7.0
Oxen-D***	7.05	$25.38	7,907	$28,445	3.23	6.5

Source: Kenneth Mulder and Benjamin Dube, 2011 LEAFS Data Analysis

Note: This table describes the kilograms of organic vegetables and their wholesale value in terms of time (hours), space (hectares), and energy (megajoules) invested.

* EROEI: energy return on energy invested

** Oxen-CS (oxen combined system): This analysis considers the oxen feeding system to be part of the system, thereby increasing the need for land and labor to maintain the animals.

*** Oxen-D (oxen direct): This analysis does not include the human labor or land needed to feed the oxen.

prices (see table 4-1). It is interesting to note that the energy efficiency (measured as energy return on energy invested, or EROEI) of all four calculations ranges from 2.3 to 7.0, which is significantly higher than the range of 0.26 to 1.6 that is typical for conventional vegetable production in the United States.[6]

Trade-offs are inevitable in farm management systems, but seldom do aspiring farmers get to test out the practicalities of different systems, much less measure them with the sophistication provided by Kenneth's expertise. The most elusive variable is energy, but it is arguably the one that currently warrants the most scrutiny.

The farm is the natural starting point for rectifying the imbalance between inputs and outputs, but if we are truly seeking balance in our food system, we must also assess the basic energy parameters that frame our daily decisions as consumers. In doing so, most of us gravitate immediately to the production and distribution aspects of our food system. Granted, those are critical components to tackle. However, food production and distribution often seem a bit beyond the scope of control for the average person, and—somewhat contrary to our recent intense focus on food miles—the transportation portion of our energy diet is actually relatively small in comparison to other parts of the food system that are based upon and driven by consumer choices and household habits.

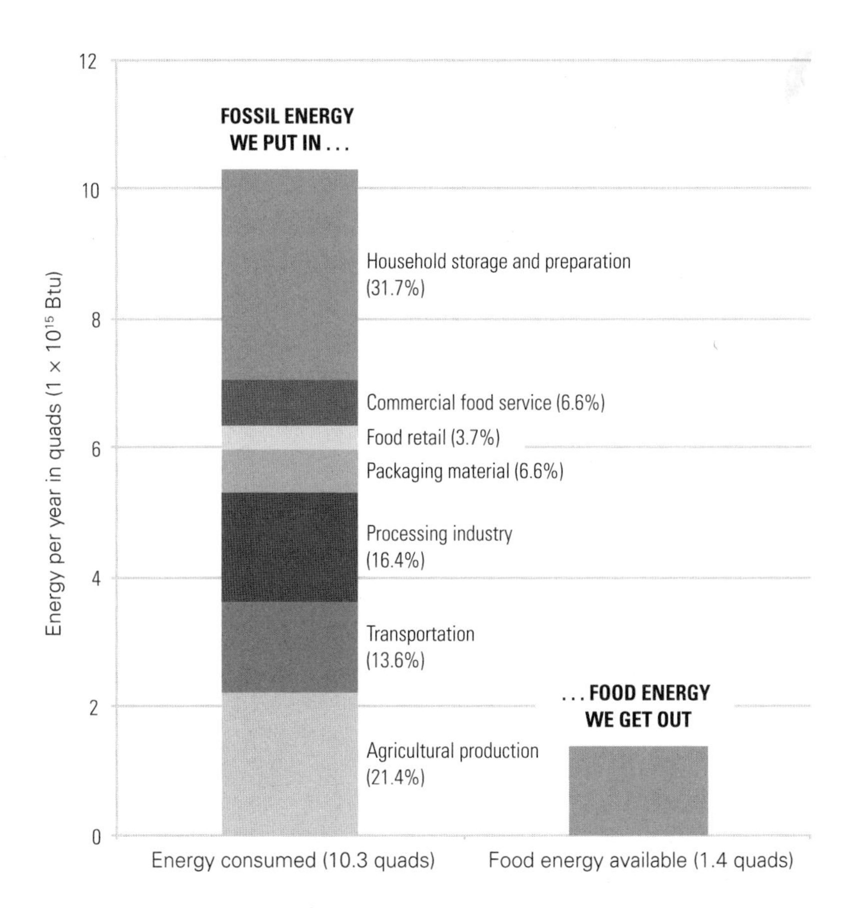

FIGURE 4-3. Energy Flow in the U.S. Food System. Source: *U.S. Food System Factsheet*, pub. no. CSS01-06 (Center for Sustainable Systems, University of Michigan, 2011).

As it turns out, the elements of the food system most within our control often tend to be those parts of the system that are closest to home, and they are also among the most energy-consumptive components found between farm and fork. The food and energy decisions we make *in* and *near* the home have the greatest impact on our personal energy-to-food ratios (see fig. 4-3).[7] Household storage and preparation represent the largest single sector of energy use in the entire food system. When it comes to energy issues and food systems, "local" starts to become quite personal.

In order for the food and energy dilemma to really hit home, so to speak, it helps to remember that every step in the farm-to-plate process increases total energy inputs, making food waste an issue that we can ill afford to toss casually aside. As we work our way through the food chain, it will become increasingly obvious why reducing waste is such a critical link in creating resilient local food systems.

Food Production . . . but Let's Call It "Farming"

Farming is about energy flows. "Food production" is about a terminal point in the act of agriculture.

As soon as we begin using the word "farming" again, all of the implicit associations with farming begin to reemerge in our shared thoughts and language—planting and harvesting seasons, the cumulative wealth of generations, the farmscape, the role of farms within the community. Suddenly, place, time, and the stewardship of inherited traditions all start to become important again. We can quantify production, but we can qualify farming. That is to say, we can instill it with values, not just interpret it through metrics.

These days, when we attribute values that we think define good ecological farming practices, we generally tend to speak of "sustainable farming." Sustainable farming is the careful management of energy flows, not just to get to one final product at the end of a season but to ensure that those energy flows allow for a regenerative use of the land far into the future.

What better way to think about food and farming than through a pie chart? Figure 4-4 clearly depicts the sticky fingers in the energy pie at this point in time. It's obvious: we are overly dependent upon photosynthesis that occurred millions of years ago to fuel the growth of our modern farms through fossil fuels. It's also worth noting that fertilizer and pesticide production is heavily dependent upon fossil fuels, not just for powering the manufacturing process but as primary components for many of those products.

These on-farm energy consumption figures are staggering in their enormity. Sadly, renewable energy sources do not even appear in the data, although there is a small renewable energy component in the electricity sector. Trying to envision how little ol' *local* can begin to confront the farming sector's entrenched dependence on fossil fuels is challenging even for the best of optimists.

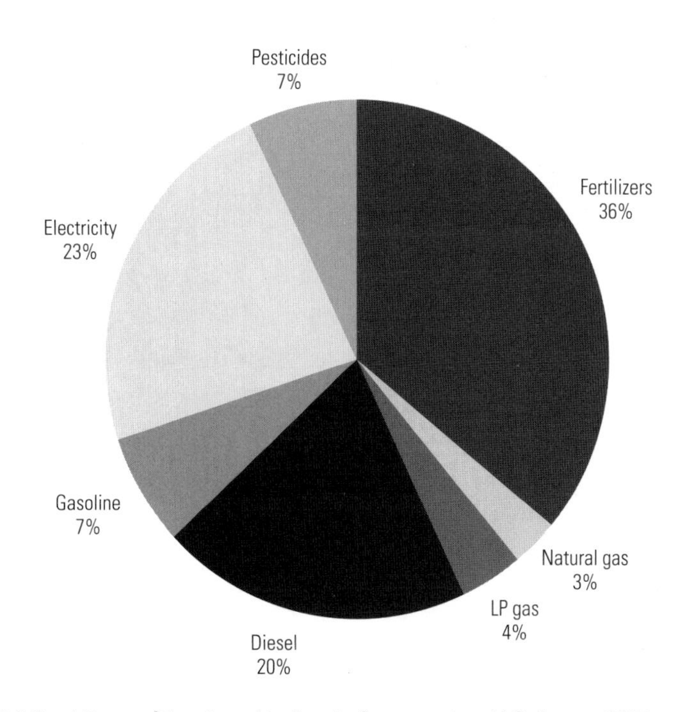

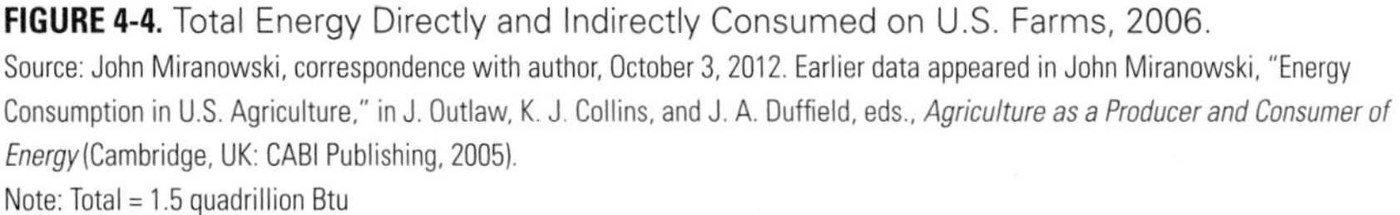

FIGURE 4-4. Total Energy Directly and Indirectly Consumed on U.S. Farms, 2006.
Source: John Miranowski, correspondence with author, October 3, 2012. Earlier data appeared in John Miranowski, "Energy Consumption in U.S. Agriculture," in J. Outlaw, K. J. Collins, and J. A. Duffield, eds., *Agriculture as a Producer and Consumer of Energy* (Cambridge, UK: CABI Publishing, 2005).
Note: Total = 1.5 quadrillion Btu

It's worth considering the growing vulnerabilities in the current system (although this is more an exploration of potential opportunity than an exercise in optimism). Most farmers in the United States are "price takers." In other words, these farmers cannot quickly change their asking price for their products when energy prices surge. Rather, they are all too often in the position of taking the going market price, perhaps with consternation but ultimately little recourse. This is particularly true in national and international wholesale and commodity markets in which individual farmers have almost no leverage; the smooth and invisible hands of supply and demand almost always trump the calloused hands of any farmers in those expansive markets. Farmers selling in direct local markets still have some control in aligning asking price with energy costs (sometimes by way of face-to-face dialogue with buyers). Local markets, therefore, can give farmers an economic edge amid volatile energy prices. Nonetheless, the energy consumption problem remains largely unaddressed. The burning question lingers: Can the rebuilding of local food systems help minimize on-farm energy usage as well as promote renewable energy sources?

In this case, "perhaps" may be a better answer than a definitive "yes." Typically, local food systems can better nurture smaller-scale agriculture than can national and global systems. And much of the energy use on farms is a direct result of scale and farming methods. Small-scale agriculture tends to require fewer fossil-fuel inputs than large-scale agriculture. When biointensive methods are utilized, small-scale agriculture also tends to be more productive on a per-acre basis.[8] Of course, any such change in shifting toward human-powered systems means that we need a dramatic increase in the number of farmers, as well as supporting infrastructure in each sector of the food system—appropriately scaled infrastructure that is economically viable.

However, before we risk becoming utterly small-minded and completely lose focus on the bigger picture, we have to think bigger about ways to maintain and rebuild what sustainable agriculture expert Fred Kirschenmann and colleagues call the "agriculture of the middle." Without it, local supply and demand for most agricultural products in virtually all regions of the country will remain out of balance. For Kirschenmann and his colleagues, agriculture of the middle "encompasses a spectrum of farms and ranches that are declining because they are too small to be served well by commodity markets and too large to be served well by direct markets. Most Agriculture of the Middle farms are characterized by: (1) their size; (2) their business organization; and (3) the production and marketing strategies they adopt to remain viable."[9] In other words, these midscale farms are concentrating on producing significant amounts of food—too much to direct-market but too little to sell as commodities—utilizing farming methods that demonstrate a sensitivity to ecological concerns, animal welfare, and community well-being. Imagine a grass-fed beef farm with several hundred cattle, an organic grain operation of several hundred acres, or a farm raising three hundred heritage-breed pigs. Each of these operations is not only midscale but also focused on certain explicit values related to its farming practices—values that have traction in the marketplace.

But can midscale agriculture help mitigate energy demands in the on-farm sector? Perhaps. Kirschenmann and his colleagues feel the primary economic solution for midsize farms is to participate in "values-based food supply chains." In other words, their marketing should be based in large part on the methods they use to produce their goods, organically and otherwise. Organic farming methods are almost always more energy efficient than conventional methods, regardless of scale, simply due to reduced fertilizer and pesticide use. The elimination of synthetic fertilizers and pesticides would shave off about one-third of American farming's energy inputs.

There can be energy trade-offs in going organic, however. Vegetable, grain, and fruit operations may eliminate many or all of their off-farm fertilizer and pesticide inputs, but a portion of those energy savings is sometimes partially compromised by the necessity of replacing these field applications of inputs with increased fieldwork. Not using herbicides often means that a farmer must cultivate the soil frequently in order to keep aggressive weeds at bay. Giving up the spraying of particularly potent fungicides and pesticides in an orchard can require the farmer to spray organically approved substances such as copper- and sulfur-based sprays with increased frequency, since the protective effects from these less harmful substances often don't last as long as their conventional counterparts. Using a tractor for additional cultivation, mowing, or spraying can diminish the anticipated energy savings, but there are also obvious resilience-building benefits with shifting to organic management strategies.[10]

What a farmer does to decrease his or her energy consumption is often quite hidden from the ordinary consumer. However, within a local market, energy-conservation efforts and investments on behalf of the farmer can be highlighted. The hallmark of the current local food push in the United States is undeniably *awareness*: awareness of the food we eat, the farms we support, the communities we inhabit, the investments we make. A farmer focused on energy conservation has some serious bragging rights. While the farmer may not have a bully pulpit in the national marketplace to convey his or her energy-conscious advances, local marketing does allow the farmer the opportunity to demonstrate forward-thinking commitments by means of farmers' market displays, farm tours, school presentations, and other such venues.

It can be a challenge to inform consumers just what measures a farm is taking to reduce its own energy consumption. Sound bites, labels, and certifications simply don't tell the tale. Part of the farmer's task is to convince consumers to invest not just in the farm but in its decision making. If consumers are willing to patronize a farm based in part on its approach to energy use, they can help support the farmer's transition to less energy-intensive farming methods. In most cases, energy-conservation efforts on the farm represent cost savings over the long term. As one farmer makes the change and can demonstrate the benefits, that decision will reverberate throughout the local farming community, encouraging other farmers to take similar measures.

The other approach to this energy and agriculture question is perhaps the simplest: the YIMBY (yes, in my backyard!) approach. There's little wasted fossil-fuel energy in home gardens, balcony vegetables, backyard livestock, community gardens, and tightly designed homesteads. While it may require some energy investment on an individual basis, it's a good reminder: the more balanced our lives, the more balanced our diets.

Food Movement: Wheels, Water, Rail, and Air

Distribution drives the food system. It's currently the fastest-shifting sector in our food system, in part because it's market-driven and steered by technological development. When we don our energy lenses, we quickly see that a local farmer bringing food to the local market in a pickup truck is efficient only if the distance between the farm and the market is very short. The more a food transportation vehicle is built for moving large amounts of food across significant expanses, the more energy-efficient that vehicle tends to be (see fig. 4-5). Ironically for the local food advocate, transportation efficiency actually peaks with oceangoing ships, declines slightly with rail and more so with diesel trucks, and finally diminishes with the iconic farmers' market pickup. It's only in compact cities and small rural towns that energy efficiency can jump back up with the farmer back in the driver's seat—or rather, the saddle. Bicycle transport wins the efficiency game in linking local farms to consumers right in the neighborhood.

Rebuilding local food systems faces two contrasting challenges with transportation:

- Underdeveloped expertise and infrastructure for moving local food from the farm to the local consumer efficiently and cost-effectively.

- The daunting influx of inexpensive foods from distant places that overwhelm the nascent local food marketplace.

Inherent in both challenges is, again, the issue of scale: scale of production, scale of processing, scale of distribution, and certainly scale of purchasing, all of which impact product availability and price. When we hit the distribution issue, we see the challenge of taking on the Goliaths of the food world with nothing but a slingshot in the form of a pickup truck.

Local food initiatives have to take the physical movement of food seriously. From both an energy perspective and a consideration of an appropriate economy of scale, distribution is one of the greatest hurdles. Distribution is fraught with complexity, and generalizations about it seldom work. At the risk of oversimplification, much of the early activist work in rebuilding local food systems was based on the premise that we needed to get people to local food by way of farmers' markets and farm-based CSAs. Now, as the reconstruction efforts mature, we are realizing that we also need to get local food to people; and not surprisingly, it turns out that it's generally much more efficient to use a delivery vehicle to get food to people rather than having a multitude of cars converging on a farmstand or a CSA operation.

The U.S. food system is increasingly dependent on the use of freight services, that is, ships, trains, and large trucks. The good news is that freight services have, in fact, achieved some important efficiencies over the past few decades, but those advances are somewhat overshadowed by the fact that shipping distances for all food commodity categories continue to grow. Poultry, egg products, fresh fruits, and fresh vegetables are all showing signs of increased energy requirements in the distribution sector.[11] Sadly, given their nutritional benefits, the perishable nature of fresh fruits and vegetables means that shipping them requires more than two times the energy necessary for transporting all other types of food.[12]

Nonetheless, local and regional food systems advocates across the country are working passionately and quickly to come up with innovative new approaches to distribution that can overcome some of these barriers. It's worth briefly exploring several of the newer approaches that seem to hold promise for building energy-efficient local food systems in the distribution sector.

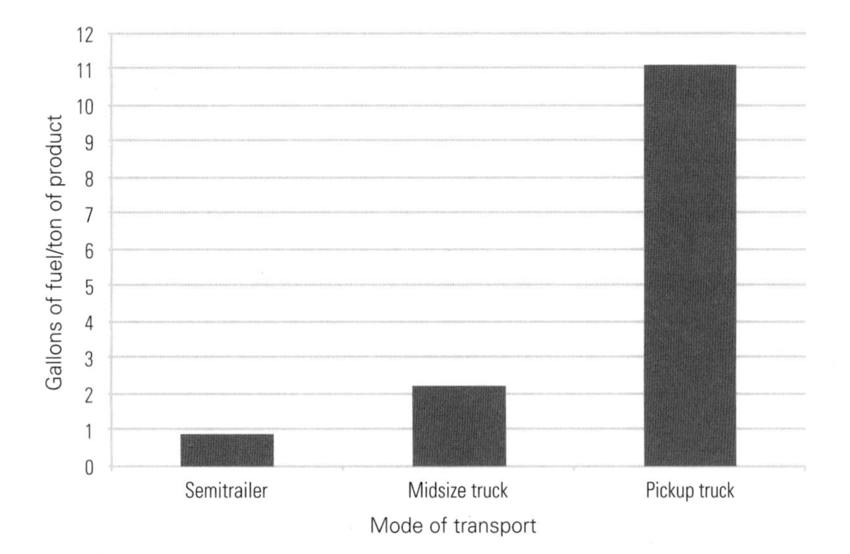

FIGURE 4-5. Transportation Fuel Use per Ton of Product. Source: R. P. King, M. I. Gómez, and G. DiGiacomo, "Can Local Food Go Mainstream?" *Choices* 25, no. 1 (2010): 4 (Agricultural and Applied Economics Association).

Efficient Vehicles and Cleaner Fuels

Fuel is a vital concern in any distribution system. Distributors are well aware of the benefits of fuel-efficient vehicles, since fuel costs are central to their ability to achieve optimal net profits. Some distributors are also working hard to find cleaner and more ecologically appropriate fuels for their delivery operations; meanwhile, tightening pollution standards in some states and cities are encouraging movement in this direction. Perhaps no other component of the U.S. food system has such a tight link between energy efficiency and profits, nor is any other sector more vulnerable to the volatility of fuel prices. When these factors are combined, more localized distribution makes sense from both a business and a marketing perspective. A few regional food distributors are tackling energy issues head-on. For example, Veritable Vegetable, a distributor in San Francisco that ships throughout the Southwest, is converting its fleet of tractors and trailers to diesel-electric hybrids that save 35 percent in fuel consumption. (Not limiting its energy conservation efforts to the wheeled part of its operation, the company is also committed to a zero-waste policy, diverting a stunning 99 percent of its waste from landfills, and it has honed in on highly efficient refrigerated storage. It has even installed 560 rooftop photovoltaic panels to serve a portion of its warehouse electrical needs.)[13]

Computer Systems for Delivery and Pickup

As the demand for local and regional foods increases, distributors are looking to reduce fuel, labor, and infrastructure costs by maximizing the efficiency of on-the-road movement. Despite the relatively limited radius of local and regional distributors, delivery routes can require sophisticated and costly software. Not only are the deliveries and pickups frequent for local distributors, but the variety of products on board is often much more diverse and perishable than long-distance truckloads of identical or similar products. Think of the difference between shipping a load of watermelons across the country versus a regional distributor picking up fruits, vegetables, and artisanal meats and cheeses from a variety of farms and processors along a much shorter route and dropping them off at multiple stops. The regional distributor has a much greater challenge in maximizing efficiency.

Entrepreneurs and local food system advocates are continually developing improved software that better collects and analyzes data to coordinate pickup and drop-off times and determine efficient routes. This enhanced data analysis helps ensure maximum capacity and efficiency throughout the entire distribution run. After all, empty trucks can drive a distributor to ruin.

Regional Distribution

Volume, supply, and consistency can pose serious challenges to distributors of local products. The sum of needed volume, steady supply, and consistent quality is efficiency—efficiency in human labor and other energy sources. Efficiency translates into cost, and cost and availability drive market potential. Many distributors, therefore, have consciously adopted a regional emphasis rather than a focus on more constrained local markets. The regional scale often offers advantages and a certain confidence to large institutional buyers that are moving toward increased local purchasing. Product shortages, as well as price points that do not work within institutional constraints, quickly stymie efforts to bring local foods into the mainstream. It is interesting to note that some of these regional distributors are run under cooperative and nonprofit models, while others are privately held companies. (For a fascinating comparison of various local and regional distribution models, see the collaborative online map titled "National Distribution Models," initiated by the Center for Integrated Agricultural Systems at the University of Wisconsin, accessible at http://bit.ly/NDM-map.)[14]

Food Hubs

"Food hub" is a relatively new term gaining credence among local food advocates. Although there's no single definition, a general point of agreement swirls around another term of interest: "aggregation." In order to address the problems of volume, supply, and consistency common in many local and regional food systems, advocates and entrepreneurs alike are developing systems for aggregating products. Aggregation simply means that an organization or a business gathers and combines local (or regional) products by sourcing from multiple farms in order to achieve the desired inventory for each product. Since scale is often an issue in creating a stable inventory of local products, aggregation becomes a means of assuring clients of a reliable local food stream. Some food hubs gather and distribute products, while others offer processing facilities. In some cases, food hubs are more about brokering relationships and building regional agricultural capacity and consumption.

Aggregation is a complex goulash. The recipe calls for entrepreneurial zest, a thickened roux of software savvy, and a medley of cooperative farms. Pallets, bins, and cases are filled with aggregated material and marketed to retailers, restaurants, institutions, and even individual consumers. Ordering and delivery are complicated enough, but efficiency often requires a pickup of products for the next cycle of aggregation, too. Regardless, aggregation is a promising avenue for increasing local food production and consumption, and careful management of the distribution system can help maximize energy efficiency.

Workplace CSAs

Community-supported agriculture (CSA) farms offer multiple benefits to farmers and consumers, with consumers generally picking up their produce from the farm or a designated pickup site. From an energy perspective, the most problematic aspect of the CSA model is the need for CSA subscribers to drive from a multitude of places to pick up their produce. Of course, these pickups can also serve as important relationship-building and educational opportunities, but too much driving by too many people has its costs in fossil-fuel consumption. Schedule conflicts, traffic concerns, or the inability to drive can all negatively impact a CSA's potential consumer base. As a partial solution, the workplace CSA model adds an element of energy efficiency and convenience: A company or institution works with one or more farmers to offer its employees CSA shares delivered to the workplace for a regularly scheduled pickup. In some cases, the employer offers its employees the option of a regular payroll deduction to cover the costs of the CSA share over time. The employer might even contribute a portion of the costs, provide coolers to employees for storing the produce while they're at work, or allow farmers to set up a booth or tent for additional sales.

Internet Orders for Home Delivery

Putting this concept on paper (or into an e-book) is perhaps unwise, since the velocity of the Internet-driven world far exceeds the capacity of the publishing world to keep up. As you read this chapter, some enterprising local food advocate or businessperson is probably sitting at a computer devising or revising a method for selling local foods online. Sometimes these entrepreneurial models are farm-based, while at other times they are linked to distributors and aggregators in various guises. In many cases, customers enter their orders online in a specific time frame, and deliveries are made to customers' homes, workplaces, or specified pickup points. Done well, these models can minimize transport distances of food products, as well as eliminate the need for consumers to jump in their car and head to the nearest (or not so near) preferred food retailer. While these home delivery services may help bring new consumers into the local food market, the convenience and flexibility of these models can impact customer bases at CSAs and farmers' markets.

Processing

It is difficult to say whether our eating habits are driven by changes in the food system or vice versa. While we may follow all of the loss-leader promotions right into the gaping mouth of the supermarket's aisles of processed food for economic reasons, the way we schedule our lives around eating also impacts our food habits.

As we continually decrease the time that we spend cooking and cleaning in the American home kitchen, the processing of our food occurs increasingly in commercial facilities. Those facilities, in turn, look for labor efficiencies that can increase net profits. In an effort to cut costs and maximize standardization, many of those processors opt for mechanization over human labor. Food processing uses energy in the transformation process and also in the movement of materials and ingredients from facility to facility. The energy impact of those decisions is quite clear, as the gross energy demands for food processing have risen more than in any other sector of the food system, as shown in figure 4-6.

It's not immediately clear just how local food systems can begin to address this rather astounding growth in energy use in food processing. Upon reflection, however, several responses emerge. The first and best is that we simply need to depend less on processed foods, both because of the high-energy inputs and the packaging with all of its embedded energy and increased waste—not to mention the health implications. Buying fresh, lightly packaged foods directly from farmers and local distributors makes sense for not only calories invested but also calories derived. That is to say, less-processed foods tend to be less energy intensive to produce and not as calorically loaded as their highly processed counterparts. (Remember that food processing also uses energy in the movement of materials and ingredients from one facility to another—often hundreds or even thousands of miles apart.)

State and federal governments can also offer financial incentives for processors willing to invest in energy-efficiency infrastructure and innovations. So another response is to advocate for such policies on behalf of businesses that comprise our own community fabric. The impact of increasing energy prices on net profits will probably be the most significant impetus for energy efficiency in the processing sector, however. At some point, processors will be forced to embrace the renewable resources available in their respective locales. Processors in sun-drenched areas will capitalize upon drying techniques, solar ovens, photovoltaics, and solar hot water systems. Food entrepreneurs in colder climes may finally adopt so-called polar power systems that utilize frigid outdoor temperatures as a common resource for enhancing refrigeration efficiency. If businesses will

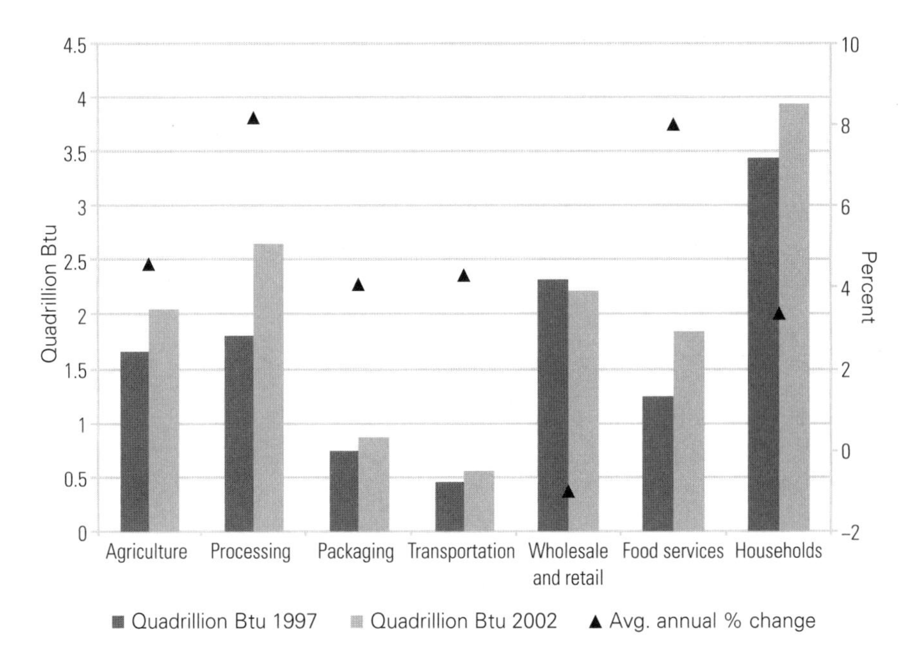

FIGURE 4-6. Change in U.S. Food System Energy Consumption by Stage of Production, 1997–2002. Source: P. Canning et al., *Energy Use in the U.S. Food System,* Economic Research Report no. 94 (Washington, D.C.: USDA Economic Research Service, March 2010), 20.

adopt these kinds of technologies prior to the inevitable spike of energy costs, not only will they be well positioned in the marketplace, but they can help bring more confidence and expertise into the marketplace.

Richard Travers is one Vermont entrepreneur who saw the potential for capitalizing upon climate in this regard. As he describes it,

> We were sitting around our dining room where we had an old restaurant refrigerator. The compressor went on, and it made so much noise, you couldn't hear the other person talking across the table. It was 20° below zero outside, one of the coldest nights of the year. I got to thinking—why not take that cold air from the outside?[15]

That winter-evening inspiration turned into a brilliant business opportunity. Travers went on to develop the Freeaire Refrigeration system, which pulls cold outside air into commercial refrigeration units. Perhaps we need to modify the motto a bit: "Think locally; act accordingly."

Food Storage and Preparation

Data sometimes hurts, especially when it hits home. Just when it seemed like we could blame the farmer, the processor, and the distributor for our food energy woes, lo and behold, our constant culinary vacillations between hot and cold have conspired to put the American kitchen in the crosshairs of our food energy hunt. But this should be no great surprise, since the kitchen is the heart of household thermodynamic transitions.

Our refrigerators, freezers, ranges, microwaves, and dishwashers make our lives more efficient, but they also expand our energy budgets and our household energy costs. Cooking, refrigeration, and dishwashing consume more than a quarter of our household energy use. Of that energy consumption, refrigeration is responsible for 64 percent of the total, while cooking and dishwashing account for the remaining 26 percent and 10 percent, respectively.[16] Add in lighting, heat, convenient kitchen gadgets, and exhaust fans, and the kitchen becomes the primary hub of household energy use—and when we can't stand the heat but don't want to leave the kitchen, we power up the AC.

So what does the household kitchen have to do with energy and relocalizing food systems? Everything. We make key energy decisions in our kitchens through a variety of short-term and long-term choices. Our choices of foods, storage techniques, preparation methods, kitchen design, and appliances all feed into our household energy footprints. And then there is an even bigger decision at hand: whether to stay at home to eat or to go out and grab a meal.

But before we run out of the house for yet more material to digest on the energy front, it's worth thinking for a moment about thermodynamics and our food system. Once food is harvested, the human concern is to ensure that the food is safe, palatable, and nutritious. That's where some of our biggest food energy questions come into play. We start to manipulate food temperatures in order to properly store and prepare foods.

Think for a moment about how many times we might alter the temperatures of a simple food product such as sweet corn. After cutting it off the cob, we heat and blanch it, freeze it for an extended period, and then thaw and cook it at some point many months later. Those are significant energy transformations. The embedded energy of the packaging needs to be considered, too, regardless of whether it is disposable or reusable.

The sweet corn example demonstrates the beauty of harvesting and preparing items fresh from the garden and preparing them as simply as possible in the kitchen. If we do it well, we've captured both taste and optimal nutritional quality by eating seasonally. As it turns out, the epitome of energy efficiency—home preparation of fresh foods—is married to the pinnacle of nutritional quality! Not only that, but the effort didn't involve driving anywhere. In fact, the entire effort might have even involved some physical activity and aesthetic enjoyment.

But these direct, fresh experiences with our food are becoming less and less common. Consider a mélange of statistics that start to provide a picture of how we relate to food—and, ultimately, energy. The time spent in preparing and cleaning for a meal at home in the United States declined from an average of 65 minutes per meal in 1965 to 31 minutes in 1995.[17] Between 1977 and 2006, the percentage of calories eaten away from home by children increased from 23.4 percent to 33.9 percent.[18] A typical American consumes about 2,200 pounds of food per year at an estimated rate of 3,747 kcal/day (the FDA recommends an average daily consumption of 2,000 to 2,500 kcal/day).[19] One researcher estimates that we Americans consume about 33 percent of our total calories in junk food.[20]

As we put the pieces of this puzzle together, we start to move toward a simple but reasonable guiding principle: every step we take away from *home*, *fresh*, and *seasonal* in the harvesting and preparation of our food increases our personal caloric intake and magnifies the energy footprint of that food. That said, we cannot afford to just chill out in a moment of self-congratulation—the heat is still on in the home kitchen.

Refrigeration

We humans have exploited *heat* in the preparation of our food for millennia, and we got really good at it. But we never had the opportunity to make extensive and controlled use of *cold* until the past century. Whereas heat drove our culinary adventures for most of human history, refrigeration has, in many ways, surpassed heat in its importance to our contemporary food systems.

The "cold chain" that binds together so many elements of our food system is a marvelous thing. In addition to minimizing food loss and waste, it provides us with enhanced nutrition and food safety as well as delectable staples of the modern diet, like ice cream (a personal favorite, so the energy realities involved are painful for me to describe in full here). The cold chain has offered us convenience and bounty from around the globe. As a result, refrigeration is probably the single most important contributing factor to the gradual melting away of robust local food systems around the world.

The use of refrigeration certainly helped some local economies flourish—we need only think of Florida and citrus as an example (although we might consider whether that growth helped to create a diverse, equitable, and healthy food system).[21] Many previously robust local food systems withered with the influx of distantly produced refrigerated goods, some exotic and some far too familiar. Perhaps the most inane example of how refrigeration has upset our local, regional, and even national economies is that of apple juice. Apples are the second most consumed fruit in the United States and our third most valuable fruit crop, but freezing technologies and economic policies have allowed China to capture over 60 percent of the U.S. frozen apple juice market.[22]

Of course, the United States has also followed similar trade strategies made possible only through the use of refrigeration technologies. As a counterexample to the apple juice dilemma, the United States has glutted the Chinese market with low-priced poultry products, leading to a series of cross-Pacific games of chicken in trade negotiations.[23] The point is not to cast blame on particular countries for various trade practices but to make it clear that refrigeration can be used both to the benefit and to the detriment of local food systems and their associated economies.

Refrigeration creates massive national and international markets, consumer convenience, and a diverse diet. It even provides locavores with the opportunity to eat more local products year-round, and it clearly has the capacity to provide us with products of optimal nutritional value. We should also bear in mind that refrigeration is the vital link in supplying safe and nutritious food from rural areas to peri-urban and urban areas—places that are growing in population at a rapid rate.

Our increasing reliance on refrigeration for convenience has also had a series of unwanted effects. It takes only one cursory circuit of a supermarket to feel the increasing encroachment of refrigerated aisles. The battalion of refrigerated and frozen goods is marching ever inward from at least three walls in most grocery stores, and that sensation is backed by data of all sorts. In the process of equipping our homes with refrigerators, freezers, and microwaves to accommodate an increased number of refrigerated products, we have also disposed of appropriate technologies and cultural traditions that helped minimize the need for refrigeration. Not only have we significantly moderated the internal temperatures of our homes in the past several decades, we've also eliminated cold pantries and root cellars from the typical house design. The common "summer kitchen" in the South was one sensible feature that kept the main areas of the house cool, but it is no longer in the contemporary architect's repertoire. Springhouses and icehouses were once ubiquitous in rural areas, but those concepts have also virtually disappeared. Even food preservation traditions that did not rely upon any type of refrigeration have waned with fading memories and flashing regulatory messages about food safety.[24] Salting, drying, and various forms of fermentation got shelved in our rush to the refrigerated aisles.

Revisiting the traditional foodways of your region can provide interesting insights into a world with minimal refrigeration and an era of fewer processed foods. At the same time, it's important to demand sensible and highly efficient refrigeration technologies. Rest assured that those worlds can come together. I'm reminded of it every time I go down into the basement on a sunny day and pull a tub of homemade ice cream out of our superefficient Sundanzer chest freezer that runs on the equivalent of an eighty-watt solar panel. High in fat, low in calories—with a capital *C*, that is.

"Away-from-Home" Food

We Americans spend almost 50 percent of our household food budgets on eating outside the confines of our homes, and we derive more than 30 percent of our caloric intake from those meals and snacks.[25] It only makes sense that researchers and market analysts would come up with a name for

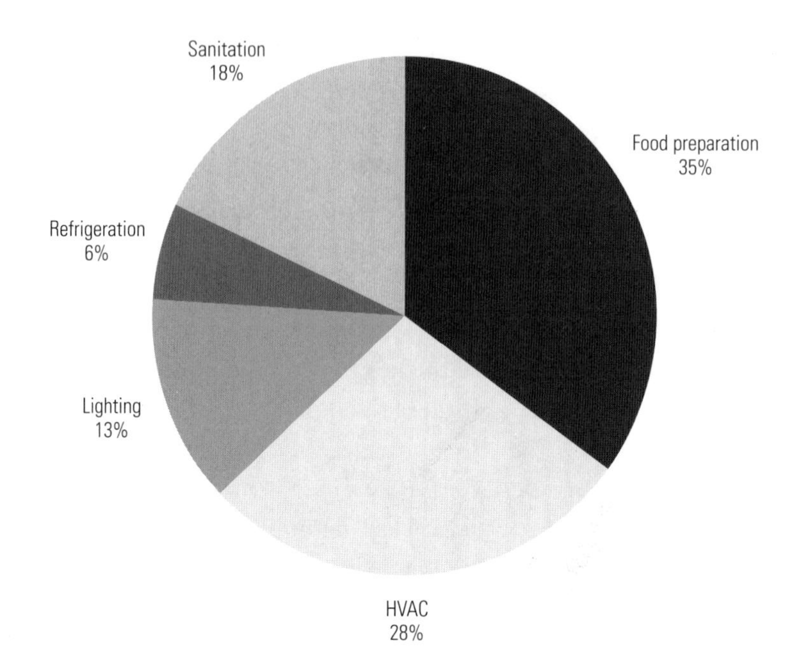

FIGURE 4-7. Energy Consumption in a Typical Full-Service Restaurant. Source: *ENERGY STAR Guide for Restaurants: Putting Energy into Profit* (Washington, D.C.: U.S. Environmental Protection Agency, November 2010), 1.

all of that dietary and economic activity.[26] "Away-from-home food" is an American passion, second only to our love affair with the car. They are, of course, in cahoots, but that's another steamy story for later.

Energy consumption in the "away-from-home food" category differs significantly from that of home food storage and preparation. Refrigeration actually becomes the smallest energy concern, with food preparation and HVAC systems becoming much more energy-intensive than in the average home. Sanitation is obviously a central concern and therefore an energy-intensive demand in a food service setting. Lighting is also a high-energy demand, as it needs to be sufficient for work, safety, and hygiene. Energy bills in a public eating establishment are an integral element in determining profit margins profit and loss, so some players in the food service industry have been quick to adopt energy-efficiency technologies and practices.

So how does energy in the food service sector relate to local food systems? The simple response is "waste." However, this simple answer is not lacking in complexity. Remember that food service is essentially a waiting game: food service personnel are trying to predict both the timing and the scope of customer demand. It's not like cooking in the home kitchen, where you generally know who is coming to dinner and when they will be arriving. Even if you do miscalculate, you can always put the leftovers in the refrigerator for the next day. In contrast, a miscalculation on the part of a restaurant manager or a grocery store deli operator typically results in food—with all of its embedded energy from farm to plate—being tossed into the garbage . . . yes, garbage. All too seldom does this food go to the charitable food system or a municipal composting operation.

And then there is the issue of disposable utensils, containers, and packaging—materials made of Styrofoam, plastic, and paper, born out of habit and not necessity. Patronizing a local full-service establishment as opposed to a fast-food restaurant or supermarket deli can result in significantly less waste of energy and packaging. Not all local establishments are necessarily conscious about the number or types of containers that they use, but places where customers sit down to eat generally generate less post-consumer waste than fast-food establishments and the increasingly ubiquitous prepared-food sections of supermarkets. Cafés and restaurants also tend to build more rapport and community interaction, as hospitality is an important part of the equation for the success of a local business. And it's also much simpler to advocate for change in how foods are served in locally owned food service establishments than it is in chain restaurants.

As soon as we step out of our homes in pursuit of food, we cross an energy threshold that is worth considering. In all cases—upper and lower—Calories and calories count.

Waste

Like everything else in the food system, food waste isn't that simple. Unlike everything else in the food system, waste knows no bounds—that is, it cuts across all components of the food system. Food is lost and wasted in every sector, from production to consumption. However, the pervasiveness of food waste also means that it's one of the biggest opportunities for rebuilding local food systems. Before making that argument, though, it is important to understand the issue of food waste in more detail.

Technically speaking, the term "food loss" is related to losses of quantity and quality of food in the initial production, processing, and distribution stages. "Food waste," in contrast, tends to refer to the loss of food in the later stages of the food chain, ranging from storage spoilage to kitchen prep scraps to unconsumed prepared foods. The distinction between these terms can be helpful, but for the sake of simplicity, most discussions opt to avoid misconstrued nuances and simply use "food waste" as an all-encompassing term.[27]

Regardless of terminology, one point is writ clear: the most technologically and economically advanced cultures in the world have the highest rates of food waste on the planet—and that's even without including the astonishing amount of packaging and carry-out containers associated with our dietary habits. You would think that nations endowed with such economic and technological capacities would lead the way in reducing and recapturing food waste, but we are far from that reality, as evidenced in figure 4-8.

Every time food is lost or wasted, all of the embedded energy that went into producing that food is also wasted. In other words, the dilemma is not just about the loss of the calories and the nutrients in the food itself. Nor is it solely about the squandered opportunity to feed the increasing number of malnourished people in our country and beyond, although that injustice alone should be reason enough to move us to act. Food lost and wasted is *energy* wasted. It also represents the arguably unnecessary dispersal of pesticides, carbon, airborne particulates, and other pollutants associated with producing foods.

Waste is generally the first element examined and redressed in an audit of any kind of energy system. It may not be the sexiest consideration, but it's the most important: minimizing waste is the best way to maximize efficiency. Consider a home energy-efficiency analysis. Perhaps the homeowner is particularly excited about installing a renewable energy system (such as solar panels) for her home. The first step is not to size, site, or install the new system; rather, it's to determine current wasted energy. Until the sources of wasted energy are addressed, it makes little sense to invest in new sources of energy, no matter how "clean" or "renewable" they might be.

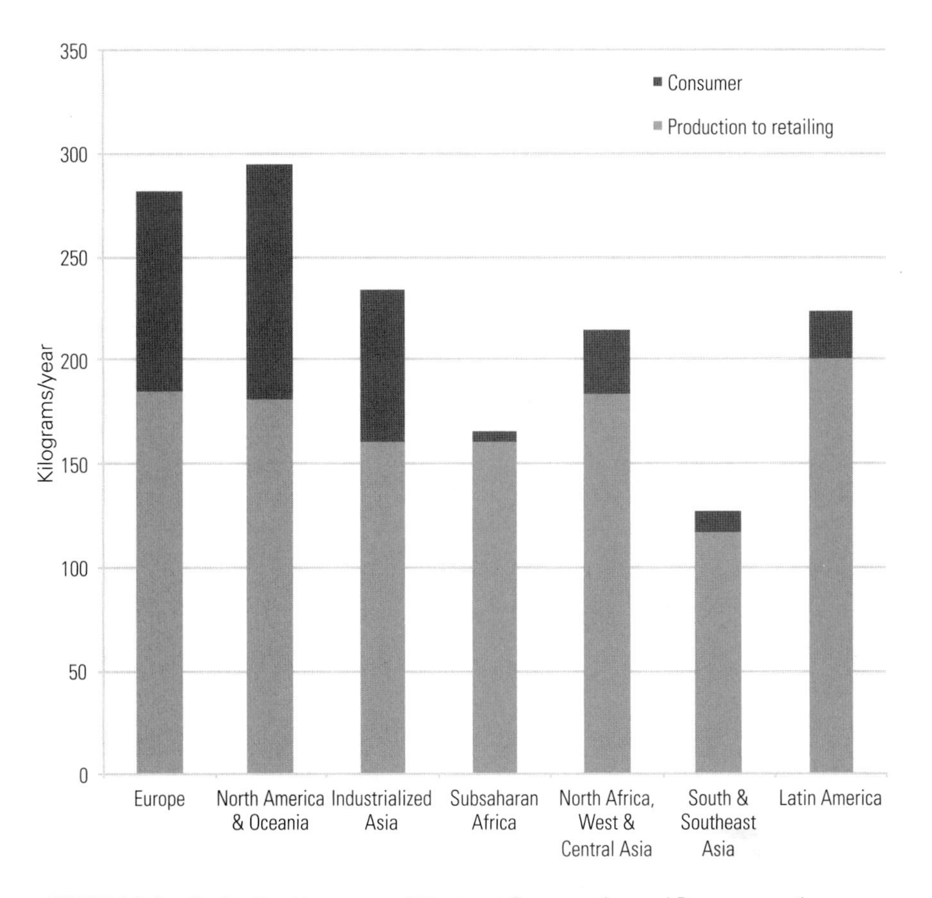

FIGURE 4-8. Per Capita Food Losses and Waste, at Consumption and Preconsumption Stages, in Different Regions. Source: Jenny Gustavsson et al., *Global Food Losses & Food Waste: Extent, Causes, & Prevention* (Rome: U.N. Food & Agriculture Organization, 2011).

Other than being much more complex than a home energy audit—by several orders of magnitude—a farm-to-plate energy audit also should focus on conservation. Only after we determine causes and potential remedies of food loss and waste can we then turn our attention to the energy systems employed in transforming seed and breed into the food on our plates.

One strong but typically ignored argument for local food systems is that we can more easily track energy use and food waste in localized food systems than in the highly dispersed and complex food systems at the national and global levels. Furthermore, when we have to contend with waste on a local level, we tend to be more cognizant of the levels and the impacts of that waste. The more distant and dispersed the waste, the less heightened our awareness and concern.

Most of us are likely unaware of how much food is wasted on a global scale: Approximately one-third of the edible foods produced worldwide are never consumed by humans. That amounts to a stunning 1.3 billion tons of food wasted annually. In the United States, the food waste percentage is closer to 40 percent. In wealthier industrialized countries, food is lost and wasted throughout the entire food chain, but a significant amount of perfectly edible food is wasted at the end of the food chain. In poorer countries, food losses tend to occur more at the earlier parts of the food chain, with minimal waste closer to the consumer end.[28]

Food loss and food waste in the United States are so enormous in terms of squandered nutrients, dollars, and energy that the overall impact is hard to fathom, in part because the results are relatively well hidden in Dumpsters and landfills relatively far removed from our daily orbits. Somehow, we Americans each account for approximately 600 to 650 pounds of lost and wasted food, most of which we barely see or consider.[29] And seldom do we connect food waste to energy waste.

Describing such food waste is an exercise in inexplicable contrasts. The immediate image is, of course, a reeking mishmash of spoiled foods compressed into an enormous metal container, oozing liquids that might even repel most vermin. But then there are utterly irrational images: Dumpsters full of perfectly intact packaged items still within their expiration dates, baked goods not even twenty-four hours old, five-star entrées that somehow never made it to the dining room. Or entire truckloads of fruit turned away from their destinations because they were too ripe and therefore had too short a shelf-life for a grocery store to accept. Regardless whether it came from the next town over or from Mexico, the entire shipment was bound for disposal. Composting has its merits, but human consumption should be the first priority. If that isn't possible, livestock certainly relish such meals.

Many people, myself included, have long congratulated ourselves for feeding our livestock and compost piles with food waste, assuming that we have closed an important ecological loop. In reality, though, we've only put lipstick on a pig (I swear that pig winked at me when I did it myself, though), since the food that we compost or feed to our livestock typically has higher energy invested in its entire "life cycle" than the physical energy it delivers to our livestock. The positive aspect is that we are at least utilizing an efficient biological process to dispose of food waste, enhance soil fertility, and perhaps even sequester carbon.

This biological approach is generally more efficient than an approach utilizing mechanization and large-scale infrastructure, but it is not a silver bullet. By incorporating food waste into our livestock systems, we are in some ways progressing toward a less energy-intensive food system. But we also need to be honest with ourselves and acknowledge that we are doing more to address waste disposal than to reduce the energy footprint of wasted foods in any significant way. The energy losses occurred long before our livestock ever smelled a good meal coming.

It's only natural for a pig or a chicken to be attracted to food waste, but should the same really be the case for a local food systems advocate? Absolutely. Not only can local communities audit their waste streams with more precision and care than a state or federal entity, but municipalities and regional agencies are already heavily involved in the management of solid waste streams. Advocates for local food systems shouldn't waste any opportunity to further their cause, and with the proper framing of the arguments, unexpected allies will flock to the cause like . . . well, like seagulls to a landfill. Careful and creative management of food waste at the local level significantly energizes the rationale for rebuilding local food systems, as a result of these economic and environmental benefits:

- Maximizing the diversion and appropriate consumption of discarded or unused foods that are still edible, particularly for food-insecure populations
- Maximizing the energy recapture of food waste through anaerobic digestion
- Minimizing the transport of heavy, water-laden food waste
- Minimizing nutrient loss from the food system by transforming food waste into engineered soils, while also reducing the potential for the leaching of these nutrients into waterways
- Highlighting local eating establishments that can document their efforts to achieve zero waste of food, eating utensils, and carryout containers

Keeping food waste local—whether in the form of charitable donations or compostable material—is both cost effective and energy efficient. Transporting, processing, distributing, and even storing food waste is energy intensive. Due to its significant water content and its bulk, food waste is heavy and expensive to transport. Landfills are essentially inefficient storage vessels that create unwanted by-products through decomposition and leaching. In fact, food waste comprises an astonishing *one-third* of all material sent to landfills, and it is estimated that landfills in the United States produce approximately 20 percent of the nation's methane emissions—energy lost and pollution unleashed.[30]

In contrast to this absurd use of landfills, good management of compostable materials can reduce leaching, build soil fertility, and capture clean-burning methane. Urban agricultural initiatives are especially well suited to reap the multiple benefits of processing food waste, often in combination with other organic wastes such as leaves and grass clippings. Enormous amounts of food and other organic wastes are exported out of many urban areas on a daily basis, at significant cost to businesses and municipalities. At the same time, urban gardens, lawns, and parks need significant soil amendments not only to rebuild fertility in existing green areas but also to remediate potential garden and park sites constrained by concrete, asphalt, and compromised soils.

Some metropolitan areas, however, are working hard both to capture those valuable natural resources and to reduce costs. The West Coast has been leading the push for diverting organics from the waste stream, in part because of high tipping fees at landfills ("tipping fees" are the charges assessed for dumping different types of waste, normally on a per ton basis). Portland, Oregon, has been demonstrating just how quickly a city can move in capturing organics by instituting a weekly pickup of household food scraps and moving to a biweekly garbage pickup, thereby encouraging good separation practices at the household level and boosting regional compost production.[31] In addition, Portland is set to activate the country's first grid-tied municipal biogas generator, a facility powered by recovered food waste.[32]

Of particular interest to urban areas is the ability to convert these compostable materials into "engineered soils" that can be designed for the special needs of rooftop gardening and farming. These soils can be engineered to be light, highly absorptive, and crop-specific in their nutrients. Just as is the case on a rural farm or in a suburban garden, optimal soils in a green roof project yield optimal results—for crops and the buildings themselves. Green roofs quickly translate into urban acres, and they hold significant agricultural potential, while also moderating summer temperatures and capturing CO_2. Keeping food waste local can significantly enhance the potential for local food security and accessibility, while also reducing the energy and climate impacts of food that is already energy-intensive.

In order to get a sense of just how much energy is contained in food waste, we need only compare it to human excreta (urine and feces, also known as biosolids). The energy from food waste has obviously not been "biologically processed" and utilized by our bodies. As a result, decomposition of an average ton of food generates approximately 376m^3 of biogas (primarily methane), more than three times the biogas produced from the same quantity of biosolids in our wastewater systems.[33]

Communities and businesses across the globe are constructing anaerobic digesters, airtight vessels that are fed organic wastes such as food, biosolids, and yard waste. As these organic wastes decompose, methane is captured and utilized as fuel for heat, electricity generation, and even clean-burning vehicle fuel. In this way, keeping food waste local expands a community's energy self-reliance, while also mitigating pollutants in waterways and the atmosphere and reducing disposal costs. Other countries around the world—industrialized and developing—are leading the way in adopting anaerobic digestion technologies, realizing savings and a sense of energy independence. In the meantime, the United States lags far behind—but leads the way in demonstrating how to waste a good opportunity and pay more to do it.

A final takeaway lesson with regard to energy, food systems, and waste stems from carryout containers and other food packaging. Our fast-paced food choices in the United States result in the disposal of approximately two hundred billion disposable cups per year. Yes, you read that correctly: two hundred *billion*. Starbucks alone sends coffee on its way out the door in three billion cups annually (and, to the company's credit, it's worked more diligently on the disposable carryout container issue than just about any grab-and-go company in the United States).[34] To be fair, this says more about our consumption habits than about any one corporation. After all, the average American chooses to eat fast food about 150 times per year, and we collectively dispose of approximately 1.8 million tons of fast-food packaging every year.[35]

Maybe it's worth going into a café, greeting other locals, sitting still for a few moments, sipping tea out of a *real* cup, tipping the waitstaff, and then waving good-bye, fully energized and steeped in local tradition.

Chapter 5

Environment

It's not hard to make local foods look scrumptious, smell even better, taste delectable, and sound like the pinnacle of sustainability. And the realities often match the hype and the buzz. But some of the most compelling arguments for local foods and their benefits are not quite so viscerally appealing at first glance—especially when we start examining the environmental benefits of local food systems, which often have to do with the messier sides of agriculture, such as dealing with heaps of decomposing plant matter and animal waste. At that point, the photographers start to shy away, and the perfectly coiffed mic-bearing journalists start to wish they were back in the safety of the sanitized studio.

The front end of the local food system is much more glamorous than the back end. We go from the romance of the farm and the enchantment of the palate to the dirty spoils of the battle for sustainable food. Regardless, the more we tighten our local ecological cycles, the better our grip on keeping healthy local food systems intact in our communities. The good news is that we have the science, the technology, and the expertise to do just that. The frustrating news is that we have had all of those things available to us for several decades, but we have yet to employ them at the large scale or extensive scope necessary.

When talking about food systems, many of us describe them as everything that happens "from farm gate to dinner plate." In certain circles, however, the whole cycle is sometimes referred to as "soil to soil," reflecting the idea that a sustainable food system is one that begins and ends with the careful management of the foundation of it all: the soil. I'm proud to say that Vermont's new ten-year strategic plan for building its statewide food system is described as a "soil to soil" initiative. In a country that is currently losing approximately 1.7 billion tons of agricultural topsoil annually, we need to think hard about conserving topsoil and rebuilding the agricultural potential of our soils. "Energy independence" is often touted as a laudable national goal by people of all political stripes. Perhaps it is also time for our nation to achieve "fertility independence." Ultimately, it doesn't matter whether you are compelled by a sense of urgency for local self-reliance or for national security. Soil fertility is key to both.

No one understood those realities better, wrote more prolifically about them, or advocated harder for both local and national solutions to the crisis underfoot than Jerry Goldstein. I don't know quite how it must feel to be a pioneer in any arena, but I can only imagine that it must be similar to being the leading edge of the wing on a high-speed aircraft. You're the point of first resistance, cutting through the surprising force of something completely invisible, encountering both the brute force of a head-on impact and the unexpected modulations of unforeseen turbulence. Yet it's that job of being the leading edge that provides the possibility of lift for the rest of the aircraft.

In many ways, that was Jerry's mantle, donned with humility and humor but worn for the entirety of a career. The founder of what is now known as *BioCycle*, the premier publication for anyone interested in the nexus between composting, renewable energy, and sustainability (not to mention local food systems), Jerry passed away as I was writing this book, on May 17, 2012. I had asked a colleague, Dan Sullivan, the managing editor at *BioCycle* at the time, to review the chapter that you are about to read for accuracy and to ask if Jerry's daughter Nora might be willing to give it a quick read as well. They provided helpful comments, as well as this general assessment: While the material may be cast somewhat differently, there's nothing particularly new here—it's what we've been saying and pushing for decades.

Point well taken, and there's the rub. What Jerry dug into (pun clearly intended) over the course of his eighty-one years set the stage for our local communities and our nation to think hard about sustainability, particularly when it comes to issues of so-called waste. Granted, we've made some headway. Nora describes Jerry's approach in her editorial tribute to her father in the June 2012 issue of *BioCycle*:

> At his core, Jerry was pragmatic, identifying the steps and initiatives to just get started, while providing encouragement and many

> models via project profiles. He had little patience for negativism (not worth the time and energy). But he had a vast amount of patience for positivism—we can do this, let's just do it, learn and improve. A good way to describe Jerry's outlook is: If one person or community is doing something, it's a pilot; if two or three are, they are models; when we hit eight to ten, it's a trend![1]

Jerry was convinced that the future of our society depends on our ability to understand waste as a resource, although his keen intellect compelled him to discriminate between appropriate and inappropriate means of utilizing waste as a resource (a prime example being whether to compost or incinerate organic wastes). Just consider the titles and the publication dates of a few of his early works:

- *Garbage As You Like It: A Plan to Stop Pollution by Using Our Nation's Wastes* (1969)
- *The New Food Chain: An Organic Link between Farm and City* (1973)
- *Sensible Sludge: A New Look at a Wasted Natural Resource* (1976)

Jerry also saw that what we do with organic wastes has more than just environmental ramifications. As he described in *The New Food Chain* from 1973:

> More and more in the past few years, some grandiose concepts have been getting mixed in with the compost heap. Though it's still the same old compost heap as it always was, in the 1970s the heap has come to mean something more. It is still simple. But now, instead of only rating the heap for its ability to make a soil fertile, we talk about it in such terms as social practice and harmony with the environment. To many people, it's still just a pile of garbage and manure—true, a controlled pile on its way to becoming humus—but nevertheless, just a pile. To others, it's a vision of a society in harmony with the environment.[2]

I spoke with Nora several weeks after Jerry's death to gain some perspective on her sense of his legacy. She described how much of his work began just when American gardeners and farmers were beginning to recognize the significant role that organic wastes play in organic gardening. "It became a guiding principle," she noted, "but not as big a principle as we'd hoped." In the early days, the emphasis was on garden and food waste at the home scale; then Jerry and his colleagues began recognizing and urging that our culture needed to do more with suburban and urban waste. "Treating and utilizing sewage sludge was obviously just something that should be done, but that all became stigmatized and that whole issue eventually suffered from a backward evolution," Nora said.

In Nora's view, the real effectiveness of the pioneering work that Jerry and his colleagues did can be seen by the growth of the trade industry shows featuring organic waste management products. "Jerry began our national conferences in 1971," she noted. "In the early years, there was only a handful of exhibitors. Today, depending on the conference, there are 60 to 70 vendors." Jerry recognized long ago that the necessary changes had to happen at the national level, not just in the backyards of a few well-informed organic gardeners. While the home scale matters, the industrial scale truly transforms.

Jerry also realized that the requisite transformations weren't going to happen simply with the growth of the organic waste management industry and the promises of clean energy through anaerobic digesters. The momentum of these initiatives also depended upon policy. In 1975, Jerry came up with the idea of a national soil fertility program that could be linked to the 1976 U.S. bicentennial year and potentially be embedded in that year's presidential race. Initially dubbing the concept the "National Humus Program," he offered the idea that "perhaps we can even get our candidates in 1976 to include a Humus Plank in their presidential platforms. . . . Two hundred years should be more than enough time to begin laying the foundation for a permanent agriculture in the U.S. By translating organic wastes into soil organic matter, a major step will have been taken."[3]

Jerry utilized his journalistic and diplomatic skills to push the issue forward to the point that presidential candidate Jimmy Carter incorporated it into his election platform (don't forget that Carter was a peanut farmer, and it is a rarity to have a farmer among candidates for national office these days). By 1977, the major provisions of Jerry's proposed program, now called the National Soil Fertility Program, made it into the form of a House bill titled the Land and Water Resource Conservation Act of 1977 (H.R. 75). These provisions focused on the need for research and analysis related to the proper utilization of organic wastes for rebuilding national soil fertility. Unfortunately, the bill didn't pass, but Jerry's family is pushing for the development of a new version of the National Soil Fertility Program in his honor.

Even though the National Soil Fertility Program hasn't (yet) become policy, the research did happen. We now have several decades of superb research on how best to manage organic waste and transform it into soil fertility, energy independence, economic development, and long-term food security. The next phase is implementation. First, however, it's important to use the rest of this chapter to remind you of what Jerry and a host of his colleagues have been saying for half a century.

It's time to listen up and get on board. We're taking an underground tour of the local food systems. Welcome to the swelling underbelly of our eating habits.

A pie gone awry. That's the essence of figure 5-1 on page 66. Before recycling is taken into account, food "scraps" are the second-largest category entering U.S. municipal solid waste (MSW) facilities. Once paper recycling is taken into account, food waste steps into the number-one position.

The fact that so much food never made its way to hungry families or even to livestock is tough enough to swallow. Then yet another unpalatable reality sets in: only 3 percent of that food waste is diverted for the recycling of nutrients through composting, anaerobic digestion, or some other means of reclamation! Meanwhile, we are absolutely dependent upon synthetic fertilizers to grow the majority of our food: fertilizers that pollute our water, increase our dependence on fossil fuels, contribute to climate change, and link food security to markets outside our control.

And it's not as if the problem is getting any smaller. Figure 5-2 demonstrates the growing girth of our throwaway society.

Approximately 50 percent of all the packaging we use globally is for food, although it's worth remembering that appropriate food packaging can actually prevent food loss, and some materials used for packaging are actually "green" products or by-products that can be recycled in some

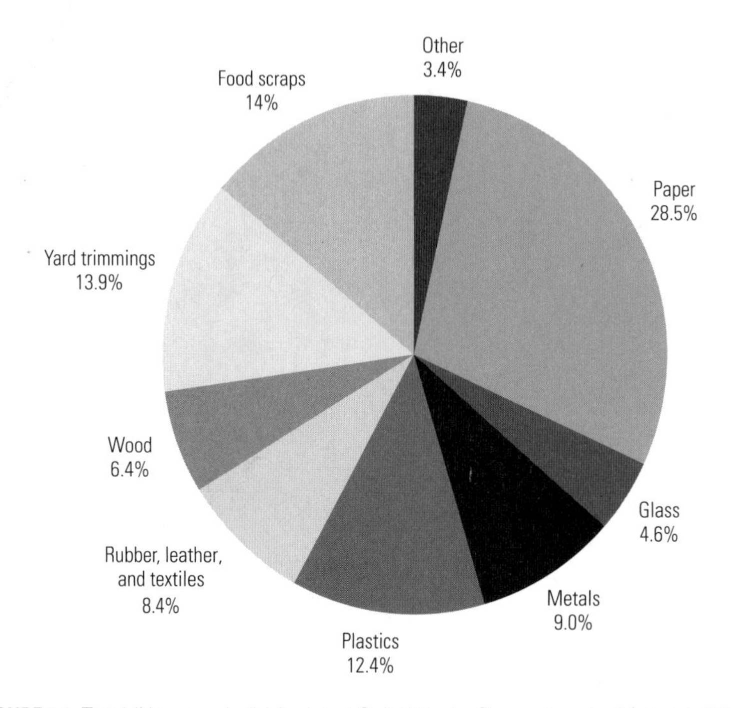

FIGURE 5-1. Total (Nonrecycled) Municipal Solid Waste Generation, by Material, 2010. Source: *Municipal Solid Waste Generation, Recycling, and Disposal in the United States: Facts and Figures for 2010* (Washington, D.C.: U.S. Environmental Protection Agency, December 2011).

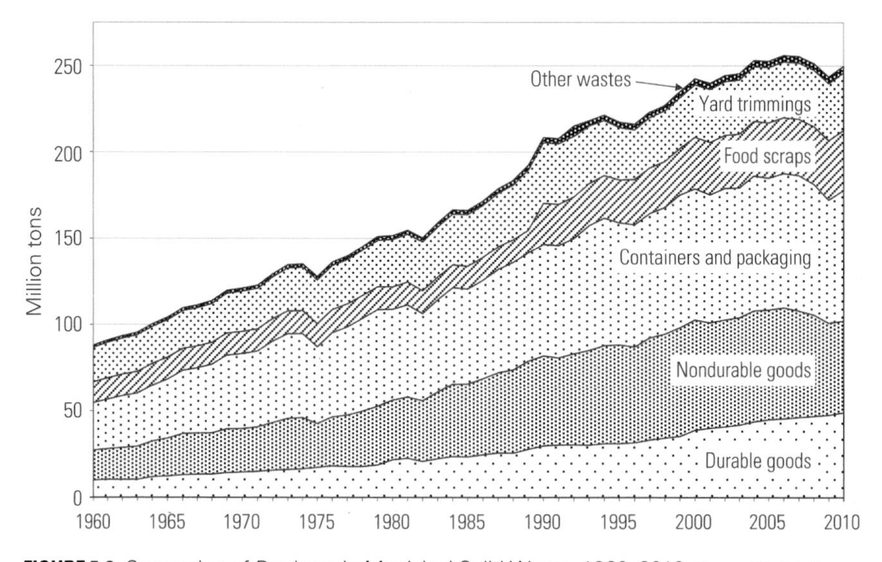

FIGURE 5-2. Generation of Products in Municipal Solid Waste, 1960–2010. Source: *Municipal Solid Waste Generation, Recycling, and Disposal in the United States: Tables and Figures for 2010* (Washington, D.C.: U.S. Environmental Protection Agency, December 2011).

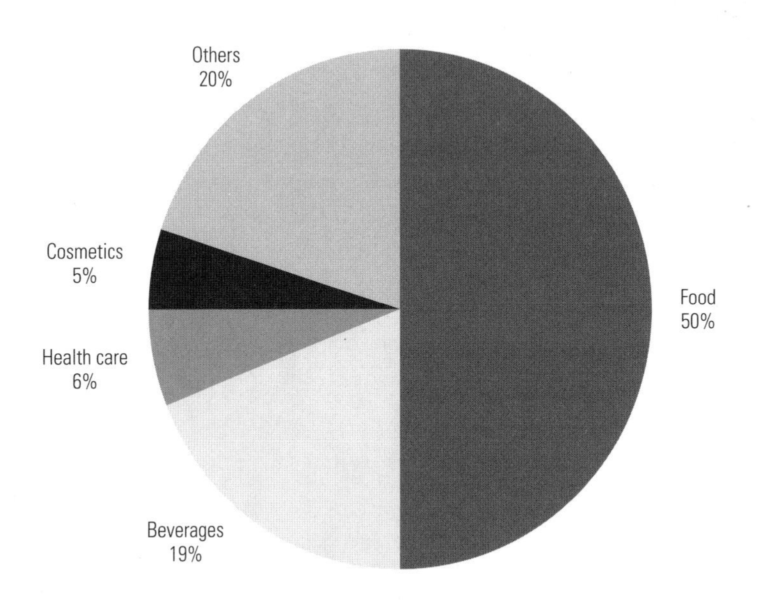

FIGURE 5-3. The Global Consumer Packaging Industry, by End Use, 2009. Source: Pira International, "Lucrative Packaging Opportunities to Be Found in the Middle East and North Africa," August 4, 2009, http://www.pirainternational.com/BusinessIntelligence/TheFutureofPackaginginMENA.aspx.

ecologically appropriate fashion. Nevertheless, as figure 5-3 shows, beverage packaging adds another 19 percent to that startling 50 percent number.[4] It's almost enough to make you long for the days of terra-cotta vessels and woven baskets.

In the grand scheme of things, simple, ecologically appropriate packaging for getting food from the farmer to the consumer makes sense. Proper packaging can reduce food waste. When it comes to the massive amount of packaging waste associated with fast food, however, our gluttony swells our landfills and decimates our forests. Forests in the southeastern United States are the most threatened in the country by logging, much of it associated with almost one hundred regional paper mills that service the packaging needs of the fast-food industry. These forests supply products for 60 percent of the U.S. paper industry and 15 percent of the global demand. The paper industry's intense and repeated harvesting has put at risk an estimated one-third of the plant communities in this biologically diverse region.[5]

Food, paper, yard waste, and manure all comprise the resource recovery category we call organics. Organics are the materials in the waste stream that have the capacity to decompose, and most of them can play a critical role in maintaining and restoring our ecological systems. In fact, they have a place in our ecosystems, and that place is *not* in landfills and incinerators.

Consider for a moment the degraded state of soils in our various human habitats. Finding urban soils adequate and safe for growing food can be a real challenge, particularly given the potential for heavy metals and other pollutants to accumulate in open urban spaces. The fate of urban agriculture ultimately resides in the health of its soils—and fortunately, there is great potential for generating plenty of healthy urban soil through the careful management of a city's own refuse. It really is like magic: a city can generate its own wealth simply by recycling its own waste and creating compost and even lighter engineered soils for rooftops and terraces. Our disposition to dispose is nothing more than a wasted ecological and economic opportunity.

Ornate green lawns often hide the reality that suburban soils are also frequently at a deficit, having been scraped, manipulated, and compacted, and then later doused with synthetic fertilizers and herbicides. Although our lawns don't tend to be particularly healthy or diverse, at least they are relatively stable ecosystems.

In contrast, we constantly manipulate our agricultural ecosystems to meet our management and marketing strategies. Agriculture is seldom benign—rarer still is it a net positive from an ecological perspective. The worst result of our food-focused manipulations is the estimated 1.73 billion tons of topsoil that we lose annually in the United States; that's two hundred thousand tons per hour![6] On most conventional farms we are not replacing this lost soil fertility; rather, we are substituting natural fertility and the loss of topsoil with synthetic chemicals. In the meantime, we bury and burn waste-stream organics as if there is no tomorrow . . . or at least not a green one.

Nothing matters for local food systems more than soils. Ultimately, *soils* feed us, not food systems. Resilient food systems are established upon resilient soils. Even food justice—how we feed each other—depends upon how we feed our soils. Yet instead of establishing food-secure futures with our soils, we've been building houses, suburbs, highways, superstores, and industrial complexes *on* the best of them. And we've created agricultural systems that have inverted the ecological hierarchy of soil stewardship and production, allowing production to supersede long-term care of the soil.

At the same time, we have created a food system that wastes approximately 40 percent of the food produced, and we divert less than 3 percent of the food casually tossed into the national waste stream from the fate of landfills and incinerators. The waste figures are an ecological travesty, but they are all the more disturbing when we consider the fact that one in five children in the United States is at risk of hunger.[7]

If it all seems daunting and depressing, then consider the exciting potential of linking waste recovery to ecosystem recovery in local food systems. Regenerative food systems depend upon nutrient recycling, so dumping the by-products and the end products of our food system into landfills and incinerators is replacing an ecological loop with an ever-tightening noose around our necks.

We should be just as concerned about "food waste recovery miles" as we are about the food miles between the farm gate and the dinner plate; transporting these materials even moderate distances is expensive. In fact, we need to be thinking not just about the food waste recovery *odometer* but also the food waste recovery *clock*. Every day we lose in recovering food waste is a lost opportunity to capture biogas, a by-product of anaerobic decomposition that can be turned into energy. Methane, the primary product of anaerobic digestion, is an important target for resource recovery, as it is twenty times more potent a greenhouse gas than carbon dioxide (over a one-hundred-year period). Methane is also the cleanest-burning and most efficient of the fossil fuels. Being able to capture methane and transform it into heat, hot water, and electricity is one more reason to keep waste recovery local (although there are potential climate drawbacks with "waste-to-energy" projects as well).

In essence, complete resource recovery in a local food system is textbook sustainable agriculture—as long as every effort has been made to avoid the "wasting" of those resources in the first place. Increasing the efficiencies of resource recovery in our food system—our *entire* food system—provides local communities with macronutrients, micronutrients, organic matter, and energy, while also minimizing the release of methane and other

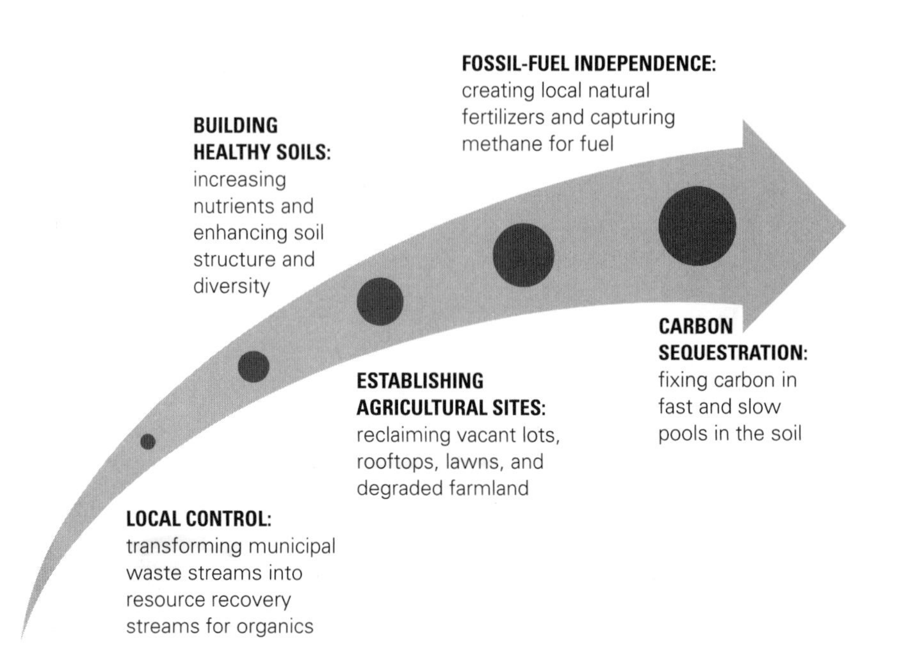

FIGURE 5-4. Linking Local Resource Recovery with Local Food Security.

pollutants into the atmosphere. As figure 5-4 illustrates, there are multiple commonsense benefits to linking the resource recovery of organics with local food security.

The ABCs of Sustainable Agriculture

To understand how critical the resource recovery of organics is both for the environment and for our local food systems, we need to work our way through part of the sustainable agriculture alphabet. Let's start at the beginning, with *C*.

Constant C

Carbon is the chemical keystone of agriculture. In our era of climate change, it is also the chief focus of our concern. The problem is not carbon itself but rather what we do with it.

No arena of human activity is more tightly intertwined with carbon than agriculture. The reason is that dirty four-letter word: "soil." All farmers manage soil; good farmers steward soil. The earth's soil is estimated to contain about 3.3 times more carbon than the atmosphere and approximately 4.5 times the carbon contained in all living things.[8] In other words, our soils are our most important carbon sink. Increased levels of carbon in our atmosphere are the driving force of climate change—and agricultural activity can both increase and decrease carbon levels in the atmosphere.

For the past several centuries, farming methods in the United States and elsewhere have increasingly contributed to the release of carbon into the atmosphere. Tillage, the cultivation of soils, opens up the earth and encourages decomposition, thereby releasing significant amounts of carbon. Think of decomposition as a fire. When we introduce oxygen, it speeds up the "burning" process (the breaking down of organic matter into various components), with a concurrent release of carbon dioxide. Cultivation, therefore, releases carbon. In contrast, a balanced natural ecosystem such as a forest or a prairie actually sequesters atmospheric carbon. The plants and the surrounding soils take up carbon dioxide and "store" it in the soil at various levels and in the dead and decaying biomass of plant and animal tissues.

It's not that sequestration actually locks carbon into the soil forever. Nor do all types of soils store carbon equally. Even the carbon in soils differs. Now we're getting into the nitty-gritty of local!

First of all, it's important to understand the distinction between organic soil carbon and inorganic soil carbon. This distinction has nothing to do with the USDA or any other certification process, but it has everything to do with Mother Nature's natural process. Inorganic soil carbon is predominately the result of the weathering of rocks and of carbonic acid. This inorganic soil carbon tends to be deeper in the soil profile than most of the organic soil carbon (although there are exceptions), and there is generally a long time frame—thousands of years, in most cases—for this inorganic carbon to find its way into the atmosphere. Due to the typical depth and relative stability of inorganic carbon, we ordinarily do not focus too much of our agricultural concern on it.[9]

In contrast, organic soil carbon is much more likely to migrate from the soil to the atmosphere, and it tends to dwell in the upper levels of the soil. Comprised of living, dead, and decaying organisms, organic soil carbon is ordinarily at its highest levels in undisturbed soils. In other words, the ecological manipulation that we call agriculture generally alters organic soil carbon levels. The more disruptive to the soil the agricultural activity, the more intensive the release of carbon from the soil and into the atmosphere. However, agricultural practices that mimic natural processes in relatively stable or undisturbed ecosystems[10] usually normalize or even enhance carbon storage in the soil. A well-managed rotationally grazed pasture is one of the best examples of a farming method that can build organic soil carbon stores.

The Unending OM

Organic matter, referred to as OM in farming circles, is central to soil health and agricultural productivity. Soil organic matter, according to soil scientist Rattan Lal, can be defined as "the sum of all organic substances in the soil comprising: (i) a mixture of plant and animal residues at various stages of decomposition, (ii) substances synthesized through microbial and chemical reactions, (iii) and biomass of live soil micro-organisms and other fauna along with their metabolic products."[11]

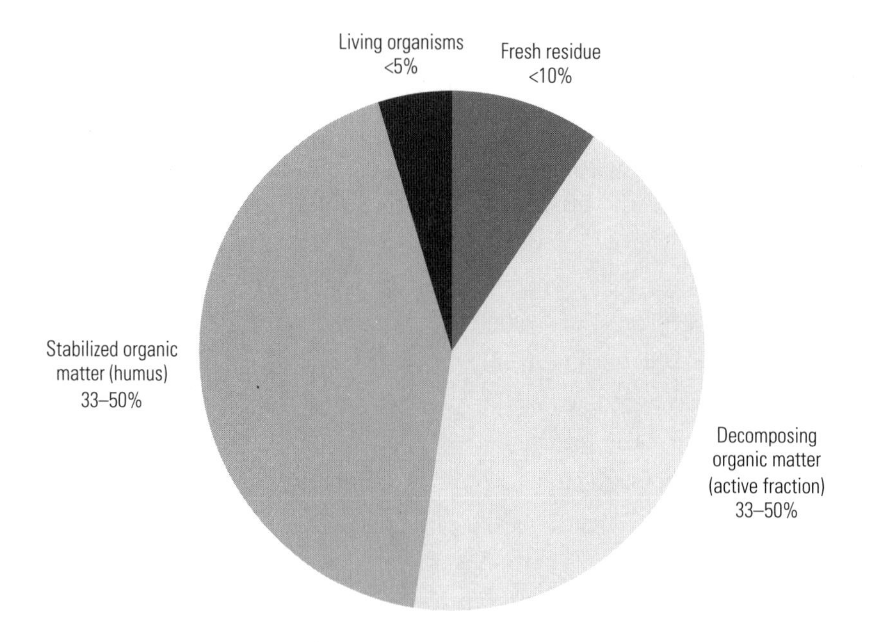

FIGURE 5-5. Components of Soil Organic Matter. Source: *Soil Biology Primer* (online), available at http://soils.usda.gov/sqi/concepts/soil_biology/biology.html (accessed September 27, 2012).

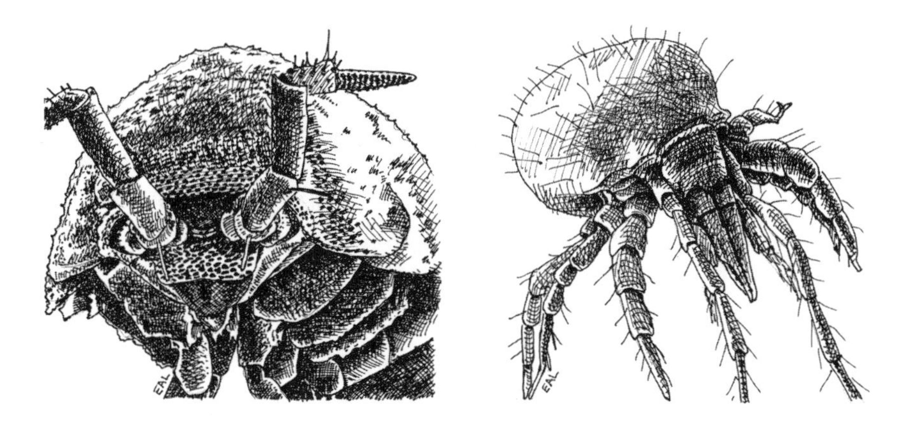

FIGURE 5-6. Microscopic Locavores. Drawings by Erin Ackerman-Leist.

OM essentially has all of the components of a B-grade Western movie: the living, the dead, and the dying. Another way of understanding soil organic matter is by means of the chart in figure 5-5.

If you prefer your B movies in the horror genre, then you can visualize the unending death and decay in the OM world by way of a whole host of microscopic images (see fig. 5-6). (No wonder we fear death.)

Scientists have grouped organic soil carbon into three basic categories, or pools. Each pool is grouped according to how long it takes for the carbon to break down and be replaced (see table 5-1). The speed of the turnover is related to the stability of the carbon in the pool.

TABLE 5-1. Carbon Pools

CARBON POOL	CARBON DURATION IN SOIL	% OF CARBON IN SOIL
Fast (*labile* or *active*)	1 year	Less than 10%
Slow (*humus* or *stable*)	10–100 years	40–80%
Passive (*recalcitrant* or *refractory*)	1,000+ years	10–50%

Source: Compiled from information in James Walcott, Sarah Bruce, and John Sims, *Soil Carbon for Carbon Sequestration and Trading: A Review of Issues for Agriculture and Forestry* (Australia Bureau of Rural Sciences, Dept. of Agriculture, Fisheries & Forestry, 2009), 4–5

The fast pool is essentially what the microscopic locavores in figure 5-6 would consider to be "farm-fresh" organics—plant and animal litter in its early stages of decay. The slow pool is composed of organic materials that are in their later and somewhat slower stages of decay—what we would generally refer to as "humus." The passive pool is the organic material breaking down over thousands of years, oftentimes in the lower horizons (i.e., levels or layers) of the soil.

The relevance here to local food systems may seem distant at first, until we begin to think about the immense capacity at the local level to manage carbon. There will always be costs to managing carbon through the municipal waste stream, but there is also potential for associated local economic benefits. The carbon-rich organics that flow into the municipal solid waste stream can be handled in multiple ways, all of which either stabilize or release their stores of carbon. Landfills and incinerators, for example, are not efficient either in capturing carbon or in generating financial or natural capital. Other resource recovery methods, namely, composting operations and anaerobic digestion (both of which will be explained in greater detail later in this chapter), can store carbon and nutrients in the OM they produce. And increasing OM levels in fields, lawns, gardens, tree farms, and restoration sites will increase the fertility of the soils.

In some cases, managing waste stream organics to produce OM can even help create soils in places where they did not previously exist: city rooftops, vacant lots, balcony containers, building courtyards, and even floating barges. Soils can also be generated to replace soils that were intentionally removed, such as in strip-mining locations. One benefit to adding OM to places where soils are highly compromised is that soils deficient in OM typically have a higher potential for stabilizing carbon.[12] Linking land reclamation to greenhouse gas mitigation is a boon for rebuilding local food systems. Urban agriculture operations immediately come to mind as prime benefactors of such efforts.

The obvious aim in transforming waste stream organics into much-coveted soil OM and natural fertilizers is to create ways to maximize the stability of carbon, i.e., the slow carbon pool. A balanced manufactured growing medium is generally a mixture of OM from the fast and slow pools, both of which play a role in healthy plant growth and carbon sequestration.

(It is worth recognizing that the fast pool is more vulnerable to losing carbon if various parameters change suddenly, including climatic conditions.)[13]

OM serves a number of other purposes, too. It greatly enhances a soil's ability to retain water and nutrients, which is important for the vitality of plants both in normal conditions and during periods of stress. OM also contributes to a healthy soil structure by adding a variety of particle sizes, beneficial organisms, and necessary chemical compounds. In sum, it not only adds to the productive capacity of the soil, but it also helps create soils that are more resilient in the face of drought, disease, and flooding.[14] Good farmers meditate daily on the benefits of OM.

The Necessity of N

Nutrient with a capital *N*. Nitrogen. The Big Daddy of the periodic table for farmers. The Green Machine. When I teach the three macronutrients for plants—nitrogen (N), phosphorus (P), and potassium (K)—I give my students a mnemonic tip for remembering the role of each member of the NPK trinity in fostering a plant's growth:

- N: "shoots"
- P: "fruits" (fruits and flowers)
- K: "roots"

When we see the splendor of green vegetative growth, we are witnessing the power of nitrogen. Nitrogen is the most influential nutrient in plant growth and crop production. Adding nitrogen to the soil generally increases plant growth even if nitrogen is already present because much of the preexisting nitrogen in the soils is temporarily stored in the humus.[15]

This point brings us to a distinguishing difference between synthetic fertilizers and fertilizers coming from the management of organics such as food waste, paper products, animal manures, and human excreta. Synthetic fertilizers make nitrogen available to plants in quick fashion, whereas fertilizers made from organics release less nitrogen immediately. These more natural fertilizers, created from organic materials, tend to hold on to the nitrogen for extended periods of time, releasing it gradually, as happens in a natural ecosystem. Over a long period of time, humus decomposes and gradually releases nitrogen and other important macronutrients and micronutrients, all of which are important for plant growth. Furthermore, these nutrients are stored and protected within the complex and diverse nature of living, dead, and decaying materials.

Composted materials create a carbon-based storehouse of nutrients that provides some short-term benefits to plant growth and numerous long-term advantages. While this tight storage of nutrients in composted material may result in slower and more subtle benefits to plants than the application of synthetic fertilizers, there are significantly fewer issues with the leaching of excess nitrogen into our waterways and the atmosphere.

The complexity and diversity of the soil food web contribute to both the comparative stability and the long-term availability of nitrogen in soils amended with these carefully managed organics. In contrast, when synthetic fertilizers are applied to cropland, only a small percentage of the nitrogen is actually taken up by the plants. The remaining nitrogen escapes down our waterways in various forms, and up into the atmosphere as nitrous oxide (N_2O). The released nitrous oxide is of concern because of its role as a more potent greenhouse gas than carbon dioxide; it lasts much longer in the atmosphere and absorbs considerably more long-wave radiation.[16]

If nitrogen is such an important driver in agriculture, then it only makes sense for us to be in constant search of local, dependable, and ecologically produced sources of the big N. Another good reason is that synthetic fertilizers are fossil-fuel dependent, and the price fluctuations can be as volatile as the substances themselves (nitrogen compounds are commonly used in explosives).

As described in the previous chapter, the production and application of synthetic fertilizer is responsible for nearly one-third of all the energy used in agriculture—and the resulting pollution has serious health and environmental consequences worldwide. In contrast, nutrients captured from the municipal waste stream help us tighten our belts and our regional resource loops. If we want resilient communities, we must ensure that we have reliable and cost-effective sources of soil fertility.

The Possibility of P

Many of us are aware of the concept of peak oil, but few people seem to be aware of peak phosphorus. Phosphorus is another critical macronutrient needed for crop production and for the maintenance of healthy livestock. Phosphorus serves numerous purposes in the growth of plants, especially in the development of genetic material and energy transfer in cells. It also contributes to root growth, stalk and stem strength, flower development, seed formation, and efficient crop maturation.[17] Phosphorus is particularly important in agricultural operations that involve removing most of a crop from the land for consumption. Replenishing the soil by replacing the store of phosphorus taken away with the plant is critical to continued fertility. Our cereal crops such as wheat and corn are prime examples of food sources highly reliant upon proper levels of phosphorus in the soil, and livestock depend upon phosphorus for multiple physiological functions. In fact, phosphorus is the second most common element found in livestock, outdone only by calcium.[18]

Phosphorus is, however, a mixed blessing in modern agriculture. In our zeal to enhance the production of crops and livestock, we have applied it in such quantities that it has become a pollutant in some regions of the United States. High phosphorus levels create algal blooms and excessive vegetative growth in our water resources, and many efforts are under way in parts of the country to mediate the long-lasting impacts of excessive phosphorus levels in our soils and waterways. One farmer friend of mine, Ed Lewis, constantly puts forward a reminder to our local conservation district board: "For years, the government paid us farmers to apply phosphorus to our fields, and it came in by the trainload. Now, the government is paying us to try and reduce phosphorus runoff created by their own policies fifty years ago. . . . Sometimes it ain't easy living life and seeing things go full circle!"

Nonetheless, phosphorus is critical to the productivity of agricultural systems, and the largely untold story is that our penchant for procuring phosphorus supplies through mining operations is about to encounter a collision. As our appetite for phosphorus continues to increase, the nonrenewable global supply of mined phosphorus is disappearing. Data make it clear that we are edging ever closer to peak phosphorous, the point at which we will face consistently dwindling quantities and diminishing quality of phosphorus available through mining.

To put it into context, the United States—historically the biggest user, producer, importer, and exporter of phosphate rock and phosphate fertilizers—has only about another twenty to twenty-five years of available phosphorus within our borders. In 2009, 67 percent of all phosphorus was mined in just three territories: China (35 percent), the United States (17 percent), and Western Sahara, a part of Morocco (15 percent). The United States halted phosphate rock exports in 2004 and has maintained increased consumption only through imports from Morocco. Meanwhile, China is protecting its domestic supplies by establishing a 135 percent tariff on phosphorus exports since 2008.[19]

If those realities seem distant to local food realities, you might want to reflect upon the drama in the global food markets when phosphorus prices rose 800 percent in the fertilizer frenzy of 2007–2008, corresponding to the disruptive rises in food costs around the globe.[20] With some scientists predicting that global food demands will likely increase somewhere between 70 and 110 percent by 2050, we all must pay careful attention to the supply of key agricultural inputs such as phosphorus and nitrogen.[21] Furthermore, with the steady (and problematic) growth in the international demand for meat and dairy products—all of which are highly dependent upon adequate levels of phosphorus in animal feed—phosphorus is an ecological lynchpin in connecting supply with demand. Just say "cheese!" Now you get the picture.

What did we do before we started mining phosphorus? Well, once humans began to realize how chemistry and plant physiology worked together, thanks to Justus von Liebig's pioneering work in this arena in the mid-1800s, they sought out sources of phosphorus. Bird guano, ground-up bones, fish remains, ash, and human excreta were predecessors to mined phosphate. Guano, harvested from the accumulated droppings at bird nesting sites around Peru and in the Pacific, was a favorite source in the mid- to late 1800s, but that finite resource quickly dwindled. In the late 1800s, flush toilets were introduced and a key phosphorus resource—human excreta—disappeared with a mighty *whoosh!* from most everyone's field of vision and from most agricultural fields. As it turns out, pee and P have more than an aural connection: human urine is particularly high in phosphorus. Scientists believe that human urine alone could produce a substantial portion of the phosphorus needed for our global agricultural production, and some studies have indicated that the use of urine as a fertilizer produces significantly higher vegetable yields than do mineral fertilizers.[22]

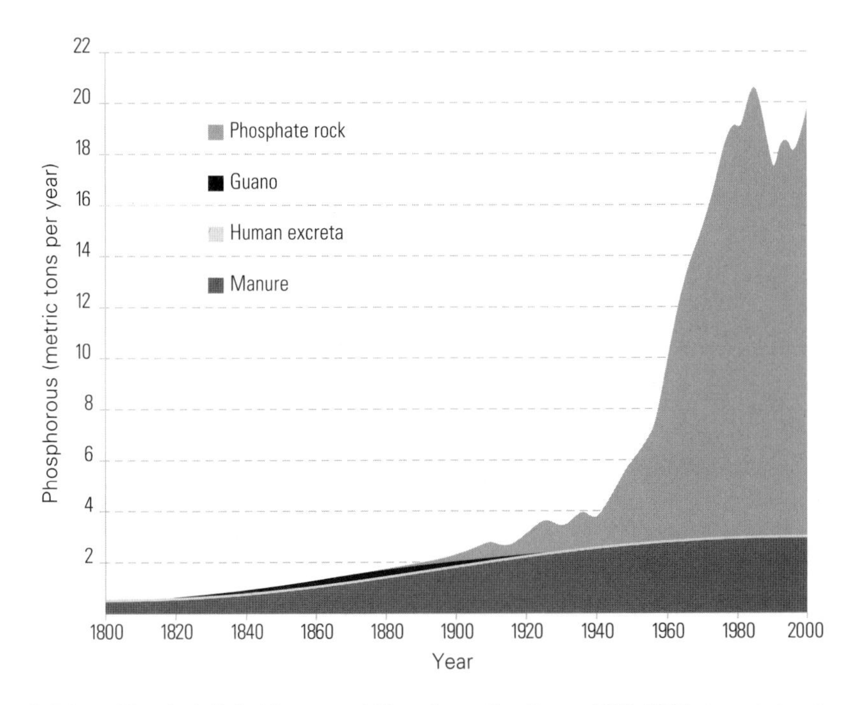

FIGURE 5-7. Historical Global Sources of Phosphorus Fertilizers, 1800–2000. Source: D. Cordell, J. O. Drangert, and S. White, "The Story of Phosphorus: Global Food Security and Food for Thought," *Global Environmental Change* 19, no. 2 (2009): 292–305.

In a natural ecosystem, phosphorus functions in a cycle. The phosphorus contained in plants is returned to the soil by means of urine, feces, and dead and decaying matter. In an organic farming system, phosphorus, nitrogen, and carbon levels are all maintained in similar fashion—nutrients are constantly returned to the soil through composting, cover crops, and careful pasture management. In a more industrialized farming system, crops and livestock are typically raised with numerous inputs from elsewhere. Inputs replace cycles. And it is those inputs that can break down local food systems . . . or make them all the more important.

Food Recovery and Microbe Management

As we face the enormous scale of our municipal waste stream dilemmas, it's rather astounding that we are so reliant upon microorganisms to solve the bulk of the problem. These tiny organisms are the initial beneficiaries of our food waste recovery efforts. Masters of decomposition, they have voracious appetites. Like us, they do have some dietary preferences and required working conditions, but they're typically too busy eating to complain much. Our job is to design systems that capture and contain the recovered organics and support optimal conditions for the feeding frenzy, and to run it as a sensible business. It gives micromanagement a whole different meaning.

Although there are multiple variations on how it's all done, the two most sensible means of capturing organic material out of our waste stream and converting it into one or more valuable products are aerobic composting and anaerobic digestion (AD). The essential difference between the two processes is that aerobic composting requires oxygen, whereas anaerobic digestion occurs in the absence of oxygen. Both make more sense than the far-too-common alternatives: putting organics into landfills or incinerating them.

The recovery rate for food waste in the United States is abysmal overall—less than 3 percent of food waste was pulled from municipal waste streams nationally in 2010.[23] While that figure shows the problem, the increasing success of a number of U.S. cities in "organics diversion" demonstrates the potential for rectifying the situation. Separation of food scraps from the rest of the waste stream—known in the resource recovery business as diversion—is the challenge. Central to the success of any such effort is creating a market-based incentive for haulers, simple and sanitary systems for households and businesses, and a supportive regulatory environment for everyone involved.

Once the diversion questions are resolved, the issue becomes what to do with the collected organic food wastes. The bulk of the answer resides in the desired final product of the process, but the by-products also need to be considered. Sometimes the by-products, such as biogas, end up being the drivers in the decision-making process. In the end, the biggest supporters of local food systems are actually microscopic.

Composting

The kitchen compost pail is becoming increasingly common around the country. Prior to 1990, the amount of food waste composted on an individual basis was not even assigned a number by the EPA; rather, it was simply listed as "negligible." Now, we're up to just over a third of a pound per person per year through solid waste collection programs.[24] That number doesn't make for great bragging rights, but at least it is finally quantifiable!

Remember that Western B movie we were talking about earlier? Composting is essentially the gunfight at the O.K. Corral. All of the hired guns are present and itching to get things started, and as soon as the newcomers pull into town, the breakdown of law and order—or at least organics—commences. Once the pileup begins, bacteria and other organisms start to work. Thermophilic ("heat-loving") bacteria get to work at the insulated center of the pile, and mesophilic ("moderate-temperature-loving") bacteria start breaking down the organics around the center of the pile, all in conjunction with a host of organisms such as actinomycetes, fungi, and protozoa.

"Microbe management" in this context involves working to ensure that the composting process is both efficient and uniform. Composting operations essentially manage the proper ratio of carbon (C) to nitrogen (N), with C:N ratios generally ranging from 25:1 to 40:1. Most food waste contains a relatively high nitrogen content, so adding carbon in the appropriate amounts is important for both the composting process and the quality of the final product. Moisture, oxygen, and temperature levels are carefully monitored to ensure peak efficiency of the composting process.

Many large-scale composting operations utilize windrows—long, high rows of compost that can be turned and churned to oxygenate the pile and ensure that all of the compost goes through a thermophilic stage. The heat in this stage kills most pathogens and even weed seeds, and it allows for the proper breakdown of the organic materials into a much-coveted soil amendment and growing medium. Rich in organic matter, nitrogen, phosphorus, potassium, and micronutrients, compost is a living soil that will only gradually release its nutrients (although too much water can leach them away if the compost is not well managed). Increasingly, large composting operations are running more of their operations under cover. Having a roof and even an enclosed facility enhances the requisite micromanagement: controlling moisture and temperature levels, as well as capturing runoff and minimizing odors.

Some composting businesses are also moving toward in-vessel composting, adding additional elements of control and more efficient mixing of materials. Traditionally, composting operations have been located on the outskirts of towns and cities—close enough that the hauling costs to the site are not exorbitant but far enough away that runoff, odors, and pests (insects, rodents, and birds) do not present a problem. However, developments in compost science and technology are enabling operators to locate their operations closer to the sources of food, yard, and paper wastes. Siting these locations nearer to town and city centers makes compost as a soil amendment and growing medium more accessible, affordable, and popular. Carefully produced compost can be a boon to gardens, parks, rooftops, container gardens, lawns, and even large civil engineering projects such as highway construction that require good soil. In essence, this compost starts to close the ecological loop for local food systems.

Compost brings to the soil numerous benefits compared to synthetic fertilizers: minimal nutrient runoff, enhanced nutrient retention, increased absorptive capacity, superior drought tolerance, improved biological diversity, and better structure. For our purposes, one other benefit stands out: compost can be locally produced under local control with local dollars, creating local jobs and local resilience. Next to good food, I cannot imagine anything that has more potential to bring together bleeding-heart liberals and die-hard conservatives than locally generated compost (in fact, the perfect composting operation uniform may well be camo pants and a tie-dye shirt). There's something for (and from) everyone in compost!

Aerobic composting is, in many ways, an exercise in ecological simplicity. However, in this era of climate change, it also has a serious drawback: the release of CO_2 and other greenhouse gases. Additionally, the loss of heat in the thermophilic process is a loss of energy if the heat does not have an end use. This heat can be captured to raise ambient temperatures within a structure, as was done for centuries in some agrarian societies. These cultures would locate a house on top of an animal stall where the animal bedding, manure, and urine combined in a deep pack that helped keep both animals and humans warm during its slow decomposition throughout the winter. More complex but versatile technologies now exist that involve transferring heat from the compost medium to another location through the circulation of a liquid that moves the heat from one point to another. Such heat recovery systems from compost are particularly useful in operations such as dairies, where hot water is in constant demand.

Municipal and farm-based composting is a biological process requiring heavy equipment for processing piles and windrows but relatively little additional infrastructure. The dividends are clear, but so are the losses: invisible greenhouse gases escape into the atmosphere instead of helping to reboot the whole system. Instead, just imagine the purr of a methane-powered truck, rumbling out of the solid waste management facility to collect more organics for feeding its anaerobic digesters. Methane never sounded so good.

Anaerobic Digestion

Think of an anaerobic digester as a giant stomach, supported by an army of breathless microbe managers ready to consume any variety of organic materials. Food waste, yard trimmings, nonrecyclable cardboard, and human and livestock excreta are all examples of potential feedstock (i.e., the materials that are put into the digester). Of course, some materials and combinations are more efficiently digested than others. The more inclined a material is to rot, or putrefy, the more easily it will be digested. Materials with high levels of lignin, the substance that provides the rigidity that we associate with "woodiness" in plants, are more resistant to anaerobic digestion, so these materials must be introduced in limited quantities.

Once the organic waste is collected, it is mechanically processed into a thick soup and then transferred into the closed system. Oxygen is eliminated from the digester, and anaerobic bacteria begin their work, breaking down the organic materials into solids and liquids, known as the digestate, and a trapped biogas that is roughly 60 percent methane and 40 percent carbon dioxide (depending upon the feedstock). The biogas is captured for use as a fuel, whereas the digestate has multiple end uses, depending upon its contents and the final stages of processing. Facility designs and management techniques determine the state and content of the final digestate products, which are either separated as solids and liquids or combined as a sludge. The digestate can be aerobically composted to optimize its use as a soil amendment; the carbon contained in the digestate is very stable and can serve as an important sequestration tool.

The amount of biogas produced varies according to the energy content in the feedstock. Plant material with minimal lignin levels will typically create more biogas than will excreta (manure) from humans and animals—up to three times as much biogas, in many cases. The digestive processes of humans and livestock are geared to capture significant energy from foods, so the excreta contain less energy for extraction through anaerobic digestion. In fact, both municipalities and farms working with excreta in anaerobic digesters often mix in food wastes to increase biogas production.

Both aerobic composting and anaerobic digestion create greenhouse gases, but anaerobic digestion provides a mechanism for capturing a significant portion of these greenhouse gases and converting them to a renewable source of energy. As mentioned previously, methane is the cleanest and most efficient of the volatile hydrocarbon fuels that are so useful to our contemporary existence, and it can be used to produce heat, electricity, or vehicular power. Some carbon is still released as a result of the anaerobic digestion process, but a thorough review of current research indicates that anaerobic digestion tends to be the most efficient process for capturing the residual energy of organic waste materials and for sequestering carbon and soil nutrients necessary for rebuilding soils.[25] Furthermore, converting biogas to electricity, in the combustion phase, can remove up to 99 percent of volatiles such as ketones, aldehydes, ammonia, and methane from the biogas. One study indicates that the anaerobic digestive process for electricity production is therefore seventeen times more efficient than aerobic composting.[26] While the precision of these numbers will always be challenging, the trend itself is clear. Anaerobic digestion technologies are probably critical to our future—and other countries are surging ahead of us in employing these technologies.

Similar to the environment in the digester, the decomposition process in landfills is primarily anaerobic. So why should local communities and solid waste facilities invest in anaerobic digestion instead of simply capturing the escaping methane from landfills? By now, it should be clear that there are several key reasons never to use landfills for the disposal of food waste (parallel arguments could be made for incineration):

- Methane capture is highly inefficient in most landfills.
- Nutrients that could be valuable for the soil become mixed with multiple pollutants and inorganic materials.

- Off-gassing from the breakdown of other materials in the landfill creates a mixture of undesirable and often dangerous gases other than methane and carbon dioxide.
- Nutrients are buried and virtually irretrievable.
- Encouraging methane capture from landfills as a solution incentivizes the disposal of food waste and other organics into landfills.

In sum, anaerobic digestion and composting are resource recovery; landfills and incinerators are resource disposal. Whether we are predominately ecologically minded or economically minded, disposing of our increasingly limited resources makes no sense whatsoever. If we have any concern for future generations, allowing food waste and other organics to enter landfills is simply unethical.

In the end, the point is not to make a definitive pitch for a particular method of food waste recovery for a given community or business, but simply to argue for *recovery* in lieu of disposal. Each municipality and entrepreneurial initiative must choose the recovery method that best fits its needs. The beauty of composting and anaerobic digestion approaches to waste recovery is that local control and investment are already embedded in how we manage these systems. We can manage the processes and the products in ways that build regional agricultural productivity and potential while also helping to confront the realities of the local waste management stream.

Ecological Sanitation

We're finally beginning to think in terms of food systems as ecological cycles. My friend Dan Sullivan (previously an editor at *New Farm* and *BioCycle* magazines) likes to describe the cycle this way: "Food, energy, and waste . . . food, energy, and waste . . . food, energy, and waste."

As an academic, I embrace this kind of analytical ecological cycle approach because the notion of a system helps tidy things up. On the other hand, farmers like myself (I live in both worlds, you'll remember) don't always get too excited about "food systems" because we experience the messy side of things almost on a daily basis—things on the farm just don't always fit into tidy theories and diagrams. And sometimes those diagrams don't tell the full story.

I confess that I like riding the fence on this one (the wooden one, *not* the electric one). I appreciate what we can learn and change by studying food systems, and yet I like to maintain a cautious arm's length from a full embrace. Studying food systems helps us unpack our contemporary dilemmas and reassemble the parts in a more sustainable manner. But the tendency toward the tidy among some theorists in the food systems world worries me. If we're going to depend on the analytical nature of food systems to take us down the right path toward more equitable and ecological practices, then we need to begin with the full slate of components that comprise a food system. When I teach students about sustainable farming systems or when a government official does a nutrient management plan for a farm, a complete ecological assessment requires us to look at all of the inputs and all of the outputs, manure included. Things coming and going on the farm aren't always pretty, but they all play a role in the realities, for better or for worse. A thorough systems analysis doesn't involve picking and choosing only the components within our comfort zone.

When I look at our study of food systems, I see two gaping holes. One is the estimated 335 million dry tons of animal manure produced by CAFOs (concentrated animal feeding operations) and AFOs (animal feeding operations).[27] That is a critical issue confronting our nation's farms and rural communities. Then there is the eight-million-ton gorilla, the one that we don't want to look at, think about, or acknowledge in any way, especially when it comes to linking it to our food system: human excreta.

Eight million dry tons: that's the general estimate of how much we produce in captured human excreta (wastewater treatment plant engineers refer to the final product as "biosolids") per year in this country. It's all a product of our food system, and whether we like it or not, it's local. Heck, it's so local that it's downright personal!

Whereas the livestock manure issue is something that farms are supposed to manage and state and federal agencies are charged with regulating and monitoring, the human biosolids dilemma has a local dimension that we seem to wish would just go away. The question of what we do with human excreta is firmly rooted in the local domain, handled all across the United States by our municipalities. It's one part of the food system that we must think about in much more detail and complexity, particularly as we think

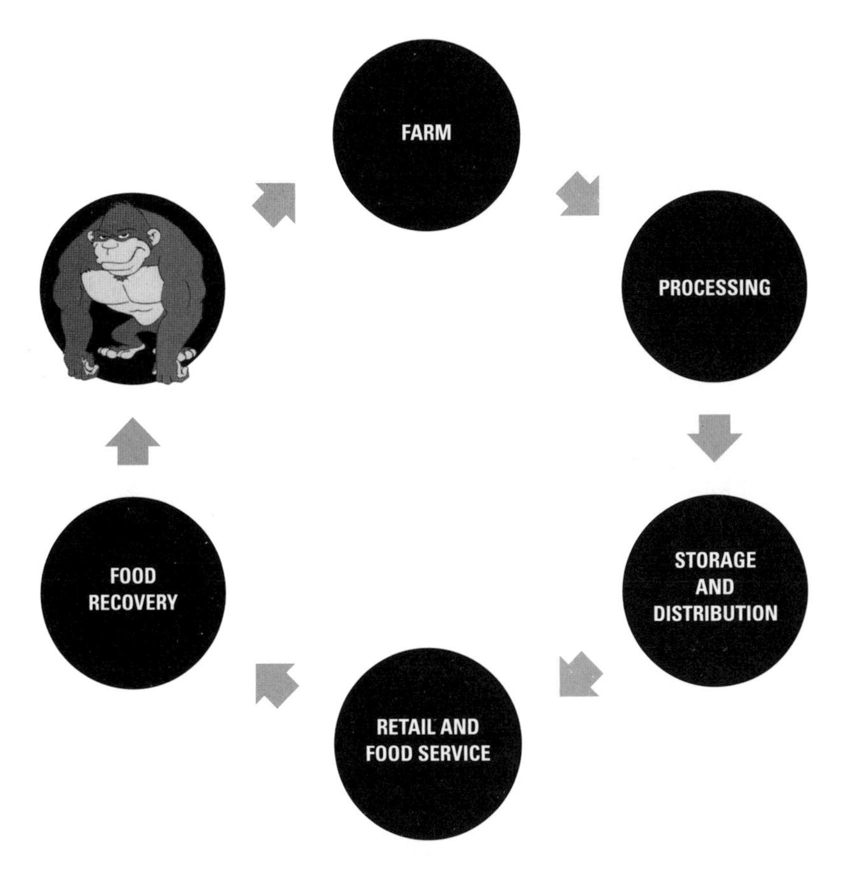

FIGURE 5-8. The Missing Eight-Million-Ton Gorilla.

about the ways in which managing nutrient cycles are linked to the importance and resilience of local food systems.

The proposal to bring human excreta back into the local food system is not that radical. Nor is it by any means a new call to ecological sanity and ecological sanitation. In 1872, Victor Hugo portrayed the dilemma bluntly but eloquently in *Les Misérables*. It's worth remembering that he wrote these words as the flush toilet was first coming into use and phosphorus hunters were setting sail to find monumental mounds of manure left by nesting birds:

> Paris casts twenty-five millions of francs annually into the sea. And this not metaphorically. How, and in what way? Day and night. For what object? For no object. With what thought? Without thinking.

What to do? Nothing. By means of what organ? By means of its intestine. What is its intestine? Its sewer.

Twenty-five millions, the most moderate of the approximative amounts given by the estimates of the special science.

Science, after groping for a long time, knows now that the most fertilizing and effective of manures is human manure. The Chinese, let us say it to our shame, knew this before us. Not a Chinese peasant, it is Eckeberg who states the fact, goes to the city without bringing back, at the ends of his bamboo stick, two buckets full of what we call filth. Thanks to human manure, the soil in China is still as young as in the days of Abraham. Chinese wheat yields one hundred and twenty fold. There is no guano comparable in fertility to the detritus of a capital. A large city is the most powerful of stercoraries. To employ the town in manuring the plain, that would be certain success. If our gold be filth, on the other hand, our filth is gold.

What is done with this filth? It is swept into the gulf.

We send at a great expense fleets of ships to collect at the southern pole the guano of petrels and penguins, and the incalculable element of wealth which we have under our hand we cast into the sea. All the human and animal manure which the earth loses, returned to the land instead of being thrown into the sea, would suffice to nourish the world.

Those piles of ordure at the corners, those carts of mud carried off at night from the streets, the frightful barrels of the department of highways, and the fetid streams of subterranean mire which the pavement conceals from you, do you know what this is? It is the flowering field, it is the green grass, it is the mint and thyme and sage, it is game, it is cattle, it is the satisfied lowing of heavy kine at night, it is the perfumed hay, it is the gilded wheat, it is the bread on your table. It is the warm blood in your veins, it is health, it is joy, it is life. So wills that mysterious creation which is transformation on earth and transfiguration in heaven.

Restore this to the great crucible; your abundance will issue from it. The nutrition of the plains produces the nourishment of men.

You are at liberty to lose this wealth, and to consider me ridiculous in the bargain. That would be the masterpiece of your ignorance.[28]

It is a fact that the vast majority of our local food systems are not self-reliant or self-sustaining in terms of fertility inputs, much less energy. Too many of our soils are eroded, compromised, covered, and sometimes just altogether missing. To rebuild soils and their fertility, we need reliable sources of organic matter, nitrogen, phosphorus, and other nutrients. When possible, those sources need to be local—in other words, within the sphere of our influence and care.

Describing and analyzing our food system objectively demands that we acknowledge the food system in its entirety, even if the realities of the full ecological cycle are not in keeping with our social norms or levels of personal comfort. We tend to keep our public discourse about food systems clean and academic. It is only when we start talking about livestock and farming systems that we begin discussing excreta. As an ecologist, I am intellectually obligated to teach about manure in discussing farm systems. Should I not have a parallel obligation to discuss human excreta when teaching about human food systems? If I were teaching human biology, I would certainly discuss urine and feces. The "end result" of our local food systems is clearly part of the ecological cycle, but most of us in the United States refuse to acknowledge the link. Apparently, it's okay to talk about these matters in agriculture and biology classes but not in food systems classes. Well, pooh.

Some will consider the previous argument to be a mere academic exercise. However, we face a much less abstract conundrum: municipalities across the country are simultaneously importing the very same energy-intensive agricultural nutrients that they are landfilling and incinerating. The charts in figure 5-9 demonstrate the clear correlations between biosolids and soil fertility. Note that the essential elements needed for plant growth are all present in biosolids, including sufficient levels of nitrogen and phosphorus. Furthermore, the addition of human excreta to soils contributes additional carbon to the soil—in sharp contrast to the petrochemical fertilizers that add no carbon to the soil but do release significant amounts of carbon in their production.

By now, I've probably made some people uncomfortable and even infuriated a few others by broaching a variety of cultural and ideological taboos, but perhaps even more so by linking the resilience of local food systems to

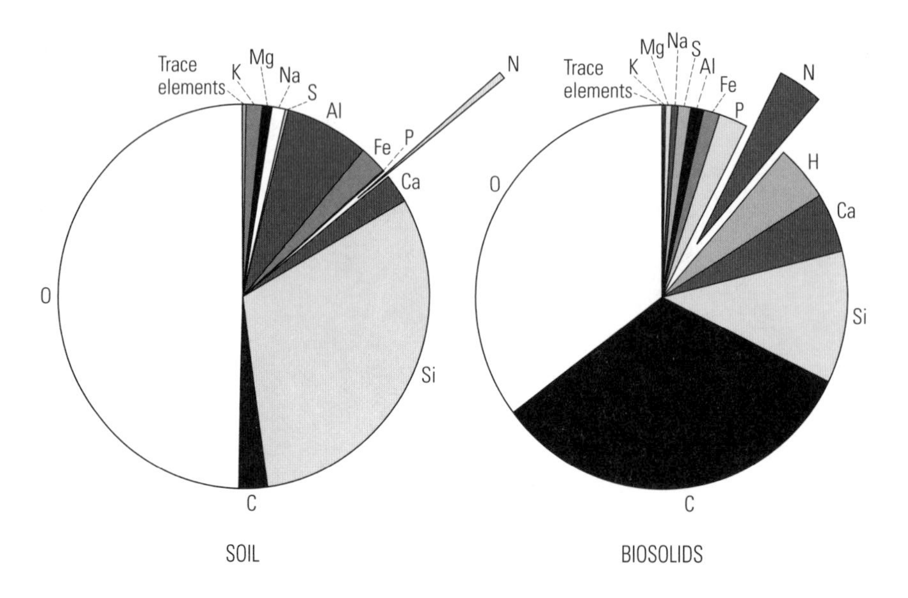

FIGURE 5-9. Comparison of the Chemical Composition of a Typical Soil and Typical Biosolids. Source: Charles Henry, Dan Sullivan, Robert Rynk, Kyle Dorsey, and Craig Cogger, *Managing Nitrogen from Biosolids* (Washington State Department of Ecology and the Northwest Biosolids Management Association, 1999).

the management of human waste. The biggest problem in this dialogue—if anyone is still speaking to me, that is—is that our American municipal wastewater systems not only combine urine and fecal matter but also mix in every other chemical and item that we wash down the sink, the toilet, and sometimes even city storm drains. As a result, this "sewage sludge" can be a cocktail of undesirable chemicals, objects, and pathogens, at least during the initial stages of treatment. The debate is made all the more complicated as government agencies and others try to determine acceptable (i.e., safe) levels of the various substances present in the final biosolids product.

Part of the reason that we have not been able to get to a point of ecological sanity in the United States with regard to including human excreta in our food system is that we have not pursued "ecological sanitation." A concept in use both in highly advanced industrialized cultures like Sweden and also in developing countries such as India, ecological sanitation is the design of hygienic systems for handling human excreta as a resource, not as a waste product.[29] Sometimes referred to as "ecosan" (for "ecosanitation"), these approaches to handling human excreta range from simple and inexpensive to high-tech and pricey. Many of the systems separate urine and feces in the collection phase of the human excretion process and then handle them separately as valuable nutrients and even sources of energy. The urine can be collected and utilized immediately or after a waiting period, whether at the residential, farm, or community level. Fecal matter, on the other hand, requires further handling, through either aerobic composting or anaerobic digestion.

The advancement of anaerobic digester technology offers some interesting possibilities for processing separated fecal matter into safe and beneficial natural fertilizers. Odors, potential pathogens, leachate, and greenhouse gases are contained, while contact with disease vectors such as insects, vermin, and heavy precipitation is minimized, if not altogether eliminated. Heat, hot water, and electricity can all be produced through the combustion of the resulting biogas, and once again, virtually all of the volatile compounds will be eliminated at this stage. Not only is fertility reclaimed, but a community can also increase energy autonomy with the capture and use of biogas for lighting, cooking, electricity, and vehicles.

We are already utilizing some anaerobic digesters for the treatment of human excreta in the United States, although our biosolids are a combination of all the materials that enter our wastewater treatment systems. If just the excreta were being treated, separate from all of the other potential contaminants that appear with some frequency in our wastewater systems, then the entire debate of where and how to use human excreta as fertilizers would be less complex. (Note that I did not say simple. Just less complex.)

Moving toward ecological sanitation requires a change in mind-sets and often an accompanying shift in technologies and wastewater infrastructures. Ideally, we would divert urine from the wastewater stream and isolate feces from all of the other materials that end up in our wastewater. Some solutions for these goals are centered at the residential scale, while others are dependent upon a community's collection and treatment facilities. Some community systems rely upon belowground storage of excreta in separated aging chambers, while high-tech approaches use vacuum systems to move fecal matter to a central location with little to no water use. Ideally, the collection and storage systems are linked to biogas capture technologies for energy applications and greenhouse gas reductions.

The common denominator of all of these systems is the driving principle that human excreta is a local resource worth utilizing. The distinction between these systems is that they need to reflect the resources and needs of the local communities in which they are developed. There is no one-size-fits-all scenario, nor is there a single perfect system available for all communities. In fact, the development of the best infrastructure for transforming the "end result" of our food systems into fertilizer and fuel is often a series of strategic and experimental steps for a community. The experimentation in technologies, infrastructure, and management of these systems in local communities across the world is tremendously exciting. Innovation through local experimentation and global cross-pollination provides us with a rich storehouse of ideas and approaches.

We are in the early stages of a transition with regard to turning human excreta into a resource. Resistance runs deep as we confront an issue we've worked hard not to deal with for more than a century. Many of my sustainable agriculture compatriots are adamantly opposed to the use of biosolids for agricultural applications. Given the current technological shortcomings of many municipal systems, I share many of their concerns about the potential for contamination and pathogens in some biosolids. However, I think it critical for us to embrace the possibilities rather than simply rejecting the status quo. We also need to recognize the incremental nature of this enormous undertaking.

One example of how we can move forward in stages is by addressing the food waste and human excreta issues simultaneously through shared anaerobic digestion facilities. Combining food waste and human excreta in an anaerobic digester—a process known as codigestion—is sometimes a transitional step forward, but it also has certain advantages over maintaining separate resource management streams. Codigestion can offer the following benefits to local communities and businesses:

- The synergy between the different feedstocks can actually heighten the efficiency of the anaerobic digestion process, in part because food waste contains as much as three times more energy than human excreta.
- Utilizing existing wastewater treatment anaerobic digestion facilities can be an economically feasible interim step, providing opportunities to experiment and plan for future operations.

- Wastewater treatment plants with anaerobic digesters often have extra capacity, and the volatile solids destruction rate for food waste is 86 to 90 percent, meaning that these extra inputs create additional energy outputs without overly taxing the existing plant's capacity.[30]

Advocates of sustainable agriculture and local food systems simply *must* participate in helping to develop locally based solutions that incorporate human excreta into our local food systems thinking and infrastructure. By stonewalling on this issue, we are ignoring a host of ecological realities. The contrast is stark, as evidenced in figure 5-10.

The first photograph shows the immense infrastructure of a synthetic fertilizer industrial complex in Indonesia, an operation requiring extensive fossil-fuel inputs. The next photo depicts a wastewater treatment facility in the United States that uses anaerobic digestion to convert food waste and human excreta into biosolids and clean-burning biogas. If we convert the visual images into verbal contrasts, the realities of our choices become even clearer (see table 5-2).

We need to be talking more about the ecological drivers than the truck drivers in our discussions about local food systems. We were too quick to allow "local" to trump "organic" as the locavore movement took hold. Our

TABLE 5-2. Comparison of Synthetic Fertilizer Production and the Management and Use of Organics through Anaerobic Digestion

FACTOR	SYNTHETIC FERTILIZER PRODUCTION	ANAEROBIC DIGESTION OF EXCRETA & OTHER ORGANICS
Function	Industrial manufacturing	Predominately biological
Infrastructure	Significant	Moderate
Feedstock	Fossil fuel	Recovered organics and microbes
Input of nonrenewable energy sources	High	Low
Pollutant and greenhouse gas contributions	Significant	Mitigated
Production location	Distant	Local
Soil health impact	Negative	Positive

FIGURE 5-10. Synthetic Fertilizer Complex and Wastewater Treatment Facility. Source: Mosista Pambudi/Shutterstock.com (top); Liviu Toader/Shutterstock.com (bottom).

FIGURE 5-11. Organic Soil Compared to Conventional Soil. Photograph courtesy of Rodale Institute from *The Farming Systems Trial: Celebrating 30 Years* (Kutztown, Penn.: Rodale Institute, 2011), http://www.rodaleinstitute.org/files/FSTbookletFINAL.pdf.

focus on local foods instead of local food systems was myopic. An exploration of local food *systems* requires us to dig deeper, to think ecologically and complexly, to pay attention to the fundamental driver of local food systems—the soil. Examine figure 5-11 for a moment. This photo of two soil samples demonstrates the difference between an identical soil maintained in two different ways for thirty years. The darker soil on the left has a high number of aggregates and clearly possesses a higher organic matter content and a better soil structure for plant growth than the conventionally managed soil on the right. These soil samples were taken from the Rodale Institute's Farming Systems Trial, a long-term study to characterize the differences between organic and conventional farming methods.

Is local "better" than organic? Can we continue to ignore the differences in production methods if we are concerned about the resilience of our local food systems? Can we continue to ignore the importance of fully reintegrating all by-products and end products of our local food system into a nutrient cycle that makes ecological sense?

Ultimately, sustainable agriculture practices and local values have to be interconnected, albeit with inevitable points of tension. There will need to be some give-and-take from both perspectives, but agriculture is essentially an extended compromise between us and the soil.

It's clear by now that no silver bullet will put to rest the whole host of environmental dilemmas posed by our modern food system. However, there is at least one loud retort loaded in the barrel of the sustainable agriculture cannon: *resource recovery drives regenerative food systems.*

The United Nations' Human Settlements Program's "Global Biosolids Proclamation"

Therefore, we urge all Earth's communities and nations to advance programs that put to their best use the resources—nutrients, organic matter, and energy—in excreta and wastewater sludge, to the extent reasonably possible, as determined by local professionals and the communities with which they work. For these goals to be reached, this proclamation will need the strong support of water quality and waste management professionals worldwide, their professional organizations, environmental and agricultural groups, research scientists and academic institutions, international development agencies, other NGOs, political leaders at all levels of government, regulatory agencies, the media, and the general public.[31]

Chapter 6

Food Security

When we started researching the potential for local flash-frozen produce in Vermont's public institutions, neither Garland nor I ever considered that the project might land her in prison. Yet there she was—surrounded by razor wire, armed guards, and male inmates—processing and freezing vegetables grown within the confines of the Northwest State Correctional Facility in Swanton, Vermont. Local frozen products, straight from the cooler. After all, what do we really need—license plates or nutritious food?

Garland Mason, the new markets research specialist at Green Mountain College, and her colleague Hans Estrin, local food network coordinator from the University of Vermont's Cooperative Extension Service, had arrived that morning, towing the state of Vermont's mobile flash-freeze unit behind their dual-wheeled pickup. The guard at the gate signaled for them to proceed slowly through the first gate, before a guard at the second gate had them stop and open up the vehicle and the flash-freeze unit for a thorough inspection. No freeloaders, no weapons, not even any knives for processing vegetables. The guard motioned them through the gate to the rear of the dining hall facility, where Garland, culinary staff, and the inmates would spend the next two days processing the vegetables grown within the walls of the facility.

The odd juxtapositions of Garland's two-day adventure—young college graduate among young prison inmates, gardens within correctional facility walls, and cutting up a half ton of vegetables with no knives—are all indicative of the creativity involved in rebuilding local food systems and the unusual alliances that must develop. If you're doing effective local food systems work, you should at some point find yourself outside of your comfort zone and potentially rubbing shoulders with some unexpected bedfellows, whether they be prisoners or corporate executives. Finding common cause means starting from different realities.

Garland had come to Green Mountain College as our "local-link coordinator," an AmeriCorps VISTA position dedicated to helping the college and the surrounding Rutland region rebuild a vibrant local food system. As she worked on food access and security collaborations with the college, the nonprofit Rutland Area Farm and Food Link (RAFFL), and Hans, we all wondered whether local frozen produce could increase opportunities for farmers, institutions, and community food pantries. Coincidentally, the Vermont Agency of Agriculture wanted an organization to pilot its newly constructed mobile flash-freeze unit, an experimental project that was designed to go from farm to farm, freezing produce for the marketplace. The college put in a proposal to operate the unit for several growing seasons, and we procured a $100,000 grant from the Jane's Trust foundation to get the unit out for a serious road test. Little did Garland or anyone else realize just where that two-season tour would take her.

As it turns out, growing and processing food within a correctional facility offers some advantages, not the least of which is that certain inmates have particular expertise at growing things in confined spaces. Fortunately, those skills are transferable and can be geared toward growing food, not just herbaceous materials that go up in smoke. Second, food processing in a correctional facility tends to be quite efficient. "You wouldn't believe how fast we got through all of those vegetables—they worked *so* fast!" Garland recounted to me upon her return from Swanton. "And that was with the inmates using just curved dough cutters with the plastic handles, since they're not allowed to have any 'sharps.'" I gave her a quizzical look. "Sharps—knives," she explained with a grin.

Garland and her cohort processed and froze cabbage, summer squash, tomatoes, peppers, and basil, all items grown in the confines of the correctional facility under the tutelage of master gardeners (people who have completed the coursework provided by the agricultural extension service and then lend their gardening expertise to nearby individuals and organizations).[1] The project was an experiment to link the garden training with the facility's food service, so the frozen produce was to be served in-house in later months. Other Vermont correctional facilities with gardens tried out the flash-freezing, too, and the Department of Corrections (DOC) is now working to develop a distribution system to move both fresh and frozen products among the different facilities. As with any project in the corrections world, there are pros and cons. One drawback was that the vegetable trimmings were thrown in the trash, not the compost: the DOC considers compost to be a potential place to stash weapons (sharpened carrots and zucchini bludgeons?) or even one's self, so they haven't closed that ecological loop quite yet. Nonetheless, creative thinkers are continuously connecting the dots, one by one, so a composting compromise is likely in the long run.

Upon hearing Garland's story, I couldn't help but wonder whether Vermont could make fresh or frozen local foods more accessible and affordable for public schools by having correctional facilities grow and process the produce. After all, it was a way of turning cons into pros! The affordability of local foods is often a stumbling block in schools and other institutions, and having the production and processing costs shared with a program providing inmates with gardening, culinary, and lifelong nutritional skills seemed like an excellent way to get the costs down. But as soon as I got really excited by the idea and began researching it further, I discovered just how behind the times I was.

Correctional facilities from California to the Carolinas are experimenting with all sorts of programs involving farming, the culinary arts, and community service. Hard agricultural labor for prisoners and even some associated skills training go far back in U.S. history, but many of the more recent culinary training efforts seem to have stemmed from the efforts of William E. Smith, a former White House chef who helped establish a much-lauded "cons to cooks" apprenticeship program back in the 1980s. Trained for five hours each week for three years, the inmates still serve first-rate meals under Smith while serving time in Lorton, Virginia. Smith's apprentices have repeatedly landed good jobs, sometimes even in lavish restaurants.[2]

In North Carolina, former inmates join recovering addicts and the homeless at the Community Culinary School of Charlotte. This intense training program begins with a pledge from participants: a commitment to the culinary profession and their future place of employment before beginning their intense training. Fifty students are accepted into the program each year, and they pay for their schooling through their work with local charities capturing fresh food from local farms that would otherwise be destined for disposal. This food rescue operation is the basis for the delivery of 2.5 million pounds of food each year to more than 125 charitable groups in the region—and it's all supported by an annual operating budget of $250,000 at the culinary school.[3] Another facility, the Northeast Correctional Center in Concord, Massachusetts, even serves the public in a restaurant, the Fife and Drum. You have to pass through security to get to a table, but that's a small price to pay for a hearty lunch costing just over three dollars![4]

Are these creative enterprises as beneficial for the inmates as they are for the communities in which they operate? A recent study done at yet another culinary training program in Massachusetts found that of the 176 inmates who participated in the Middlesex House of Correction's inmate culinary arts program, only 10 percent returned to prison[5]—quite impressive considering the relatively high recidivism rate at conventional Massachusetts prisons. Looking at the variety of such programs across the country and their successes, it's hard not to come to the conclusion that the table is indeed the place where we find our way to civility and common interests. Indeed, inmates commonly describe the sense of purpose and gratification they discover in watching customers' faces while partaking of the inmates' culinary creations. There need to be some rewards in this punishment business if we expect to alter career paths and life choices.

Minimal Security and the Search for a Way Out

I realize that most people don't pick up a book about local food systems and expect to read a lot about prisons and food waste. It's more fun to think about beautiful displays at farmers' markets, white-tablecloth presentations, and overflowing pantries and root cellars. But if we ignore the less obvious and more disconcerting aspects of our food systems, then we certainly cannot begin to understand the full scope of realities we face. In the end, rebuilding local food systems requires us to connect with the neighbors we've never known as much as it does to share the bounty with our comfortable acquaintances. This refocusing of our interests and priorities starts to transform

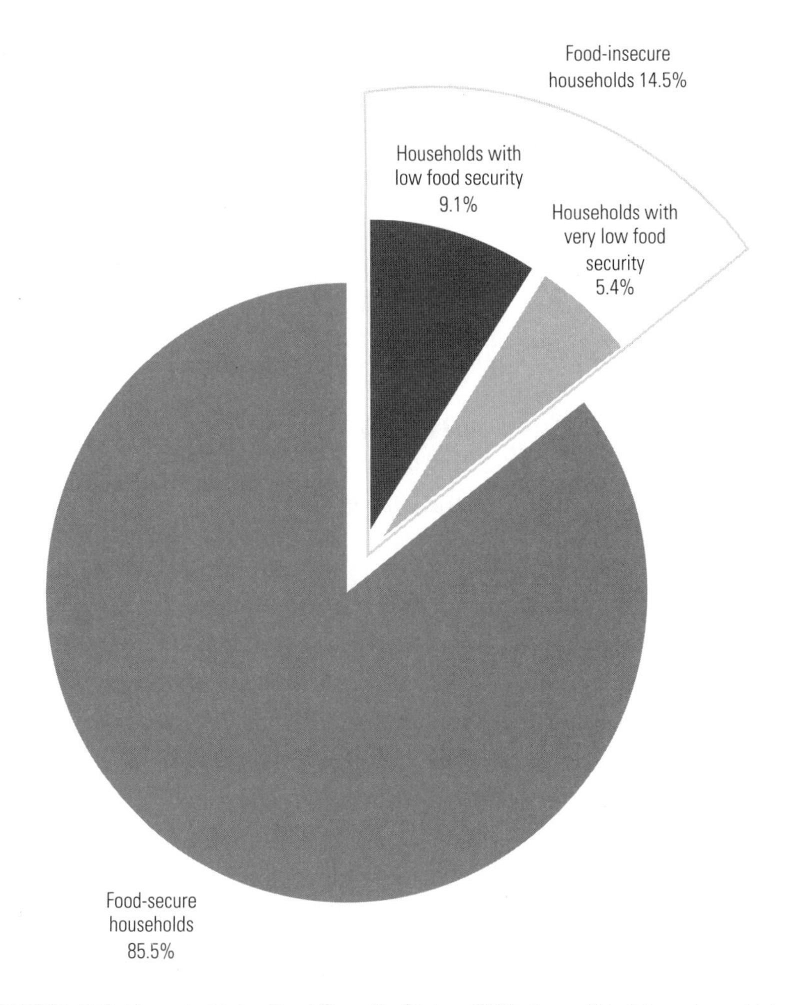

FIGURE 6-1. U.S. Households by Food Security Status, 2010. Source: Alisha Coleman-Jensen, Mark Nord, Margaret Andrews, and Steven Carlson, *Household Food Security in the United States in 2010,* Economic Research Report no. 125 (Washington D.C.: USDA Economic Research Service, September 2011).

statistics into neighbors, consumers into customers, producers into farmers, and prejudices into dissolved perceptions. In the process of peeling back the layers that comprise any community food system, we finally arrive at the hollow center around which it is all based: the threat of hunger. Just as hunger is the ultimate driver of our food systems, it is also the cruel jester that satirizes our seeming inability to create just and equitable food systems. Obviously, hunger and malnutrition are merely two prominent symptoms of deeper-rooted societal ills, but the necessity for triage means that we have to address these symptoms before taking the time to perform a full diagnosis and treat the full scope of the malady. Otherwise, the diagnosis risks becoming an autopsy—and that is unacceptable.

Unfortunately, food security in the United States is, by all measures, a privilege, not a right. Approximately one out of seven American households is food insecure. And an estimated 5.4 percent of those households experience very low food security.

However, those figures only begin to tell the realities of food security in our country. If you focus only on American households with children, then more than 20 percent of those households fall under the government's definition of "food insecure." A deeper examination reveals that over one-quarter of black and Hispanic households are categorized as food insecure[6] (bear this figure in mind as we look at the demographics of persons doing the majority of the field work and food processing labor in the following chapter).

Such are the inequities—and then come the ironies. One might surmise—and rationally so—that living in a region with a more hospitable growing climate would help ensure food access. In reality, however, the southern tier of the United States is the most heavily impacted region, as indicated by figure 6-2.

You might also assume that the inhabitants of nonurban areas would be less likely to experience food insecurity, given the seemingly logical conclusion that it would be easier to find both growing spaces and less expensive local food products in suburban and rural environments. Yet food insecurity remains quite high in many rural areas (14.7 percent in "nonmetro" areas), particularly across the South. Nevertheless, the most dramatic rates of food insecurity, as you might expect, are in the heart of "principal cities of metropolitan areas," impacting an estimated 17 percent of the population in those areas.[7]

Sadly, logic begins to seem somewhat elusive in the search for answers to food security issues in the United States. For example, how is it that we were unable to get the overall national rate of food insecurity under 10 percent even in the midst of the era of tremendous economic growth prior to the 2008 economic downturn, as indicated in figure 6-3 on page 104? (So much for trickle-down economics.) The only "logical" part of this data set is that food insecurity levels increased rapidly in response to the bursting of that economic bubble in 2008.

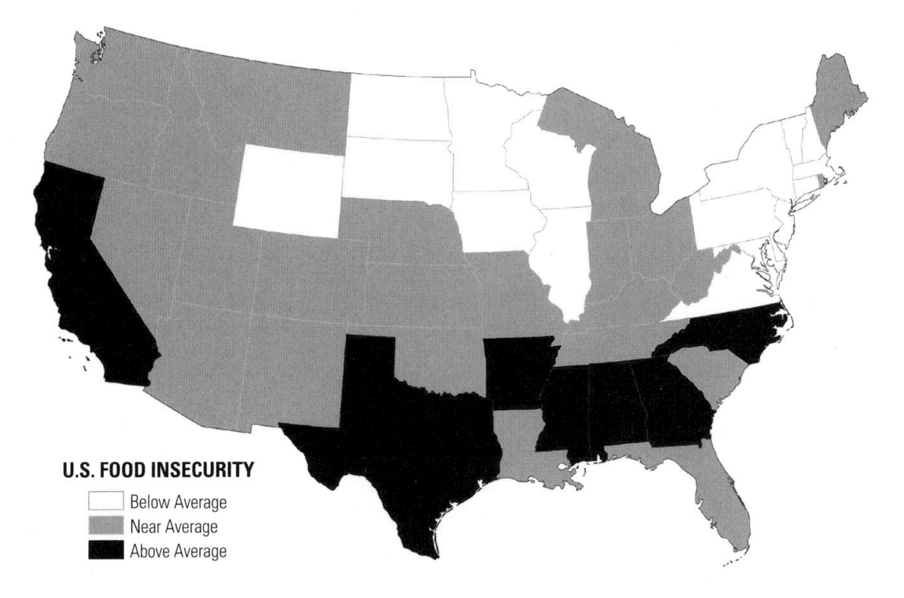

FIGURE 6-2. State-Level Prevalence of Food Insecurity in the United States, 2009–2011.
Source: Calculated by the USDA Economic Research Service based on Current Population Survey Food Security Supplement data, as shown on the web page "Food Security in the U.S.: Key Statistics & Graphics," USDA Economic Research Service, accessed September 4, 2012, http://www.ers.usda.gov/topics/food-nutrition-assistance/food-security-in-the-us/key-statistics-graphics.aspx#map.

We Americans are well versed in volunteerism, supporting nonprofits, and transforming religious ideals into action. In sum, we do a pretty good job in *responding* to problems. But we don't always seem to be so good at *fixing* problems—not even one as basic as ensuring that everyone has access to affordable, nutritious, culturally appropriate foods. The first step in addressing a problem is defining it. And the more complex the problem, the more challenging it can be to define it. Such is certainly the case with hunger: can we define it as a physiological condition that has no direct link to economic security? Malnutrition is equally complex and elusive. Despite being on opposite ends of the physiological spectrum, both obesity and emaciation are now recognized as conditions related to malnutrition. Why? Because not all calories are created equal. As a result, "food insecure" is, in some situations, a more apt term than "hunger" to describe the lack of access to foods of appropriate nutritional value.

Developing a common useful vocabulary is vital to advancing any community-based food system. This vocabulary can help us define

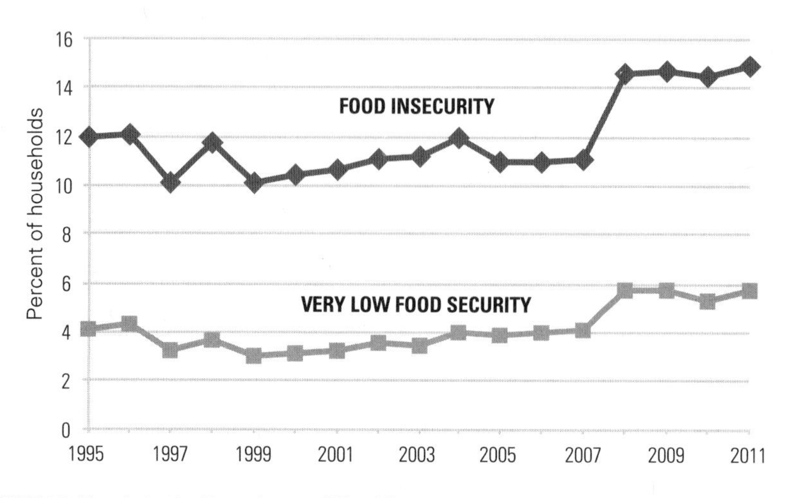

FIGURE 6-3. Trends in the Prevalence of Food Insecurity and Very Low Food Security in U.S. Households, 1995–2011. Source: Calculated by the USDA Economic Research Service based on Current Population Survey Food Security Supplement data, as shown on the web page "Food Security in the U.S.: Key Statistics & Graphics," USDA Economic Research Service, accessed September 4, 2012, http://www.ers.usda.gov/topics/food-nutrition-assistance/food-security-in-the-us/key-statistics-graphics.aspx#map.

common ground and create community-based strategies that can complement, and sometimes even provide models for, state and federal food security programs.

Food for All: Defining Moments

We are once more reminded that words matter and morph—yet another defining moment in the local food world! And what better way to clarify the issues than a food lexicon?[8] Since the terms and definitions are in a constant state of evolution, perhaps we should just abbreviate and call it a "flexicon." At any rate, coming to a common understanding of the following terms is important for understanding the challenges and approaches involved in rebuilding local food systems.

Food Security

Per the United Nations: "Food security exists when all people, at all times, have physical, social, and economic access to sufficient, safe, and nutritious food which meets their dietary needs and food preferences for an active and healthy life. Household food security is the application of this concept to the family level, with individuals within households as the focus of concern."[9]

According to the U.S. Department of Agriculture (USDA), "Food security for a household means access by all members at all times to enough food for an active, healthy life. Food security includes at a minimum the ready availability of nutritionally adequate and safe foods [and the] assured ability to acquire acceptable foods in socially acceptable ways (that is, without resorting to emergency food supplies, scavenging, stealing, or other coping strategies)."[10] The USDA has begun to describe individuals and families as "food secure" or "food insecure" based on a continuum that is to some degree quantifiable. Simple black-and-white categories can be problematic, as a family's food security can vary during the course of a year, or a family may have to consume undesirable foods to get by.[11]

Food Access

Although the definitions vary somewhat, most organizations and governments define food access as having the resources and the physical ability to procure adequate and appropriate foods necessary for a nutritious diet. "Appropriate" is a key word here, as some foods are culturally appropriate or inappropriate for certain populations, and it is also important that "access" does not include food sources that involve activities such as sorting through garbage.

Charitable Food System

As expansive in scope as it is simple in definition, the charitable food system includes the full spectrum of organizations, institutions, and agencies working on issues related to food insecurity. Spanning from neighborhood food pantries to city soup kitchens to international relief efforts, the charitable food system is the safety net for addressing short-term and chronic food insecurity issues. The groups that provide these services are often called "emergency food providers" (EFPs).

Community Food Security

Per food system researchers Michael Hamm and Anne Bellows, community food security is "a condition in which all community residents obtain a safe, culturally appropriate, nutritionally sound diet through an economically and environmentally sustainable food system that promotes community self-reliance and social justice."[12]

Per the USDA, "In the broadest terms, community food security is a prevention-oriented concept that supports the development and enhancement of sustainable, community-based strategies: to improve access of low-income households to healthful, nutritious food supplies; to increase the self-reliance of communities in providing for their own food needs; [and] to promote comprehensive responses to local food, farm, and nutrition issues."[13]

Food Justice

Maybe you'd better define this one for yourself. Or at least choose for yourself from among the options, of which there are many. That's the spirit of food justice—the right to self-determination in food access and choices. Below are two definitions to chew on for the time being, both coming from organizations well respected in the food justice world.

From the Growing Food and Justice for All Initiative, a collaborative initiative focused on "dismantling racism and empowering low-income and communities of color through sustainable and local agriculture":

> **Food Justice:** Asserts that food is a right and no one should live without enough food because of economic constraints or social inequalities. Food justice reframes the lack of healthy food sources in poor communities as a human rights issue. Food justice is inspired by historical grassroots movements and organizing traditions such as those developed by the civil rights movement and the environmental justice movement. The food justice movement advances self-reliance and social justice by acknowledging that community leadership is the way to authentic solutions.[14]

From Just Food, an organization dedicated to connecting local farms and communities with the residents of New York City:

> Food Justice is communities exercising their right to grow, sell, and eat healthy food. Healthy food is fresh, nutritious, affordable,

> culturally-appropriate and grown locally with care for the well-being of the land, workers and animals. People practicing food justice leads to a strong local food system, self-reliant communities and a healthy environment.[15]

Although "food justice" is the newest term in our "flexicon," it is perhaps the most rapidly developing concept in the food systems world, and none is more important. After all, the ultimate intent of food systems is to feed people—no exceptions allowed. The question is, how can we do it better through more comprehensive local food systems? Better yet, how is it already being done? We will explore food justice in more depth in the next chapter, after getting a better handle on what "community food security" looks like in different settings across the country.

Local Solutions: Going the Extra Mile

Food access and food security come in many forms. Wading through a litany of those forms might seem rather dry, but driving across the United States and actually *visiting* an array of food access organizations sounds fascinating to me. Perhaps it's because I'm around college students so much, but it seems like a much better use of a road trip than, say, following a jam band across the country. At any rate, that's just what Theresa Snow chose to do. She had a vision for dramatically increasing community food security in Vermont, but before implementing it, she needed to temper her ideas with lessons already learned by community food security initiatives across the country, so she went on tour.

Theresa is a dynamic icon in the Vermont food world. As a student at Sterling College—a food-savvy college virtually next door to "the town that food saved" (Hardwick, Vermont, whose story was chronicled by Ben Hewitt in an intriguing book by that title)—she cofounded Salvation Farms in 2005 with an eye toward capturing surplus fruits and vegetables in northern Vermont through organized gleaning efforts. Although gleaning can be informal and individualized, when the term is used in the context of coordinated food security efforts, it typically refers to the process of harvesting and collecting unmarketable food from farms and feeding it into an array of food access points, thereby serving a community's most vulnerable citizens and primarily through the efforts of directed volunteers.

In the first three years of Salvation Farms' work, Theresa and her colleagues gleaned eighty-eight thousand pounds of produce that never would have made it to anyone's plate, much less someone in need of a nutritious meal. Theresa then pursued an opportunity to link efforts with the Vermont Food Bank, an emergency food provider dependent primarily upon foods available through the federal government. This collaborative venture with the state food bank magnified the scope of Theresa's work, resulting in a vast statewide gleaning effort involving more than 120 farms and bringing more than one million pounds of fresh local products into the state's charitable food system. Despite that success, she was looking for ways to have even more impact in linking local "food rescue" to community food security.

Her efforts with the Vermont Food Bank were somewhat unusual in the food security world in that she was working to confront food insecurity from two somewhat contrasting approaches. The gleaning work she did with Salvation Farms contributed to what is typically referred to as "community food security," while her recent work at the Vermont Food Bank put her in the service of an EFP. These two worlds don't always come together. EFPs are dedicated to providing food to people who are food insecure, generally on a temporary basis. Community food security initiatives, on the other hand, are designed with a goal of making healthy, fresh food available to all people within a particular community by creating coalitions between individuals, organizations, businesses, and government entities. In other words, EFPs serve a critical safety-net function that ideally meets short-term needs, while community food security organizations try to take the long view by working to create resilient community-based solutions that can, over time, minimize the need for such a safety net. Many community food security organizations also emphasize the need for a more democratic food system in which all members of a society have a voice in reformation efforts. One of the pioneering leaders in this approach is WhyHunger, a national nonprofit organization founded in 1975. WhyHunger is working to build "the movement to end hunger and poverty by connecting people to nutritious, affordable food and by supporting grassroots solutions that inspire self-reliance and community empowerment."[16]

Theresa and some other creative thinkers in groups such as WhyHunger are also interested in retooling our national charitable food system so that it is more tightly linked to the local food systems of the specific communities served and therefore less vulnerable to volatile economic and environmental conditions. To illustrate the need for change, Theresa often points to the fact that the Vermont charitable food system brings in most of its beef from out of state—a real head-scratcher, since the state probably exports over ten million pounds of beef annually, much of it through dairy-farm cull animals being shipped out of state for slaughter. The dilemma is how to connect, or at least mediate between, the local agricultural marketplace and the charitable food system.

The local agricultural marketplace often has a lot to offer. Our state and national charitable food systems tend to be based predominately on federal commodities, while local farmers and processors often struggle to find a way to market or at least avoid wasting their agricultural surplus. Consumers' disdain for imperfect produce and desire for uniformity sometimes means that farmers may leave as much as half of a potential harvest in the field or destined for the compost pile—sometimes more if weather, labor, or economics are problematic.

Theresa wanted to find a way to expand the scope and impact of Salvation Farms, so she decided to leave the Vermont Food Bank and seek nonprofit status for her organization. She and her supporters then homed in on a focused mission for the organization. As a 501(c)3 nonprofit, Salvation Farms would "build increased resilience in Vermont's food system through agricultural surplus management." They aim to achieve this mission by creating an innovative management system for capturing and moving Vermont's surplus agricultural bounty, including fruits, vegetables, and meat.

Theresa currently has her eye on those fruits and vegetables; she estimates that there is more than two million pounds of surplus waiting to be captured in those sectors. The ultimate goal in this challenging undertaking is not just to make sure that these locally raised products are available within the charitable food system but also to ensure that they are processed and distributed in efficient and appropriate ways for the institutions and individuals that will use them. Not a task for the weak of heart . . . but Theresa is known for her drive as much as her knowledge. You could say it was her drive that got her on the road across the United States, a road trip designed to firm up her nonprofit ideas for increasing Salvation Farms' capacity.

Her first stop was the national Community Food Security Coalition (CFSC) conference in Oakland, California, in November 2011, where she presented her perspectives on linking fresh food and local food systems to the national charitable food system. The CFSC, as its name attests, was long the national nexus point for individuals and organizations working toward community food security, although it ceased its operations at the end of 2012.[17] Many persons, myself included, have found that attending a CFSC conference could significantly transform your thinking, if not your vocation. CFSC gatherings have always brought together a diverse array of people from all walks of life and virtually every corner of the United States. Corporate suits rub shoulders with urban ag tattoos while directors of senior centers intermingle with elementary-school food service personnel. People at those conferences were neither like-minded nor similarly attired, and that's the sign of a healthy coalition. CFSC conference days are spent learning from a variety of models around the country and even touring the models within proximity of the conference site. Whether you needed a primer or a recharge, a CFSC conference could convey the horizon of emerging ideas in solving food security and food access issues at the community level, as well as important perspectives on what individuals and organization can do at state, national, and international levels.

After the CFSC conference, Theresa spent the next month visiting a diverse array of organizations and businesses dedicated to solving food security and nutrition issues. She visited food banks, gleaning operations, prison-based programs, light processing facilities, freezing facilities, dehydrating facilities, agricultural surplus management organizations, meat preservation programs, gardening and culinary training initiatives, and innovative distributors.[18] Inspired innovation inspires more innovation, and Theresa was in search of the next step for Salvation Farms.

By the end of her trip, her vision was becoming crisper. She would develop the Vermont Commodities Program, an initiative geared toward lightly processing food products that can be provided to institutional kitchens serving the young, sick, and elderly as well as to emergency food providers. Dismissing the notion that two or more wrongs can't eventually help make things right, Theresa has begun to develop a partnership with the state Department of Corrections, the Vermont Correctional Industries (a work program for inmates), and the Community High School of Vermont (dedicated to providing persons under the custody of the Department of Corrections with a high school education) to capture and process the state's agricultural surplus through a cost-effective program that generates an affordable end product.[19]

No matter how good a nonprofit's vision, it can be tough to get started. And no matter how much a college professor thinks he knows, he always needs an expert grounded in the realities beyond the gates of the academy. So as I read Theresa's blog about her journeys across the United States and realized the common interests and questions we had in finding ways to process local foods and transform them into products available to Vermonters of all income levels, it only made sense to collaborate with her on the flash-freezing research I was doing at Green Mountain College with Garland Mason. As is the case with so much research, we didn't realize just how complicated the questions we were asking actually were.

The 2012 harvest season was one of researching and testing crops, freezing methods, bag sizes, labor requirements, and processing equipment, with Theresa and Garland working their way through gleaning strategies, processing protocols, food safety regulations, and the economics of it all. To an outsider, those first few packages of vacuum-packed frozen vegetables may not have looked like a big answer to food security. But they represented the bottom-up way in which so many of these local responses to food security evolve, fed by networks that are expanding beyond the usual suspects and even beyond early local food norms and goals, and becoming, in the process, something entirely new.

As Theresa puts it, "Sometimes people don't understand things until they exist." Or until they see the label.

Agriculture-Supported Community

Community food security—it's hard to understand what it looks like, smells like, or tastes like until you start cooking it up. That's what happened when Josh Slotnick and several other members of the Missoula, Montana, community faced the stark local realities of Newt Gingrich's 1994 ten-point Contract with America and President Bill Clinton's promise to "end welfare as we know it" (perhaps a better promise would have been to end *poverty* as we know it). As Josh tells it, "The 1996 Farm Bill did its best to end welfare as we knew it. The Missoula Food Bank and the federal WIC [Women, Infants, and Children] food assistance program were soon maxed out, and we were just three people with only a vague sense of the problem and no real solution." Nonetheless, Josh and his colleagues, Mary Pittaway and Caitlin Desilvey, started brainstorming and making community connections. Mary was the director of the WIC program in Missoula, and Caitlin was the twenty-four-year-old executive director of a neighborhood nonprofit focused on urban sustainability issues. Those early casual conversations soon transformed into strategic action.

With fewer than one hundred frost-free days a year, and one in five of its seventy thousand citizens living in poverty, Missoula is a particularly challenging place to grow local food solutions. Josh and his colleagues saw that Missoulans had a hierarchy of food choices: "'real' food, industrial food, and charitable food system options." The charitable food system options tended to be "calorically dense and nutritionally thin," as Josh puts it, and high-quality food is not generally at the top of the list of priorities for those living in poverty. In fact, Josh says, food always follows housing, child care, transportation, and clothing in terms of priorities, and "poverty dictates what you identify as food."

The solution was an evolution—a homegrown collaborative that soon became known as Garden City Harvest, a project linking what is now known as the PEAS (Program in Ecological Agriculture and Society) Farm, a youth farm, the University of Montana, the local food bank, senior citizens, a homeless shelter, and a network of neighborhood and community gardens. These gardens produce over fifty tons of produce annually, but the education that accompanies all of these efforts is just as important. The PEAS Farm is an equal partnership between the Environmental Studies Program (EVST) at the University of Montana and Garden City Harvest, with Garden City Harvest covering the operating expenses and EVST providing the salary for Josh's position as a farm manager and instructor, along with a labor pool of university students who earn credit through their work. Students from the university participating in this academic program link with area youth to power the farms and gardens. Participating youth from the surrounding community include several teenagers with drug violations. They are placed among the university work crews, creating an important educational interaction for both groups.

More than 2,500 children from the area visit the group's gardens every year, and the various sites scattered across the city host art and music events on a regular basis. Josh describes the group's evolution not only as a transformation of individuals and the community, but also a shift from the common concept of "community-supported agriculture" to an "agriculture-supported community." He notes, "It's clear that the education is working when the pronouns begin changing—when the lettuce becomes 'our lettuce,' the tractor becomes 'our tractor,' and we all begin wearing each other's sweatshirts."

As Josh sees it, all of this activity around food and agriculture is a remedy to one of the ills of modern society. "We've created a world in which we aren't necessary to our own lives," whereas growing food in community brings meaning and beauty, he says. "The farm reminds you of exactly where you are. When you feel an allegiance to your place, then you care for it."[20]

Turning Caring into Doing

If there is any one reason to grab a wooden spoon and cast-iron skillet and jump on the community-based food system bandwagon, it's because local and regional food systems provide us the best opportunity to achieve a marriage between nutrition and food security. We need fewer invisible (and therefore unaccountable) hands in the marketplace, and more calloused and caring hands that we can see and touch.

Let's face it: rebuilding local food systems is hard work, and the primary motive that drives many of us is not necessarily profit. More often than not, it's caring that energizes and guides us. We care about place, community, and nutrition. National and international food systems, on the other hand, are driven in large part by profit. Set aside for a moment whether that's good or bad. Instead, consider what happens when a food system is built around community concerns, with economic well-being for all supplanting a winner-take-all profit motive. The food system's objectives are vetted, attempted, reevaluated, and revised time and again, potentially based as much on community needs and concerns as on financial reward. Motives and outcomes may not always be transparent in an appropriately scaled food system, but they are ultimately accessible and transformable.

In contrast, try transforming the national or international food system—give yourself a decade or two, maybe even a lifetime. The penultimate example is the U.S. Farm Bill. Many food activists are relatively sanguine about making significant change in the Farm Bill, one of the largest congressional appropriations we make as a country. Not that we should stop trying, but attempting to effect change in the Farm Bill is a bit like trying to move a thick-skinned elephant with a toothpick as a lever (poking a donkey with the toothpick is only slightly more effective—it occasionally registers at least a twitch).

We don't have a lever with enough length or a clock with enough hours to adequately resolve food justice and nutrition issues working simply at the national and international food levels. Nor do we really want a one-size-fits-all response. Every community has different food justice and nutrition issues, and the needs vary dramatically even among subsets of the population in a single community, no matter how large or small. Perhaps our two-party system can agree on one thing: grassroots solutions are more practical than partisan solutions, and homegrown innovations often offer the best path forward for a community. Of course, it would be nice if even the most partisan-minded and thick-skinned of politicians would also agree to support the best of those local solutions from the federal level.

That leads us to the next point: communities simply have to lead the way in finding solutions to food security and nutrition dilemmas. If our local and regional food system leadership efforts are strong enough and the proposed solutions actually work, then the governments and markets will eventually follow suit. Policy always lags behind demonstrated need—after all, policy is more often a function of response than foresight. It's the wag that tails the dog.

And let's be clear that our work in this regard will never be done. Rather, it will require constant reinvention and vigilance. Food justice and nutrition issues have dogged human civilization from day one. We have made progress, but the job is far from over. Unfortunately, the constraints of energy, natural resources, climate change, and overpopulation are only going to make the task more difficult. And these constraints will not allow us to huddle up in circumscribed enclaves and hoard resources, ignoring others who were not as resource-rich or as "forward thinking" as we think ourselves. Such insular and callous responses also do little to promote food security or food justice. However, that kind of provincialism is in some ways a problem for later, since we've barely begun to conceive of reasonable foodshed boundaries, much less find adequate ways to feed our local communities predominately with local foods.

The statistics are daunting. And they differ dramatically, not just from region to region but also from community to community. In part, that is why community-based responses to food security and nutrition are so crucial. Federal and state programs are also critically important, but grassroots solutions with strong local commitment tend to identify the key dilemmas and develop long-term solutions. Local knowledge helps communities ask the right questions, while it is local investment in the future that tends to have staying power. The challenges differ from area to area, which is why community-based food security really matters—responses must be tailored to the specifics of the community and the region.

Homing in on Nutrition

In our push for local foods over the past few years, I think that we have made a mistake, albeit one that we are beginning to rectify. In addition to touting the energy and security advantages of local foods, we wanted to believe and say that local foods are categorically more *nutritious* than foods coming from afar—and that can be the case, for some foods some of the time. However, nutritional quality depends upon the health of the soils, how the product was raised and processed, the methods of preservation and storage, and even the safety of the containers and packaging. Local does not necessarily mean better or more nutritious if care is not paid to these factors.

For example, vitamins in fresh vegetables are highly volatile, so the time between harvest and processing can be more important to the retention of vitamins than the distance traveled or even the time from harvest to consumption. Broccoli frozen direct from a field several thousand miles away can have more vitamins than a head of broccoli picked thirty-six hours earlier in a nearby field but never refrigerated. By the same token, tomatoes canned almost immediately after harvest in California may have superior nutritional quality to that of day-old farmers' market tomatoes in New York that sat out on the display for most of a hot day, only to sit on a customer's counter for several days before being eaten.

That doesn't mean the canned tomato is the better choice. Many of those metal tomato cans are coated with bisphenol A (BPA), a chemical currently under review by the U.S. Environmental Protection Agency, the Centers for Disease Control, the National Institutes of Health, and the Food and Drug Administration. BPA is prevalent in the U.S. population; it has been found in 93 percent of urine samples taken in a study of 2,517 people conducted by the CDC.[21] The concerns surrounding BPA center around potential harmful effects on the endocrine, reproductive, cardiovascular, and neurological systems, as well as possible links to cancer.[22] Researchers are particularly focused on the potential health impacts to infants and children.[23] The chemical and food industries are both heavily invested in the use of BPA, so their views readily conflict with those of consumer advocacy groups such as the Environmental Working Group (EWG).[24] EWG has pushed hard to keep BPA concerns in the public spotlight, and some companies have responded to these pressures, including Eden Foods (as early as 1999) and Muir Glen, among others.[25]

These kinds of questions—the loss of nutrients over time, packaging concerns, pesticide residues, and so on—do mean that local food advocates have to be more specific about nutritional claims, and anyone interested in creating sound foodsheds has to be strategic about building regional soil fertility, promoting sustainable agricultural methods, and emphasizing processing options that preserve nutrients.

Ultimately, it all depends upon what you are comparing. Nutritional superiority is not based on geography; but deep-seated concerns for community nutrition definitely have correlations to proximity. Proximity increases the potential for understanding and common cause between community members of different socioeconomic and cultural backgrounds; distance makes mutual understanding and abiding concern more challenging. After all, it's easier to fix a leaky spigot at home than it is to fix one two thousand miles away; at home, you know where to find the shutoff valve, where to find the replacement parts, and whom to call if the DIY method fails. (It's worth noting that in both scenarios the strategy least likely to be helpful would be to call a federal agency in Washington for help in resolving the problem.)

The Triple Whammy

Diet, health, and income. Plain and simple. If those three words don't work for you, then look at figures 6-4 to 6-6 below. See any correlations?

Economic insecurity drives food insecurity that then creates health issues, which in turn generate additional local economic burdens. These feedback loops suffer the consequences of a poor diet, but they also demonstrate the relationship between dollars and diet. As I stated above, nutritional

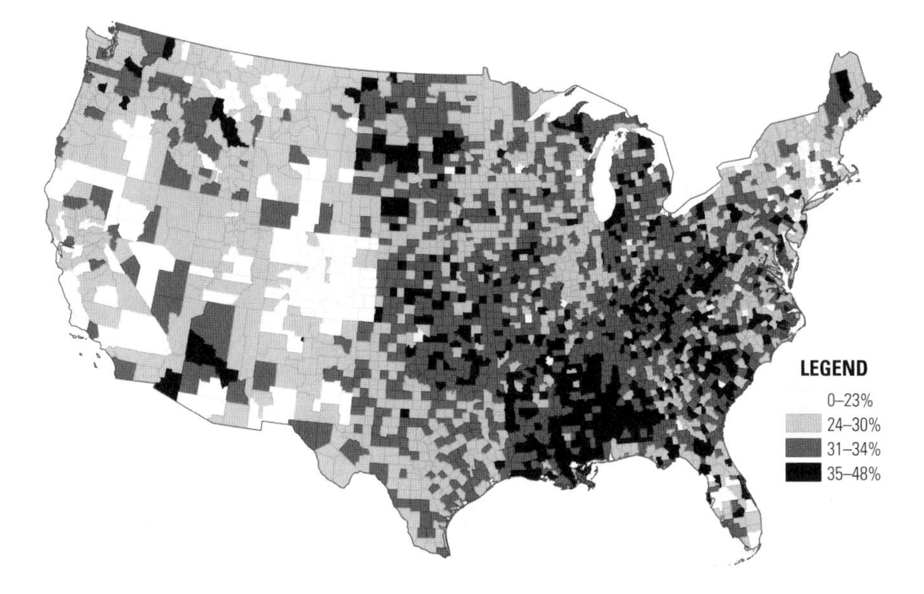

FIGURE 6-4. U.S. Adult Obesity Rates, 2009. Source: USDA Economic Research Service Food Environment Atlas, accessed July 5, 2012, http://www.ers.usda.gov/data-products/food-environment-atlas.aspx.

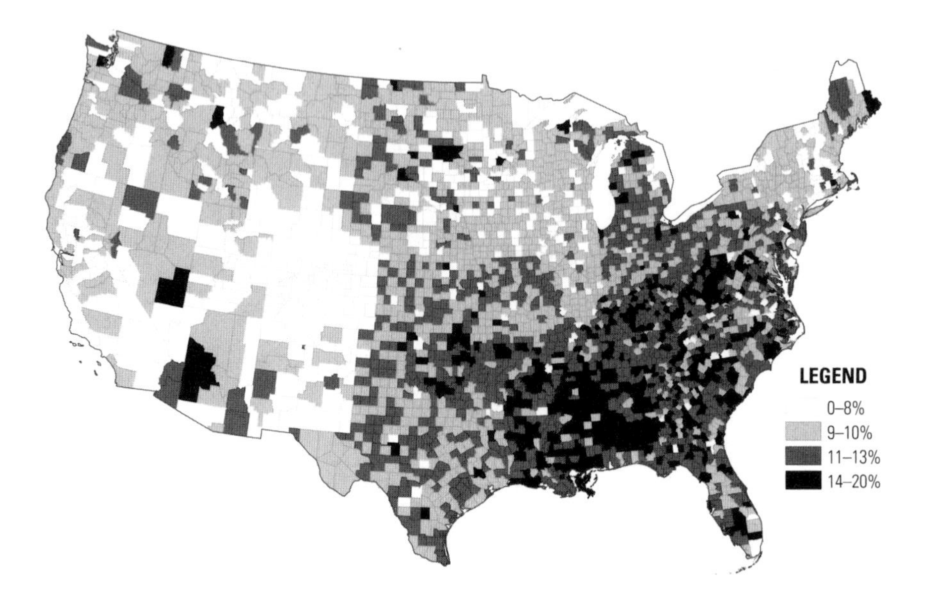

FIGURE 6-5. U.S. Adult Diabetes Rates, 2009. Source: USDA Economic Research Service Food Environment Atlas, accessed July 5, 2012, http://www.ers.usda.gov/data-products/food-environment-atlas.aspx.

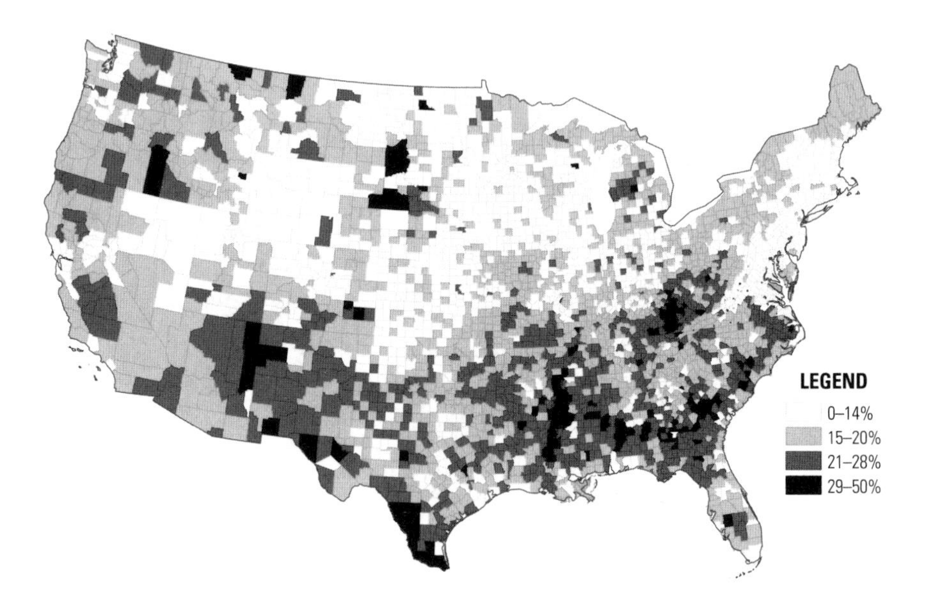

FIGURE 6-6. U.S. Poverty Rate, 2010. Source: USDA Economic Research Service Food Environment Atlas, accessed July 5, 2012, http://www.ers.usda.gov/data-products/food-environment-atlas.aspx.

superiority of foods is not necessarily based on geography. However, *access* to superior nutrition has much to do with intensely localized economic conditions. What I mean by "intensely localized" is the difference between one side of the tracks and the other, or one side of the freeway, or perhaps the county. And the ultimate solutions in these scenarios are also intensely local. A coalescing sense of urgency from within can spark the change in ways that outside assistance seldom can.

There is a vicious cycle at work in the areas on the maps where we can so visibly see the correlations. The temptation for some (perhaps the "foodie cognescenti") is to pin the blame all on diet. Despite the good intentions in such an assertion, it fails to recognize the complexity of the situation. The predominant force behind these negative feedback loops is almost always economic insecurity. Put more bluntly—and in a fashion that puts the responsibility more squarely on everyone's shoulders (this is a democracy, is it not?)—the bulk of these nutritional and health issues are ultimately about economic injustice. That economic injustice is likely to increase in intensely localized ways unless local citizens begin to work toward community-based solutions. The health care concerns and costs are only going to swell, and community health and social services will bear the associated financial burdens.

Triage is already under way for adult populations, but the biggest potential for change is among the youth. Schools are therefore a critical component of the solution, but not just because students are a captive audience. Rather, it is because schools serve such a vital function in feeding children, particularly in these beleaguered environments. For example, the average American child receives the majority of his or her fruit and calcium intake from school food, and students participating in the subsidized lunch, breakfast, and summer food programs offered through the public schools are utterly dependent upon the meals served at schools for the bulk of their nutritional needs. Of course, schools provide a vital role not only in providing nourishment for youth but also in establishing healthy eating habits. And, of course, as centerpieces of the community, schools can serve as critical beacons of positive change.

While attending the National Farm to Cafeteria Conference in August 2012, I ran into one of those change agents, a school chef named Robert Rusan from Maplewood, Missouri, an aging inner-ring suburb of St. Louis.

While eating breakfast with me, Robert recounted the transformation of his school district under the leadership of Dr. Linda Henke, who had just retired at the end of the 2011–2012 school year. "The Maplewood Richmond Heights School District was crumbling when Dr. Henke arrived—things were in terrible shape," Robert said. "In fact, it was known as 'Maplehood.' But Dr. Henke came in from another affluent school district, and she had a vision. She was dynamically strong-willed, and 'no' is not an answer for her. That's because she grew up on an Iowa farm, and you don't say 'no' to a tough Iowa farmwoman!"

The *St. Louis Post-Dispatch* described the condition of the district when Dr. Henke arrived in 2000 as follows: "The middle school was on the state's academic watch list. The district was flirting with losing full accreditation. The high school was surrounded by barbed wire. Doors with broken locks were chained shut or jammed with broom sticks. Science labs had no running water. Gangs used school walls for graffiti."[26]

Dr. Henke hired Robert, a chef in his family's catering business, to help change the students' diets and heighten their nutritional education. "When I first heard about the job opening, my first reaction was, 'They don't hire chefs in schools.' But Dr. Henke always did things the way she thought they needed to be done." And she didn't mince words when she first met Robert. She told him, "I want somebody who can cook. I'm not going to invest in food that's going to go into the waste can." It was a change from the norm for many of the people working with him. Robert laughed as he described the shift in approach: "Sometimes other food service employees would see me coming, and I know what they were saying: 'Oh no, here he comes—the cook-from-scratch chef. No cans, no bags—he wants it all fresh.'"

The motto for the transformation of the school district was "We're investing in our students." Those investments transformed the buildings and infrastructure, including some nontraditional infrastructure. District staff and teachers created four gardens, an aquaponics facility, a beekeeping operation, and a petting zoo (the district rents different animals for the children to encounter).

The district also established collaborations with the community, including a cooperative arrangement with the Salus Processing Center at St. Louis University.[27] This processing facility has been freezing local produce for the district as part of Missouri's HELP (Healthy Eating with Local Produce) program.[28] These collaborative efforts are a pilot project for the entire state, testing the potential for increasing the processing and distribution of local foods at prices that work for school cafeterias. The result thus far? Twenty-five percent of the produce Robert and his colleagues are utilizing comes from local farms, thanks to a woman who dared to bring her knowledge and values from an Iowa farm into a community about which she cared deeply.

Nature, Nurture, and Nutrition

Women are the backbone of the significant advances we are making in the re-envisioning of food systems at every scale: local, regional, national, and international. I don't think that's a coincidence. When it comes to nature, nurture, and nutrition, women have been the nucleus of so many of the most progressive initiatives. I realize the dangers of promoting any kind of gender-based stereotypes. However, I make this point for two reasons. First, I believe that our society should do more to recognize more fully and publicly the critical roles that so many women are playing in food security and food justice issues. Second, I think that we men would benefit greatly from paying closer attention to women's views and approaches to solving problems in food security and food justice.

As we begin to examine the intertwined nature of food security and food justice in this chapter and engage with it more fully in the next chapter, it makes sense to begin this inquiry with a topic universal to us all: gender. As soon as we lift the lid on the stewing mélange of gender and food systems, the food justice issues start to bubble up with such intensity that there's no way to put the lid back on the pot. I find it tremendously exciting, and one of the best prospects for a new era in food systems work. The sea change in our international discourse about food and agriculture over the past few decades has been the powerful emergence of women's voices and perspectives. Vandana Shiva, Wangari Maathai, Alice Waters, Erika Allen, Joan Dye Gussow, Marion Nestle, Kathleen Merrigan—the list goes on and on. Women have comprised the majority of students in my agriculture and food systems courses for years, and I've long attributed this fact to women's historical roles and perspectives in safeguarding nature, nurture, and nutrition. Not that we males should by any means abdicate our roles in these areas, but women have in many ways been the sentinels of sanity in those conversations.

To get a glimpse of how women are reshaping the food system landscape, it's interesting to consider the shifting realities of women's roles in U.S. food production. The USDA's 2007 census data estimated that women are the principal operators of 14 percent of the nation's farms, nearly a third more than five years earlier. While the growth of women in farming is interesting, it is even more intriguing to note the ways in which women are farming in the United States. A recent study by the Organic Farming Research Foundation indicated some key differences in how male and female organic farmers in the United States run their farms:

> Women are much less likely than men to manage organic and conventional systems on the same farm (m = 25%, f = 14%) and they are far more likely to allocate land to vegetables and herbs (m = 33%, f = 47%). They are likewise far less likely than men to devote land to field crops (m = 44%, f = 28%). The average farm size, 40 acres, held by women is far less than the average farm size held by men, 149 acres. The average income difference in favor of men is $85,000 (m = 42% <$15,000, f = 70% <$15,000).[29]

Remember those definitions of "community food security" and "food justice," and their references to growing nutritional and culturally appropriate foods in ways that demonstrate care for land, workers, and animals? The data above suggest an abiding interest among many women farmers in food justice, nutrition, and careful stewardship of resources. However, there are also disturbing differences in income and equity between female farmers and their male counterparts. Perhaps these discrepancies can be attributed to the different ways in which women choose to farm and to market their goods, but we are left to wonder whether these apparent inequities in income and ownership are parallel to those faced by women in other sectors of the workforce. Part of the food justice conversation, therefore, needs to include how local food systems can further empower women, from the farming sector all the way to the final goal of food security in the household.

In light of such data, it's tempting to see food justice simply in two sex-defined dimensions. But food justice is multifaceted, and all of the angles that we can take in viewing the world from a food justice perspective can yield a kaleidoscopic perspective. Of course, the dazzling array of shapes in a kaleidoscope is enticing, but the real point of fascination is color. Unfortunately, I have to travel outside of Vermont to find much of it.

An Urban Legend to Believe In

I knew I didn't understand urban agriculture. I knew it existed, and I knew it was critically important. I knew it was where more and more of my students were coming from and where many of them would head. But my entire agricultural experience was rural, and I am, at the core, a rural citizen and a lover of open and wild spaces. I'm also a white middle-class male academic—which makes four excellent reasons for me to take a pilgrimage to Detroit. So in 2010, when an invitation came to join a group of people from around the country to discuss the formation of a national FoodCorps—linked to a Community Food Security Coalition conference happening in Motor Town at the same time—I leaped at the opportunity.

It's not that I thought urban agricultural production techniques would be all that different from those of the diversified small-scale rural agriculture I had taught and practiced for years. Rather, I felt like urban agriculture would somehow combine food production with food security and food justice with more intensity than my rural farming experiences in Vermont and the Alps (the situation in North Carolina was different, but that's the next story). I simply needed to have a better grasp of the food justice drivers and responses in a beleaguered urban environment. Detroit certainly fit the bill in terms of a highly challenged urban population, but I'll confess that I wasn't quite prepared for the steely gumption that I encountered . . . and, in every instance, it came by way of women whom I was lucky enough to meet during my short adventure.

Nor was I adequately prepared for the intensity of what I would see in Detroit. The drive into the center of the city from the airport caught me off guard, despite Detroit's reputation as one of the most challenged cities in the country. The gaping intensity of a vacated industrial legacy shocked me: enormous factories and warehouses laid to waste by neglect, the elements, and vandalism. Shattered windows and battered doors stitched together with graffiti, the one art form that could bridge anger and justice with the pressurized hype of a spray can. I pulled off the interstate to drive through some of the neighborhoods and city streets to see things for myself. Neighborhoods were gutted, although it was easy to see that many of the abandoned houses would be upper-middle-class homes in most any other setting. I was especially shocked to see two fourteen-story brick apartment buildings standing side by side, empty and windowless. The incongruity of so many homes that were people-less and people that appeared homeless added an element of absurdity to the unjust nature of it all.

With at least one-third (some say it is actually closer to one-half) of the city unemployed and approximately one-fifth without personal transportation, the city was reeling from the decline of the manufacturing sector and the collapse of the housing market. Supermarkets had long since disappeared with the exodus of businesses and residents, with the predominately black (>80 percent) population left holding the bag. It was a mindful setting for a meeting intended to address community food security issues on a national scale.

The gathering to flesh out the possibilities for the nation's first FoodCorps included a diversity of players from around the country, including a strong delegation from Detroit. As two intense days of conversations unfolded, consensus seemed to be gathering around the need to address food access and nutrition issues for America's youth. Opinions as to why, how, and where FoodCorps volunteers should serve varied, but no one objected to the need to focus on food-insecure regions of the country. The pioneers of the FoodCorps concept represented an interesting combination of organizations—Occidental College, the National Farm to School Network, Slow Food USA, Wicked Delicate, and the National Center for Appropriate Technology.[30] In particular, Crissie McMullan and her colleagues had provided a national model through their development of the Montana FoodCorps, a program to which many of us had looked for inspiration and ideas.[31]

The first evening together for the group appropriately featured a work party at a newly conceived school garden site in the city. As we used our spades to cut into the perimeters of the demarcated outline for the garden, we started conversations, many of them cut short by sharply pivoting wheelbarrows going in different directions. Sod went to one location; excavated garbage went to another spot. We dodged glass, poison ivy, scrap metal, and each other in the excited mayhem. People of all colors, ages, and economic backgrounds mixed. The young leaders of the proposed garden gave us clear directions and kept us on task as we moved fast to leave them with an unearthed vision of good things to come.

Gradually, soil lay prone to the elements and the whimsy of school-aged gardeners, and banter turned into deeper shovel-propped conversations on the sidelines. Throughout my time with this group, I found myself weaving in and out of conversation with Aba Ifeoma, a founding member of the Detroit Black Community Food Security Network (DBCFSN).[32] She described how she and other concerned community leaders such as Malik Yakini, now the executive director of the DBCFSN, finally got to a point of desperation about the devastated economy and its tremendous social fallout and realized they were going to have to fix their own problems at a local level. "The city is doing what it can, the state of Michigan is busted, and the feds hardly seem to have a clue about how bad things are here," she said.

But she also didn't see it all as doom and gloom. An avid urban beekeeper, Aba passionately described the complexity of honey in the city, rich with unexpected taste and color from the enormous variety of plants found in an urban environment. She was also working with a group of volunteers to build D-Town Farm, at that time the largest urban farm in Detroit, with two acres in production and growing.

Aba invited me to visit D-Town and set up a tour for me with Jackie Hunt, who was on the board of the DBCFSN and a pivotal volunteer at D-Town Farm. The next morning, I followed Aba's directions and drove my rental car to Rouge Park on the outskirts of the city to get an introduction to my first urban farm. When I got to the site, I was a bit surprised to see a suburban development on my right, not unlike those of my childhood, except that the depressed state of the Detroit economy had clearly taken its toll. The lawns appeared to tell part of the story for each household, and the presence of plywood in the windows seemed to be the final exclamation point for several of those tales.

To the left, I saw an enormous fence surrounding what I knew had to be D-Town Farm, with the front gate boasting a chain that seemed more a casual dare than an outright challenge. Surrounding the perimeter of the fenced garden area was what looked to be a park unleashed to nature's whimsy, with fields of high brush interspersed with trees of all kinds, some in straight lines and others apparently much more serendipitous in their location. Beside the gate was the sign Aba had told me to look for: the inside of a tire sidewall had been transformed into a colorful piece of art. "D-Town Farm," the sign said; I'd made it.

Moments later, Jackie pulled up and stepped out of her car, straightening a colorful baseball cap perched atop beautiful graying braids. We introduced ourselves and she began unlocking the gate. "Sure is a big fence we have to have here," she said.

"People break in and steal things sometimes?" I asked.

"Sure, every now and then," she replied, flexing the padlock open. "But the fence is more to keep the deer out than the people. Those deer will rob you blind in a minute. You ever seen a fence this high for a garden?" She chuckled, as I looked at the thick poles supporting the fence, which appeared to be twice my height. "We had to get a grant for that one. The fence might be the most expensive thing you see out here!"

Crossing the threshold into the sanctuary of the garden was a step into transcendence. It was palpable. Suddenly, everything seemed orderly and under control. Anything that was trash outside—pallets, tires, old wood, furniture, buckets—apparently got a creative splash of color as soon as it passed the garden threshold, and the whimsical brushstrokes looked to be the work of young, deft hands. Jackie started walking me through the garden along the route of the eleven educational stations, explaining things as we went. The first station put the whole project into context, and its sign succinctly summarized the reason so many of us are watching Detroit with great interest:

> In 2007, the last major supermarket chain left Detroit, leaving a void that has caused many to characterize much of the city as a "food desert." However, the vast amounts of vacant land in the city of Detroit provide the opportunity for its residents to grow significant amounts of the food that we consume and to contribute to revitalizing the

> local economy. D-Town Farm is a model of how unused and underutilized land in the city of Detroit can be used to address the lack of access to high-quality produce that many Detroiters experience.

As we walked further into the gardens, I could see that "appropriate technology" in this urban scenario fit with my own homesteading philosophy—reduce, reuse, recycle—but the pieces here came more from the urban wilds than the surrounding woods of our homestead. Ingenuity was clearly at play, manifested in particular by the compost sifter they were using, apparently based on a design used elsewhere in the city. Hardware cloth was mounted to a horizontal sifter framed by two-by-fours. That sifter was placed on rolling casters that followed a wooden track at about waist height so that compost could be tossed onto the sifter and shaken by means of an easy and controlled motion. "This city depends on compost for projects like this," Jackie noted. "In the places where people want to garden, there's either no soil or soil you don't really want to trust for growing food."

We continued on the tour around the garden, looking at the season extension tunnel that protects crops during colder weather, the raised beds planted by a group of elementary-school students, the pallet-framed compost bins, and the beehives tucked away just beyond the easy access of unwary kids. We then entered a shaded woody area, where I could see black plastic pots with logs sticking straight up out of them. The bottoms of the logs were buried in soil, and I could see the telltale mushroom spore plugs and their wax coating around the protruding portions of the logs.

"Shiitakes?" I asked. I know a little about mushrooms, but I'd never seen this version of shiitake logs.

"Yes, this setup is something people can do on balconies and porches or in their backyards," Jackie replied.

"I'm starting to get the picture," I nodded. "But what's that upholstered chair for right there? A place for you to sit in the shade on a hot summer day?"

"You haven't read the book, have you?" I glanced over, and Jackie was looking at me with amusement. "What book?" I asked.

"The mushroom book!" she exclaimed. "If you'd read it, you'd recognize that right away. You can grow mushrooms right in a piece of furniture just like that one."

I was clearly out of my league. "Well, I came here to learn, Jackie, but that's not something I was expecting to run into!" We both laughed and made our way back out toward the gate.

As we got closer to the entrance, I saw an interpretive sign that I'd missed on the way in. I went over to read it:

> **Africans and Agriculture**
>
> For thousands of years African people have relied on the earth for physical and spiritual sustenance. They began cultivating plant foods over 7,000 years ago. The development of agriculture enabled early Africans to settle in one place for extended periods of time, thus opening the way for the development of civilization. Ancient and current staple crops include barley, sorghum, millet, tef, rice, groundnuts and yams. African people's extensive knowledge of agriculture was one of the factors that caused Europeans to target them as enslaved workers for their plantations in the western hemisphere.

That was it. That one little sign said it all, as far as I was concerned. Pride, history, culture, autonomy. It was the "aha" moment about urban agriculture that I'd been looking for. D-Town Farm wasn't just about food security. It was about food justice—bringing forward the depth of history and the breadth of culture. There was indeed something going on here that was much different from any farm or garden that I had visited in the past three decades.

I called Aba at the beginning of the growing season two years later to see how things were going.[33] The changes she described were nothing short of extraordinary. As of June 2012, D-Town Farm was cultivating seven acres; had shifted from a volunteer base to having a farm manager, a compost manager, an office assistant, an education and outreach director, and some contract personnel; and was offering a popular "What's for Dinner?" lecture series. The farm had also become a Regional Outreach Training Center (ROTC) for Growing Power, the Milwaukee-based beacon of inspiration for so many urban agriculture projects across the country, founded and fostered by Will Allen and his daughter Erika.[34] As an ROTC, D-Town Farm was able to integrate training sessions with the construction of new infrastructure.

The farm had outgrown Jackie's compost sifter by leaps and bounds with a new quarter-acre composting site filled with windrows. Forgotten Harvest, a Detroit-based hunger relief organization that collects surplus, prepared, and perishable foods for redistribution, delivers unusable food products to the farm. This food waste is mixed with wood chips from Wayne County's landscaping operations. All of this compost and added acreage necessitated the addition of a tractor.

The small demonstration hoophouse that I had seen in 2010 was now one of four season extension structures, with the latest addition being a thirty-by-ninety-six-foot mobile greenhouse dubbed "Rolling Thunder" by its manufacturer, Rimol. Following Dutch designs made popular in North America by organic pioneer Eliot Coleman, this greenhouse comes with three hundred feet of track so that it can be moved back and forth between three planting locations, allowing a farmer to plant "in succession," moving the greenhouse from plantings with differing tolerances for cold and heat. This mobility also allows the grower to "freeze out" pests that might overwinter inside a greenhouse, thus minimizing some of the pest situations common to greenhouse operations. Michigan State University professor John Biernbaum and others who have brought the university's resources and expertise to Detroit provided four days of technical support for the construction of the first of these new season extension structures.

Another initiative made possible through the ROTC collaboration was the installation of ten photovoltaic panels by a local utility company. And the Midwest Renewable Energy Association taught a weekend introductory course on renewable energy for D-Town Farm as part of Growing Power's focus on developing renewable on-farm energy sources.

D-Town Farm was no longer a large market garden—it had truly become a significant farm operation. When I asked Aba about the farm's connection with rural black farmers outside of Detroit, she replied, "We *need* the rural farmers. We need the alliance because of their expertise. We city people only garden at best. But D-Town actually has more in common with rural farmers because of our size—*no one* has seven acres in Detroit. Because of our size, we have different challenges compared to other places: different cultivation methods, pests, everything. It's all different. We need to put more effort into building those relationships with rural farmers."

All of this also comes amid tremendous political and economic uncertainty. In fact, the momentum at D-Town Farm is an empowered response to this instability in many ways, except for the fact that Rouge Park, home to the farm, is owned by the city. "Detroit is in such an emergency state, it has trumped everything. No one knows what is going to happen next due to the threat of a takeover of the city by Governor Snyder," Aba lamented. "In Detroit, it's particularly ominous because it's taking away the power of our local elected officials—our link to local democracy. . . . Being located in a city-owned park, we don't know what's going to happen."

Thinking more about the obvious desire among members of the black community to achieve some sense of self-reliance amid such uncertainty, I asked if some of the needed farming resources and training opportunities might come by way of someplace like Beulah Land Farms in South Carolina, a four-thousand-acre property envisioned as a place where blacks can discover self-reliance and empowerment through working in a community of black farmers. The working farm there has cattle, chickens, horses, vegetables, a fish pond, and "the most delicious artesian well water you can imagine," according to Aba. "But many of the people there are the same city people we are," she said, noting the importance of having highly experienced farmers teaching up-and-coming farmers the skills necessary for successful larger-scale farming ventures.[35]

As we talked about ways to educate more Detroiters about agriculture, I mentioned that everyone I had met associated with D-Town Farms in 2010 was female, and I had seen reference to the DBCFSN women as the "Queens Council" on Dr. Monica White's "Soil2Soul" website.[36] White sees Aba and her female colleagues as key leaders in an important form of resistance in Detroit. As White, a professor at Wayne State University, describes it, "Farming and gardening are not directly confrontational with the power structure; however, freedom farmers define gardening as a resistance strategy. Their work is internally transformative not only for themselves, but for their families and their communities in three ways: healthy living and the production of healthy food, building community through the food system, and cooking as resistance."[37]

When I asked about women's role in community food security initiatives like D-Town Farms, Aba replied, "Women really get it more than men. No slight or slam intended there—it's just that we work *with* the land a lot more.

We give to the land, and the land gives to us. It's just a nurturing thing to work the land and to work together. That's just what we women do—we nurture. The symbiotic relationship that I feel and see with women at the farm—maybe it's my bias, but I think it's women that really get it."

As we discussed the role of nurturing and nutrition in her work, I had to ask her a question that had nagged at me for two years. On my tour of the farm in 2010, Jackie had laughingly remarked that anytime someone took something from the gardens, the first thing they seemed to go for were the collards. Anytime I told the story of my visit, I had been unsure of whether sharing that anecdote would convey a stereotype or the idea of "culturally appropriate foods," a phrase used on D-Town Farm's website. So I asked Aba for some help in understanding her idea of how to define that term. "Culturally appropriate foods are familiar to the population you're targeting," she said. "For example, Hispanic foods are obviously completely different from the foods of our traditions. It seems that black people expect greens of any kind: collards, turnips, okra. It just makes us comfortable. For example, not a lot of people in our community know what to do with an eggplant. It's not a part of our past, so we have samples and recipes of these foods new to some of our customers."

I needed a little more explanation as to why some blacks embraced the notion of culturally appropriate foods but, as in the Civil Rights era, rejected so-called soul food. Aba made a quick distinction, based on history and nutrition: "For the people at D-Town Farm, the part of soul food that we reject is the unhealthy inner organ, what slaves had to eat—'the innards'—not the skeletal meats that went to the slave owners. Also, we reject the traditional way so many of the foods had to be cooked with pork fat to make them palatable. In fact, a lot of us at D-Town are vegetarians at this point. But it's not just a rejection—more importantly, it's the idea of understanding what we're eating and why and eating healthier as a result." It was a good reminder that trying to link local food security to nutrition requires as much history as it does imagination.

And it is the imaginative nature of projects like D-Town Farm that is so vital: one person's imaginings can quickly become a transformational touchstone for hundreds, and collaborative dreams can impact thousands. Take, for example, Detroit's Garden Resource Program Collaborative. In 2011, this coalition of organizations across Detroit created and maintained 1,351 vegetable gardens, yielding 73 types of fruits and vegetables within 382 community gardens, 48 market gardens, 64 school gardens, and 857 family gardens.[38]

Given the vast amount of vacant land in Detroit, an obvious but rather complex question arises: How much of Detroit's food could be raised if a significant portion of these open areas could be put into agricultural production? Such a vision begins to turn Detroit's beleaguered landscape into a field—okay, perhaps "lot" is more apt—of dreams. Estimates vary on just how much vacant land is available and appropriate for urban agricultural endeavors in Detroit, but the authors of a Michigan State University 2010 report titled *Growing Food in the City: The Production Potential of Detroit's Vacant Land* estimated that 44,085 vacant parcels provided approximately 4,848 acres of possible cultivation area.[39] The next challenge for these researchers was to determine a method of estimating potential productivity within these areas. As figure 6-7 below indicates, the authors of this study considered several alternatives in production methods. They contrasted both high-yield and low-yield biointensive approaches, as well as a more conventional commercial approach to production. "Biointensive" agriculture involves maximizing production while minimizing space requirements through a careful process of soil building through proper management, often emulating methods used for millennia in long-standing agricultural societies throughout the world, and combining those techniques with modern "appropriate technologies" such as efficient hand tools and disease-resistant crops. Commercial production, in contrast, involves a larger land base and more machinery.

The researchers also took into consideration the added benefits of crop storage and season extension tools. The results of their study indicate significant potential for any of these scenarios, at least on paper. As any farmer knows, every growing site has its own nuances, and productivity on any scale takes years of experience. Nonetheless, every venture in this direction matters, not just for the food that it produces. Malik Yakini's experience at D-Town Farm is a reminder of the importance of this kind of work: "We're changing the way Detroiters think about food, where it comes from, and who profits from its sale."[40]

Motor Town may be shifting gears, but its residents are finding new ways to drive home some old wisdom.

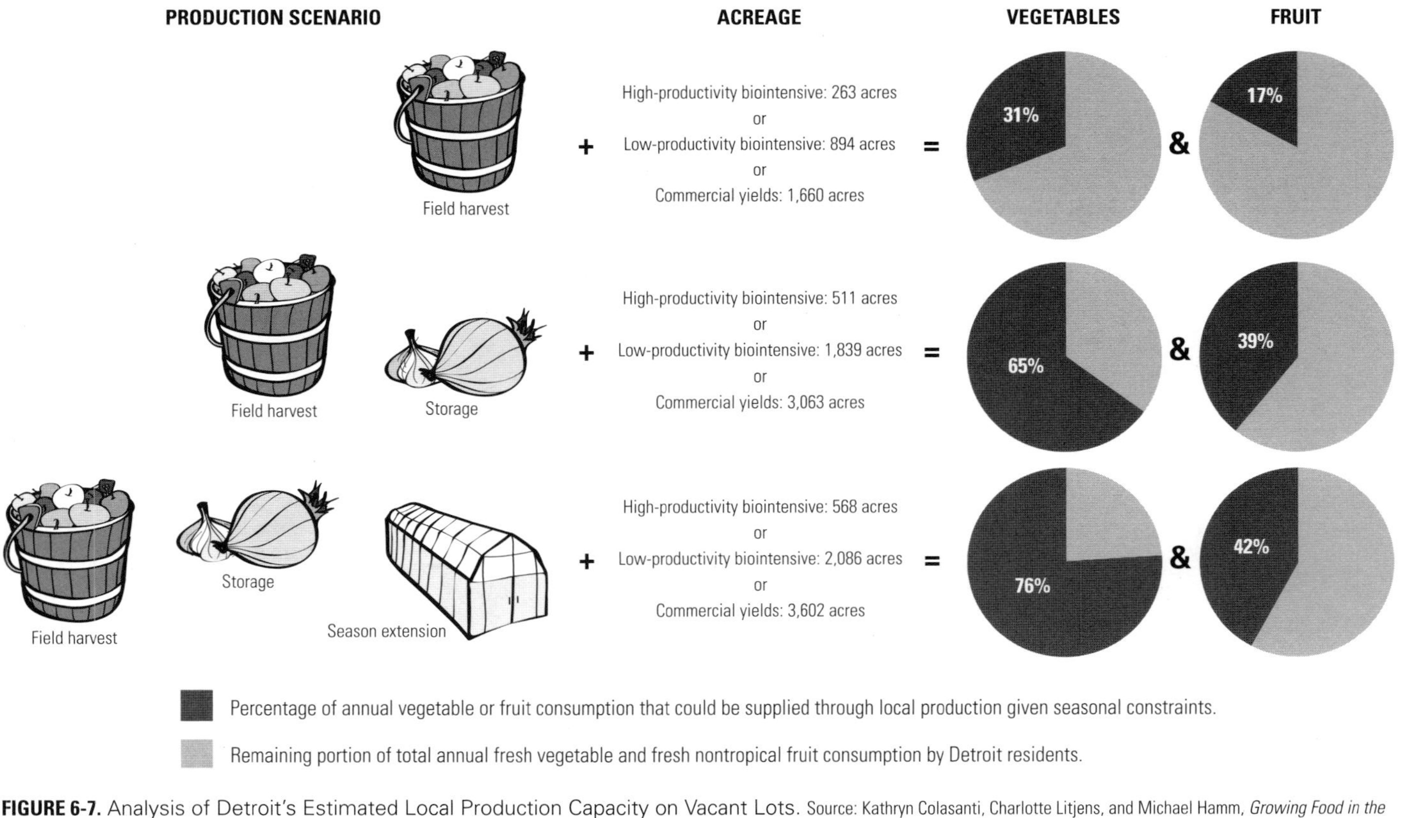

FIGURE 6-7. Analysis of Detroit's Estimated Local Production Capacity on Vacant Lots. Source: Kathryn Colasanti, Charlotte Litjens, and Michael Hamm, *Growing Food in the City: The Production Potential of Detroit's Vacant Land* (C.S. Mott Group for Sustainable Food Systems at Michigan State University, June 2010).

CHAPTER 7

Food Justice

An airplane ride is a great way to gain perspective on agriculture's role in a particular landscape. Fields unfold and roll up to the edge of our other uses of flat and fertile terrain. The warp and weft of pastures and plantings crisscross the terrain, interspersed with the dwellings and doings of our lives below. Altitude provides a sense of what we're up against in maintaining—much less rebuilding—our local food systems. The challenges become apparent with even a little altitude gain, sometimes in unexpected ways.

It was 1976, and I was a gangly eighth-grader. Despite the lack of bulk on my skinny frame, I felt like an adult as I helped my father and our neighbor Joe push the Cessna 172 by its tail and two wing struts out into the open field where we could more safely start the engine, with its roaring front-mounted prop.

Our family had just moved to eastern North Carolina from the Piedmont region of the state. We had moved from the land of textiles to the land of tobacco, and my dad's role as a progressive Presbyterian minister was shifting from dealing with the economic fallout of textile workers to confronting the entrenched racism of a rural landscape dominated by a big golden leaf—"baccer," as it was often called by the locals there in Johnston County.

A career built on compassion and community service takes its toll, and flying was Dad's best way to relieve stress. He'd grab one of us three kids and take a flight for an hour or two. But since settling into our new home, Dad hadn't found access to a plane, much less an airport. So we were both excited to accept Joe's invitation to make our first flight and see our new territory from above.

It wasn't quite what I'd expected, opening up a farm gate to get to a "hangar" that looked like so many of the other low-slung farm buildings in the region. I wasn't sure I could make out anything that would be considered a legitimate runway, but there was a plane—a nice one, with four seats. Nothing made me happier than flying with my dad, looking at the shifting patterns of generations of farmers on the landscape, imagining a life spent crisscrossing cultivated fields on a tractor—a life lived on the continuum between mud and dust.

We got the plane out into the open with a minimum of exertion and began circling the aircraft to check all of the required inspection points. Finally, Joe motioned for us to climb inside by way of the tiny step on the passenger's side wing strut. Once the doors were shut, the engine cranked with a roar, and the prop whipped into motion and transformed from two curved blades into a circular blur.

Joe eased off the brake and the plane lurched forward, pulled into the wash of the prop, with the meadow grasses whipped into submission behind us. We bumped along until we got to what finally started to resemble an airstrip. Even more than the worn grass, the parallel lines of old smudge pots on both sides of the runway indicated the trajectory of the intended runway. Joe and Dad checked the windsock, the most visible evidence of the dual nature of this farm, and we taxied to the far end of the runway, ready for takeoff.

We rumbled forward with more vibration than I was prepared for. We approached takeoff speed, and suddenly I could feel the lift of the warm humid air gather under the wings and hoist us upward, slowly at first and then with a whoosh—less a sound than a sensation welling up in the pit of my stomach.

As we lifted away from the ground, my dad and I simultaneously pointed to a big wooden cross at the end of the runway. The plane pulled up, over, and banked hard to the left, the cross slipping under us. We let Joe level off and throttle down the roaring engine before my dad pointed back and asked, "What's that cross, Joe?"

"I guess they light it up sometimes," Joe replied, his eyes on the altimeter. "The guy that owns this farm lets the Klan come here for their gatherings. Some of them fly in, sometimes at night—that's why the smudge pots are lining the runway."

I could feel the chill creep up my spine, pushing aside the usual thrill of breaking from the earth's tug. Suddenly, everything in the plane felt very heavy. I could see the gravity of it all settle into my father's expression, too.

Not that it was a complete surprise. The billboard that overshadowed the city-limit sign in our new home of Smithfield depicted a figure of a hooded Klansman on horseback. "Welcome to Smithfield," the sign read. "Help fight communism and intergration." At least the misspelling conveyed some of the ignorance behind it all.

I never did feel at ease in that agricultural landscape, and a lot of other people felt inordinately more uncomfortable than I ever did—in fact, many of them lived in justifiable fear. Whenever I feel tempted to make broad-brushed assumptions about the inevitable benefits of all things local, I remind myself of the perspective gained from that plane ride. There is a cautionary point to be made here about local control—and even about local food systems. And it is not exclusive to the South.

Disturbing stories of oppression and injustice can be told from any region in the country, from any era in our history, including the present. Some local farmers and politicians brazenly squelch both the enforcement and reform of pesticide regulations designed to protect field laborers in California; "guestworker recruiters" sometimes lure migrant agricultural and seafood laborers to the United States only to hold them in conditions ranging from misery to modern slavery; and the "disassembly" lines in meat and seafood processing plants are sites for grisly accidents and grueling working conditions for the minority populations that often comprise the bulk of their workforce, while public officials entrusted with the well-being of these workers look the other way. Local control.

A Poor Aftertaste

It is important to remember that rebuilding local food systems is not necessarily the same thing as *restoring* local food systems. What existed previously may not be a system any of us care to revisit, at least in reality. The historical diversity of crops and livestock may well be worth replicating, but the socioeconomic divisions and hierarchies of the past often left much to be desired from the perspective of impoverished or oppressed communities. "Local" can be the provincial guardian of the status quo—or it can be ground zero for food justice and empowerment.

Erika Allen, Chicago projects director and national outreach manager for Growing Power (founded by her father, Will Allen) and a Post Carbon Institute fellow, put it to me this way:

> Food is really a litmus. When there is a lack of healthy food, there is also usually a lack of employment, good schools, safe neighborhoods, etcetera. You have to put racism out there as a means of understanding it. The PC [politically correct] movement looks at those pieces, for sure, but it's far more critical to look at large economic and social structures and how they impede progress. It's a long, long history—but now with fast-paced technological and cultural changes under way, people are talking . . . and the corporate ladder is falling down! It all boils down to looking at the roots of what happened and how we got here to these inequities. The historical perspective is critical—the Good Food Movement has to examine it. I'm certainly a student of it.[1]

It's critical in this work to include the perspectives of people who have long been subjugated to the unjust economic and social structures inherent in many regional agricultural histories. While we cannot undo the injustices of the past, we can begin to unravel the conditions that supported those injustices. At the same time, local communities can work to reframe food systems that embrace a diversity of cultural and economic perspectives. Yet another approach is for self-identified groups ("communities of color,"[2] "coalitions of the oppressed") that have suffered injustice to frame their own food systems. Some Native American and black communities explicitly prefer independence from a food system controlled by those with power and privilege—a food system that has not responded to their needs and desires. In response, some of these groups are working toward self-reliance by creating their own food systems that reflect their collective history, culture, and future aspirations.

Both economic privilege and race play critical roles in food justice, currently and historically. There are times in these conversations when the distinct issues of race and class should be kept separate; on the other hand, race and class also frequently intersect in problematic and unjust ways. Revitalizing local food economies and traditions can provide communities with a unique opportunity to ensure that disenfranchised and oppressed groups have strong and respected voices. No one should be under any illusions, however, that such a process will be simple or even completely successful. Nonetheless, the effort must be made, because a just and equitable local food system cannot be built from the top down, from one end of the socioeconomic continuum, or from an unchallenged white middle-to-upper-class perspective.

Communities of color and impoverished populations suffer from the injustices and inequities of our current food system at highly disproportionate rates—as farmers, field workers, meatpacking and seafood processing laborers, pesticide applicators, and ultimately consumers. Indigenous peoples lost land and livelihoods to European agrarian dreams. Blacks, Hispanics, Asians, poor whites, and others have all suffered as the laborers behind so many of those dreams. Today, economically disadvantaged communities of color comprise the labor keystone that supports our contemporary food system. Despite dramatic changes in the technologies and economic systems that have driven the food system over the past century or more (did someone say "efficiencies"?), the inequities and the injustices persist, with the same racial groups often continuing to bear the burdens, albeit some different ones.

Ironically, it is precisely these inequities and injustices that make opting out of the so-called industrial food system a complex if not impossible option for the underprivileged, the disenfranchised, the unemployed, the homeless, the oppressed. In fact, economic constraints make those people all the more dependent upon the commodities, the transnational corporations, the cutthroat supermarket pricing, the subsidized agricultural production—the international industrialized food system. Oh, and lest we forget, the jobs that drive the system. That's where so many of the jobs are: right in the belly of the beast, the beast that feeds our bellies. And they're largely jobs that many Americans would not want (and few could actually endure) at such low wages, even in the most dire of circumstances.

Think Harder, Dig Deeper

All those activities so many activists rail against—spraying pesticides, butchering animals on an assembly line, trawling the oceans, driving big trucks across the country, stocking shelves in the Walmart food aisle, inseminating turkeys that can't do it themselves—are the jobs held by many of those who can barely afford to buy food at any cost. "Food choice" for them is not about "local," "organic," or "animal-welfare approved," but whether they can feed the kids even just one meal a day . . . and how.

It is here that the push for local food systems hits a paradox in the shape of something like an enormous wall. How can more just, inclusive local food systems be built upon a fair and equitable representation of all community members when some of the most oppressed members are so embedded in and reliant upon a much bigger and more powerful food system? In fact, they may well be jeopardizing their jobs or even their personal well-being by advocating for change. And then there is the elephant in the room: the vast majority of our national anti-hunger and nutrition programs rely on commodity foods purchased by the U.S. government for distribution to food banks, senior centers, schools, emergency shelters, Indian tribal governments, soup kitchens, and other alphabet-soup programs (TEFAP, CSFP, FDPIR, WIC, etc.).[3] The amount of food purchased and distributed, as well as the number of recipients of these low- to no-cost foods, is staggering. It all goes well beyond the scope of our too few scattered and seasonal community gardens.

As a country, we've been struggling with the effectiveness of federally supported food security programs versus community food security initiatives for decades. In fact, since the 1980s anti-hunger advocates have repeatedly fought federal government efforts to push food security initiatives out of the purview of the Beltway and onto the already bent backs of local communities. They have worked assiduously to make it clear that a train wreck is inevitable if those efforts go too far; neither the resources nor the systems have been in place for such a shift.

In general, though, much of the debate has been on the periphery of the mainstream news. For example, in 1999—well before local foods became such media celebrities—Patricia Allen at the University of California's Center for Agroecology and Sustainable Food Systems in Santa Cruz wrote a seminal paper titled "Reweaving the Food Security Net: Mediating Entitlement and Entrepreneurship" in the journal *Agriculture and Human Values* (an excellent place to watch these sorts of evolving dialogues unfold, often well before they hit the popular press). In that forward-thinking article, Allen pointed out that community food security efforts to tackle the issue of low-income food access and local food systems "are not necessarily compatible and may even be contradictory."[4] She was warning community food security activists not to assume that their efforts (such as urban gardens, CSAs, and so on) would supplant the need for a federally supported safety net, particularly for the most vulnerable populations. I think that Allen was correct to bring forward the contradictions at the time, and many of the cautions that she put forward then remain relevant. I also think that we can continue to find ways to overcome the potential contradiction between local food systems and food security concerns. After all, a contradiction is often an opportunity for creativity. And the increasing ecological constraints facing our food system will inevitably force us to span the contradictions with commonsense bridges.

A number of community food security initiatives have seized upon this creativity. The primary challenges for these locally rooted efforts are limited resources and a restricted scale, whereas the federally supported programs are challenged by the enormity of their scope and scale, as well as the sluggishness of innovation at the federal policy level. In healthy agricultural tradition, hybridization is happening nonetheless, with the feds allowing Supplemental Nutrition Assistance Program (SNAP) benefits (formerly known as food stamps) to be used at farmers' markets and some food banks, and other emergency food providers supplementing their federal commodity supplies with local and regional products.

Perhaps the most pervasive breakthrough bringing more fresh, local food to food-insecure populations comes by way of the farm-to-school efforts happening across the United States. Not only do these education-based local food initiatives build awareness among the next generation and across class-based gaps, but the shifting school-age demographic bodes well for bridging cultural and racial divides. U.S. Census Bureau estimates now show that "minorities" make up more than 50 percent of children younger than age one.[5] Given this data and the astounding increase in farm-to-school initiatives across the United States, it seems likely that awareness and consumption of sustainably produced local foods is growing among an ever-broader cross-section of Americans.

The exceptional leadership of the National Farm to School Network (NFSN) and its fast-paced growth over the past several years have helped accelerate the number of these programs and their effectiveness. One of the beauties of the NFSN approach has been its emphasis on ensuring that school initiatives be adapted to the school's resources and interests.[6] This openness to diverse approaches is critical.

But here's where we run into another dilemma. If you put your ear to the ground to listen to the rumblings in the local food world, you can hear a mixed message: a call for a national food movement and a call for more diversity in efforts to rebuild local food systems. The problem is that despite the ability of a national movement to generate momentum, it can just as easily steamroll disparate perspectives and disenfranchised whispers.

Julian Agyeman and Alison Hope Alkon are the editors of a superb recent book titled *Cultivating Food Justice: Race, Class, and Sustainability.*[7] It's a book that should be read by anyone advocating for change in our food systems. The goal of the book is to "highlight the alternative imaginings eclipsed by the pervasiveness of the food movement narrative, to offer other voices and visions of how past and present human communities relate food, agriculture, and ecology."[8] There is a tendency among many food activists and writers to espouse strong claims about what foods are imbued with all of the right values and which ones are somehow bad choices. Agyeman and Alkon consider the potential for exclusiveness in a food movement that judges people in part by the foods that they are able to access and eat and thereby help frame the importance of cultural and racial diversity in all food systems thinking.[9]

Their seminal work reminds the reader that food justice does not simply mean *access* to food; more complexly, food justice means having the opportunity to make *food choices* without being judged. Organic, local, sustainable, fair-trade—many of these kinds of food choices are far beyond the range (geographically and economically) of many people. The rhetoric espoused by some food activists risks creating a black-and-white world with some elitist overtones. Rebuilding local and sustainable food systems in the midst of a highly industrialized food system can enhance the diversity and health of food choices, but there are clear challenges in actually making those choices available and affordable to a vast number of people within communities of color and among food-insecure populations.

Food Sovereignty

One way that some racial and cultural groups have chosen to maintain and increase their food choices within the global industrialized food system is to promote the concept of food sovereignty. Although the variety of food options may seem vast in the grocery-store aisle, true food *choice* is challenged by the regulatory environment and the hardball economics played out on Wall Street and Main Street. Market concentration and one-size-fits-all regulation threaten food traditions and consumer choice.

"Food sovereignty" is another one of those terms that is in flux. The term was first put forward in April 1996 by Via Campesina ("the peasant's way"), originally an international organization for small rural farmers, and the concept was conceived as a response to the threats to small food producers resulting from international trade policies and multinational corporations. Simply stated, Via Campesina viewed food sovereignty as "the right of each nation to maintain and develop their own capacity to produce foods that are crucial to national and community food security, respecting cultural diversity and diversity of production methods."[10]

The concept has taken on additional meaning and been somewhat transformed in its recent use by some small food producers, municipalities, and even states here in the United States. In the spirit of the Via Campesina

Defining "Food Sovereignty"

The International Planning Committee for Food Sovereignty created the following definition of "food sovereignty" in 2007. It is thoughtful and well worth contemplating:

> Food sovereignty is the right of peoples to healthy and culturally appropriate food produced through ecologically sound and sustainable methods, and their right to define their own food and agriculture systems. It puts the aspirations and needs of those who produce, distribute and consume food

> at the heart of food systems and policies rather than the demands of markets and corporations. It defends the interests and inclusion of the next generation. It offers a strategy to resist and dismantle the current corporate trade and food regime, and directions for food, farming, pastoral and fisheries systems determined by local producers and users. Food sovereignty prioritises local and national economies and markets and empowers peasant and family farmer-driven agriculture, artisanal fishing, pastoralist-led grazing, and food production, distribution and consumption based on environmental, social and economic sustainability. Food sovereignty promotes transparent trade that guarantees just incomes to all peoples as well as the rights of consumers to control their food and nutrition. It ensures that the rights to use and manage lands, territories, waters, seeds, livestock and biodiversity are in the hands of those of us who produce food. Food sovereignty implies new social relations free of oppression and inequality between men and women, peoples, racial groups, social and economic classes and generations.[11]

The third of the committee's six principles for food sovereignty describes how food sovereignty depends upon localizing food systems:

> Food sovereignty brings food providers and consumers closer together; puts providers and consumers at the centre of decision-making on food issues; protects food providers from the dumping of food and food aid in local markets; protects consumers from poor quality and unhealthy food, inappropriate food aid and food tainted with genetically modified organisms; *and rejects* governance structures, agreements and practices that depend on and promote unsustainable and inequitable international trade and give power to remote and unaccountable corporations.[12]

concept, these farmers and communities are reacting against business and regulatory policies that threaten the livelihoods and traditions of farmers. Farmers who wish to sell seeds, raw milk, value-added products, and various other goods are falling under increasingly stringent regulations, regulations often created with much larger businesses in mind.

While this relatively new amplification of the phrase "food sovereignty" maintains a focus on the plight of the small farmer in a highly globalized economy, we could argue that this new iteration is more about "food rights" than Via Campesina's original thrust of "the right to food." One current hot spot for this fiery reshaping of the term is Sedgwick, Maine. In 2011, Sedgwick citizens unanimously adopted the "Ordinance to Protect the Health and Integrity of the Local Food System" at their annual town meeting. This ordinance pushes for a community's self-determination in food choices, noting, "We have faith in our citizens' ability to educate themselves and make informed decisions. We hold that federal and state regulations impede local food production and constitute a usurpation of our citizens' right to foods of their choice."[13]

The state of Maine did not appreciate the town's declaration of self-determination, so both sides have engaged in a shoving match of sorts since the passage of the ordinance. Despite the town's size—or precisely because of its tiny population of 1,102—the Sedgwick story has become a national beacon for towns across the country that view some portion of food safety regulations as a major hindrance to consumer choice and local economic development. It has also put a bit of a twist on the evolutionary spiral of food sovereignty. "Goliath, meet David."

Food sovereignty is clearly a challenging ideal for any group that is resource-thin or marginalized by virtue of prevailing power structures or stereotypes. Yet food sovereignty provides a framework for continually assessing the predominant food systems from a particular cultural or racial vantage point. It also encourages appropriate and fair representation in the economic and grassroots decision making about local food systems initiatives.

Traditional Ways, Traditional Lands, and Traditional Waters

Nowhere does food sovereignty find a more natural niche than in the claims made by indigenous peoples to their traditional lands and foodways

("foodways" referring to a group's diets, traditions, and cultural associations surrounding food). We sometimes forget that "local food" is not always about gardening and farming. If anything, agriculture sometimes gets in the way of the most visceral possibilities for relocalizing our diets and truly engaging with the region where we live. Hunting, fishing, and gathering were the local mainstays well before any European plow ever ripped open the first furrow in the New World. A tongue-in-cheek summary of American agricultural history might cast it as a history of displacement: displaced immigrant farmers in turn displacing native populations, with foodways and ecosystems gradually overwhelmed by the fast-folding wake of those displacements.

Local food activists should be careful not to lose sight of the Native American foodways in their regions. Sadly, in most of the United States, both the descendants and the food traditions of these groups have long since been absorbed into the powerful undercurrents of the mainstream. In some areas, however, such as the Pacific Northwest and the Great Lakes region, not only do many of the traditions continue, but the ideas surrounding Native American rights to food and traditions still stir up fierce controversy. Many Native American groups continue to assert their rights regarding specific hunting and fishing methods and seasons, as well as where they are allowed to pursue their traditional foodways—on reservations, private property, state land, and federal holdings. Water rights are also a critical component for the survival of Native American agriculture, and numerous disputes about historic water rights are in various stages of arbitration.[14]

When appropriate, local food systems initiatives should include Native American perspectives in everything from regional planning to market development, nutrition education, farming practices, and agricultural regulation. The idea, of course, is to include Native American voices and ideas about efforts to preserve native foodways and food rights—not to mention native traditional methods of managing ecosystems—without intruding upon tribal affairs when autonomy is preferred.

One of the best models for integrating Native American and other cultural perspectives into the rethinking of local food systems comes in the form of a report on the status of Native American and Hispano[15] farmers in the "Dreaming New Mexico" project. Commissioned by the environmental organization Bioneers, this report by the Southwest Cultural Center takes a hard, unromanticized look at the challenges facing these two longstanding agriculturally focused cultures and how they might fare as a result of emerging food relocalization efforts in New Mexico. Unexpected contradictions are revealed, such as the trend toward organic foods in northern New Mexico as a result of the gentrification of the area. Hispano farmers are seldom certified organic, and the hurdles to achieving organic status are not insignificant for many of them.[16] The report concludes that Hispano farms are shrinking due to the division of land between heirs and rising property prices associated with the gentrification of northern New Mexico.[17]

The current trends for both groups are disturbing, but the report's identification of the issues does provide an opportunity for these groups and the collaborators of their choosing to generate creative responses that might link them to a brighter agricultural future with traditional foundations. At the very least, it's rewarding to see that their views have been taken into serious consideration early on in the state's relocalization efforts.

Cultivating a Diversity of Farmers

Urban farms and gardens play a critical role in redressing some of the ills and injustices of various food systems. These projects can provide a variety of services to communities of color and food-insecure individuals—on their own terms—and the gardens and farms serve as national beacons of awareness and hope. But it is important to acknowledge that these initiatives are only one part of the answer. Even in Oakland, a location with a hospitable growing climate and a wealth of vacant lots, researchers Nathan McClintock and Jenny Cooper estimated in 2009 that maximizing production on all available urban agricultural lands would provide for only about 5 to 10 percent of the city's vegetable consumption.[18] Such numbers are by no means trivial, particularly if those food-based activities enhance job opportunities, nutrition, education, and the local economy. However, these figures make it clear that we have to think about the entire system that brings the bulk of food into cities. Therefore, the architects of newly conceived local food systems should not ignore the elements of discrimination and injustice that remain embedded in our current food system. In fact, it is critical that the system be not just examined in this light but also changed through new business models that bring more diversity and justice into the mainstream.

To what extent can the organization of our local, regional, national, and international food systems be recast to provide business opportunities for a more diverse array of players throughout all sectors of the systems? And how can we ensure that these shifts also happen on the farm? If the long view simply involves minority participation in small-scale enterprises, then the boundaries, structures, and players in our food systems have not been altered in significant ways. Currently, the national focus on projects of smaller scales in urban settings—projects that so many of us really believe in—risks losing perspective on other realities of a larger scale, namely, the plight of nonwhite farmers in rural America.

The statistic for the average age of the American farmer is almost a mantra among sustainable-agriculture wonks: 57.1 years old. But there is not nearly as much focus on the discrepancies among farmers in different racial/ethnic categories, and therefore far too little discussion about the challenges facing these different groups. Without a sense of these realities, many local food systems initiatives are unlikely to include mechanisms for addressing the inequities and hurdles faced by nonwhite farmers.

TABLE 7-1. Comparison of Data for Minority-Operated Farms (2007)

	ALL U.S. FARMS	BLACK	HISPANIC	ASIAN	AMERICAN INDIAN/ ALASKA NATIVE
Average Size of Farm (Acres)	418	104	307	124	1,431
Average Value of Sales	$134,807	$21,340	$119,634	$327,621	$40,331
Total Sales and Government Payments <$10,000	58%	79%	67%	42%	78%
Internet Access on Farm	57%	34%	44%	48%	42%
Average Age of Farmer	57.1	60.3	56.0	56.4	56.6

Source: USDA 2007 Census of Agriculture

As table 7-1 illustrates, black farmers have the least acreage, the lowest sales, the highest percent of receipts under $10,000, and the lowest rate of Internet access of any farmer racial group measured by the 2007 U.S. census. On top of that, the average age of black farmers is higher than that of any other group—by three years. American Indian and Alaska Natives are also resource-thin overall. (Bear in mind that these figures are showing gross sales receipts, with no reference to farm inputs and infrastructure.) Farming is challenging enough economically, without the additional hurdles (many of them caused by "structural racism" over the course of several generations) faced by these minority groups.

It is far too easy to focus on increasing cultural diversity among consumers in local food systems while completely ignoring the inequities in farming opportunities for nonwhite individuals. These inequities are particularly well hidden in remote rural areas. Nonwhite farmers in these areas too often face discrimination in markets, banks, and USDA offices, as was clearly documented by the famous *Pigford v. Glickman* class-action lawsuit brought against the USDA by black farmers seeking retribution for decades of discrimination in securing loans; the case was settled in the farmers' favor, resulting in billions of dollars in damages being paid out to them and their descendants.

Finally, it is important to remember that losing farmers of diverse backgrounds does more than lessen the diversity of entrepreneurs in the marketplace. These farmers also possess knowledge, skills, and products that offer alternatives to conventional farming and marketing practices—alternatives that may well prove to be of increasing importance. Keeping farmers of various cultural backgrounds in business is an important form of enhancing food system resilience. The success of U.S. agriculture for well over two centuries can be attributed in large part to the cultural fusion of ideas and technologies manifested on the American farm—a place where a borrowed idea is the seed of innovation.

Farmworkers: Borderline Realities

The enormity of the U.S. food system means that certain realities are easily lost to consumers, policymakers, and even advocates for change. Perhaps no issue is more hidden—because the people themselves so often have to hide or are hidden by their employers—than the plight of farmworkers. Any substantive inquiry into food security or food justice must take into account the outlandish contradiction that those who do the hardest and most dangerous work in our food system often find themselves on the grim end of the food security spectrum. That said, trying to unveil the realities of this vital aspect of our food system is an extremely complex undertaking.

The first challenge comes simply in classifying who is a farmworker and what type of a farmworker he or she might be. About three times as many farmworkers in the United States work in crop agriculture than in livestock agriculture. They are hired in one of essentially three ways.[19] First, they can be employed directly by a farmer on a short- or long-term arrangement and compensated on an hourly, daily, or piece-rate basis. Second, they may find employment through a farm labor contractor (FLC), which operates by providing workers to a number of farms. While having an intermediary who is responsible for the oversight of the workers may serve the farmers well, this arrangement obscures the realities from public view and is ripe for corruption and abuse. The third employment option involves the grower or FLC utilizing the H-2A guest worker visa program to bring in foreign labor for a specified type of work of a limited duration (usually less than one year).[20] Of those H-2A guest workers, 94 percent come from Mexico.[21]

Farmworkers vary in their mobility, and the National Agricultural Workers Survey (NAWS) divides them into four categories:

- Settled: Not migrating
- Shuttle: Traveling internationally between their home country and their work or over a distance of more than 75 miles from home to work
- Follow the Crop: Moving between two or more work locations spread over a distance of 75 miles or more
- Newcomer: Entering the United States for the first time and for less than one year[22]

A critical document in helping to bring these issues to light is the *Inventory of Farmworker Issues and Protections in the United States*, a report based on research conducted collaboratively between the United Farm Workers and

TABLE 7-2. Percentage of Farmworkers According to Migrant Type and Work Status

MIGRANT TYPE	HIRED	CONTRACT	ALL
Settled	72%	53%	70%
Shuttle	14%	11%	13%
Follow the Crop	4%	11%	5%
Newcomer	10%	25%	12%

Source: NAWS data (2005–2009), from *Inventory of Farmworker Issues and Protections in the United States* (United Farm Workers and Bon Appétit Management Company Foundation, March 2011)

the Bon Appétit Management Company Foundation, with support from Oxfam America. This report from 2011 represents an unusual collaboration between a for-profit food service company and several nonprofit advocacy organizations in order to clearly present the data and the issues to the public, with some innovative suggestions for change. Using the 2005–2009 data, the report helps profile farmworkers in new ways. Table 7-2 demonstrates the overall percentages for each category of farmworker, as well as the estimated differences between farmworkers hired directly by farmers and farmworkers working for farm labor contractors.

This report also provides some insight into the types of crop agriculture that are dependent upon farmworker labor in states heavily dependent upon this labor force. In reviewing table 7-3, consider the hand labor involved with the cultivation and harvesting of the crops listed. To understand why farmworkers are so critical to the U.S. food system, we need only reflect upon one of the main produce standards of the American consumer: we want blemish-free fruits and vegetables, and therefore the harvesting and handling of this produce must be done by hand to maintain the quality expected. It's estimated that more than 85 percent of fruits and vegetables grown in the United States are harvested by hand.[23]

I think that the American public grossly underestimates the amount of nonmechanized work required in crop agriculture, how hard it is, and just how much skill it requires. While working on my grandparents' farm in North Carolina—a peach orchard and nursery operation with a variety of other fruits and vegetables—I was humbled on more than one occasion by the Mexican labor we brought in when there was a task too big for us

TABLE 7-3. Top Five Labor-Intensive Crops (in Terms of Commodity Value)

CALIFORNIA	FLORIDA	NORTH CAROLINA	OREGON	TEXAS	WASHINGTON
Grapes	Greenhouse/ nursery	Greenhouse/ nursery	Greenhouse/ nursery	Greenhouse/ nursery	Apples
Almonds	Oranges	Tobacco	Pears	Onions	Greenhouse/ nursery
Nursery products	Tomatoes	Blueberries	Cherries	Pecans	Cherries
Lettuce	Strawberries	Tomatoes	Grapes	Watermelon	Grapes
Berries	Grapefruit	Cucumbers	Hazelnuts	Cabbages	Pears

Source: *Inventory of Farmworker Issues and Protections in the United States* (United Farm Workers and Bon Appétit Management Company Foundation, March 2011)

to complete on time. I consider myself a good farm laborer—thoughtful, quick, and upbeat. But when we called the farm labor contractor to bring in a crew of Mexicans to help us out with pruning or harvesting from several hundred to several thousand peach trees, I watched in awe as nimble fingers and strong backs combined to get the job done and done right. The only unsettling aspect was the unloading and loading of the farm labor contractor's van, with the contractor trading paperwork with my grandfather. I tried to imagine what lives these men were leading during their off-hours, and the relationship between the workers and contractor who delivered them and picked them up never seemed to be one of camaraderie.

The last few years have made clear just how vital these workers are to U.S. agriculture. A shortage in the national agricultural labor pool has both sides of the aisle in Congress scratching their heads. A hearing titled "America's Agricultural Labor Crisis: Enacting a Practical Solution" in the U.S. Senate on October 4th, 2011, yielded a number of enlightening testimonies from concerned citizens and business representatives from across the country.[24] One after another, the speakers that day described the importance of these workers to the U.S. agricultural economy and how immigration policies are taking their toll on the United States' ability to harvest crops and maintain livestock operations. That same week, the Georgia Fruit and Vegetable Growers Association cited a $391 million loss related to the shortage of agricultural workers. With 5,244 seasonal farmworker jobs unfilled in that state during the 2011 harvest, crops were lost at a staggering rate. In South Carolina, many farmers are concerned that their state's recent immigration laws—on parallel with Arizona's draconian policies—will create serious labor difficulties and ensuing financial losses.[25]

Are these jobs that aren't going to American citizens who might want them? That seems to be far from the case, as state after state has tried to recruit workers from its traditional workforce, to no avail. The conditions are cited as too tough, the labor too physically demanding, and the pay too minimal. There's the rub. The U.S. food system as it currently exists simply cannot function without a strong farmworker labor pool. But it is not a life to which many would aspire. In fact, the life of a farmworker is far too often a case study in injustice, from multiple angles: economic insecurity, family malnutrition, racial and cultural bias, dangerous working conditions, and sexual harassment and violence. And because so many of these workers feel threatened by any system of documentation, the data on their lives is difficult to acquire or analyze with full confidence.[26] Add to that the fact that some of these migrant farmworkers may move eleven to thirteen times during the course of a year.[27] Then toss in the fact that farmers with fewer than eleven employees are not subject to many of the same safety regulations and reporting structures that apply to other farms, and much of the data regarding working conditions simply doesn't exist.[28] The mixing of categories and definitions simply adds to the confusion, such that we might hear that there are anywhere between two and five million farmworkers in the United States (depending on how researchers categorize things and then count without actually counting). How's that for a high margin of error?[29]

The challenges in identifying and tracking these farmworkers with any degree of certainty mean that the conditions facing farmworkers are neglected far too often in local food system initiatives and planning. They are, at best, "under-considered" instead of "under consideration." So how do these workers fit within local food systems? There is not a quick and easy answer. After all, they may work for small farming operations or large multinational corporations. They may be established in local communities, or they may be transient. It seems obvious that they support local food systems through their labors, but some people claim that they undermine local economies (although these claims are made without defensible data, in my view). One thing is certain: in most regions of the country, the loss of farmworkers would unravel the entire region's food system. No pruning, no spraying, no thinning, no mowing, no harvest. No feeding, no cleaning, no milking. Yet as vital as they are to our food system, the majority of farmworkers are underpaid, underprotected, and overworked.

Table 7-4 provides an overview of the estimated income levels of farmworkers, sourced from 2005–2009 NAWS data. It is particularly important to note the discrepancies between farmworkers hired directly by farmers and those working for farm labor contractors. Again, such data are a challenge to gather and interpret, but these numbers indicate both the economic injustices and the need for more transparency and accountability in this shadowed world of our national food system.

Studies have been conducted to answer the question at the state level in various places, and the prevailing conclusion seems to be that farmworkers do, in fact, generate profits for the state's agricultural economy (and not just for corporate bank accounts). Much more research is warranted, including focused inquiries into where those profits actually end up—in the hands of the farmer, the processor, the distributor, or the kingpins of vertical integration. (Regardless, it's clear that the farmworkers themselves are probably not the primary benefactors of their own labors in most cases.) Interestingly, other research indicates that replacing labor-intensive crops with other crops not requiring as much labor actually *reduces* net profits for farmers.[30]

TABLE 7-4. Farmworkers' Annual Personal Income Levels by Employment Type

INCOME LEVEL	HIRED	CONTRACT	ALL
Up to $9,999	20%	33%	22%
$10,000 to $19,999	46%	58%	47%
$20,000 to $29,999	23%	9%	21%
$30,000 or more	12%	0%	10%

Source: *Inventory of Farmworker Issues and Protections in the United States* (United Farm Workers and Bon Appétit Management Company Foundation, March 2011)

Note: Percentages are from the total number of farmworkers with reported income data. Farmworkers who had not worked in the United States for a full year were excluded from this question.

In the end, it is evident that farmworkers not only enhance the agricultural economy but, in fact, shoulder it—and with little to show for it on their end. The knee-jerk response is to blame farmers for the hardships endured by farmworkers, but there is enormous consumer and corporate pressure to keep food prices low, and farmers' profit margins tend to be narrow, with labor costs always being a challenge on the farm.

Nonetheless, it is tragic that our food system and agricultural economies are completely dependent upon these farmworkers, and yet they suffer a variety of injustices. The *Inventory of Farmworker Issues and Protections in the United States* provides one of the most succinct and well-researched reports on these injustices, and anyone interested in this issue should read it. The authors summarize just how little regulation and oversight there is for these workers: "Few people, for example, are aware that farmworkers are excluded from the basic labor and safety standards firmly established in other employment sectors. Likewise, many people would be shocked to learn that farm work has little or no overtime limits, child labor restrictions, collective bargaining rights, or workers' compensation insurance, although agriculture is considered to be one of the most hazardous industries in the U.S." Not only that, but there are only twenty-two full-time inspectors dedicated to enforcing the federal Migrant and Seasonal Agricultural Worker Protection Act for agricultural employees (farms and FLCs) for *the entire country*![31]

One of the greatest concerns for farmworkers is one of the most difficult to regulate and monitor: chemical exposure. Farmworkers have the highest rate of chemical exposure and number of chemical-related poisonings of any labor sector in our society.[32] The issue of pesticides should be a more prominent and complex part of the local food systems dialogue. Many local food system advocates who embrace even the most basic tenets of food justice and/or community food security still perpetuate the ill-conceived notions that "local always trumps organic" and "small is always better than industrial organic." The issue of farmworker pesticide exposure highlights the weakness of these positions.

Farmworkers receive the bulk of human exposure to pesticides in the U.S. food system. They are the people in the fields and orchards, in constant contact with plants, soil, and livestock that have been treated with a variety of chemicals. Farmworkers tend to have little control over or even knowledge of their exposure, nor are they often in a position in which they can ask questions, much less express concerns. Language barriers make it all the more difficult. I sprayed pesticides for three years in Italy, all the while struggling to read the highly technical pesticide labels in either German or Italian. I remember well the frustration and occasional anxiety of not being certain whether my language deficiencies would result in improper applications of the spray or jeopardize my own safety or someone else's.

That experience working with pesticides (there is no better impetus for teaching sustainable agriculture) leads me to be highly skeptical that any amount of quantitative academic research will ever tell a complete story of how many challenges farmworkers face with regard to pesticide exposure. Who sprayed? When? Were the application rates correct? Who is keeping the records? Is the farmer abiding by the no-entry time period? How many different pesticides and herbicides have been sprayed in a given area during the course of the season? Is there a threat of pesticide drift from adjacent areas that are being sprayed? Are children playing in the area while the adults are working? Is the well providing the drinking water located in a high-spray zone? These are all basic management questions that are typically unanswered for most field workers, and they don't even begin to touch on the scientific questions that are just as important and even more complex. What's not understood off the farm is that it's often taboo to question the specifics of pesticide management and exposure, much less the whole paradigm of pesticide use. These conversations are all the more difficult in male circles, as an air of machismo and feigned indifference tends to squelch expression of the real fears. Even in the best of circumstances, if a worker does question any of it too much, the result is likely to be teasing, scorn, or termination.

All of us who advocate for change in food systems need to pull back from some of the overused mantras and assumptions and consider things from the farmworker perspective more frequently. Here's a question to begin the process: When it comes to pesticide exposure, what do you think a farmworker would choose—a season of pesticide exposure or a season of organic-based agriculture on huge farms? "Industrial organic"—even if it is organic on a huge scale—*is* in fact an alternative to something much grimmer.

The Raw Deal: Food Processing

If farmworkers are hidden, workers in the food processing sector are virtually invisible. These workers—documented and undocumented—face numerous challenges. They process fruits, vegetables, freshwater fish, seafood, poultry, swine, and beef, often in oppressive conditions. Management positions are typically dominated by white males, while the processing laborers are predominately from nonwhite racial groups. Women comprise a larger component of the labor force in this sector than in the field worker arena. Not only are women typically in positions with little power in this male-dominated industry, but they are also economically vulnerable. Unsurprisingly, this imbalance of power can readily result in cases of sexual harassment and violence.[33]

Just as farmworkers suffer from pesticide exposure, dangerous heavy equipment, heat stress, and repetitive motion injuries, laborers in the low-wage food processing sector also face numerous health risks. Workers handling fruit and vegetables in conventionally managed operations frequently encounter concentrated levels of pesticides on the products coming straight from the field, as well as repetitive and often heavy tasks that tax a body working long hours in the same job for weeks, months, and years. The likelihood of illness and injury increases at a staggering rate for food processing workers when slaughtering and butchering are involved. U.S. Bureau of Labor statistics consistently show the inherent dangers in slaughtering facilities, with the rates of reported injuries at nearly three times the average rate of injury in all of private industry.

The frequency and severity of injuries in meatpacking plants should come as no surprise when we look at the daily capacity of some of these plants. In 2007, the five largest U.S. plants for pork had capacities ranging from 18,000 to 32,000 hogs per day.[34] Poultry processing plants require workers to work at tremendous line speeds that are still not regulated by the U.S. Occupational Safety and Health Administration (OSHA). These employees also find themselves working at times with dull knives, making the work all the more dangerous.[35] The horrendous working conditions and problematic regulatory practices are well documented, and they are even featured in a number of films hitting the mainstream media, such as *Food Chain* and *Harvest of Dignity.*

The seafood and freshwater fish packing industries have some parallels in terms of dangerous work conditions, repetitive task injuries, and harassment. They also rely heavily upon migrant workers classified as H-2B "guest workers," a governmental status that grants even fewer rights and less oversight than agricultural H-2A status. Whodathunk? H-2A agricultural workers, who typically work in the fields, receive—at least in theory—some benefits that help compensate for their otherwise difficult working conditions: the possibility of free (albeit often substandard) housing on some farms, social security tax exemption, access to free legal counsel (presuming they dare or even have the necessary language skills and transportation), "reasonable" reimbursement for transportation costs to and from the United States, and at least three-quarters of the employment time spelled out in their contracts (but who's counting?). (In the end, much of it depends upon the farmers involved, and there are certainly some who have enduring and even endearing relationships with the H2-A workers who work with them every year.) H-2B workers in the seafood sector—along with their compatriots in sectors such as tourism and forestry—get none of this.

The catfish processing industry is yet another commonly overlooked sector, but it's of real concern in this discussion. In the cotton-dependent Delta states, catfish ponds and processing infrastructure were built in the 1970s and '80s to replace the waning cotton industry. Although the goal was to restore the economic vitality of the region's agriculture, not much changed in the way of socioeconomics as a result, so the historic patterns of poverty and poor working conditions continued while the benefactors of this new industry generally garnered the bulk of the financial rewards. The workers in these processing facilities are predominately black females. The working conditions in many of these facilities can be poor at best, and the lack of empowerment among these women can lead to harassment in various forms.[36]

Many of the food processing facilities described above are geared to serving national or international markets. They may be on a scale that dwarfs or even drowns local and regional food initiatives, creating the sense that they are the nemesis. But there are people inside those facilities who are some of the most vulnerable and downtrodden people in a community. In fact, chances are that many if not most of the laborers themselves fall somewhere within the USDA's food-insecure spectrum.

How do we even begin to tackle the issue of caring for the farm and food processing workers who are fundamental to such a large portion of our food system? One way involves the certification approach that has helped transform some aspects of our food system. There is a pioneering effort under way in this regard, born of a consortium of major food buyers, farmers, farmworker groups, and consumer advocates. The Equitable Food Initiative (EFI) is developing guidelines for a quality assurance certification process for fruits and vegetables that are produced with high standards for labor, environmental stewardship, and food safety. The certification process is supported by an accompanying workforce education program aimed at having growers and workers work together to develop protocols for maximizing worker and food safety. The intent of this initiative is to link the well-being of workers in the food system with quality assurance and environmental sustainability.

In the end, certifications are a process of turning conversations and concerns into heightened expectations. The same should be true of food justice at the local level. It goes far beyond talking local. It begins with local listening.

Chapter 8

Biodiversity

It's hard to drive past a general store in Vermont and not stop in, especially when there's a stop sign at the same corner as the store, the parking is just adjacent to the stop sign, and the smells of fresh-baked goods, made from scratch by the owner, Julie, are somehow sneaking out the front door. Plus, it's a moral obligation to support any local business that is also a community center. At the Wells Country Store in Wells, Vermont, you get the news about the local car crash five minutes before it happens and some insights into how a few hard-core Vermonters ended up electing a Socialist to the U.S. Senate. You also get some of the best sandwiches in the state.

I stopped to grab one of those sandwiches on my way to the college a few weeks ago, and they were out of my usual cranberry-walnut chicken salad on a hard roll with Dijon mustard. Joanie, Julie's sister, had a suggestion. "Have you tried our prosciutto? It's fabulous—and it's made from small farms in Iowa by an amazing prosciutto company there called La Quercia. La Quercia means 'the oak' in Italian, I think." There was a cultural clash going on in my head at that point—Iowa and Italy just don't strike me as likely culinary co-conspirators. I didn't realize at the time that the oak is Iowa's state tree, although I immediately made the connection to the tradition of acorn-finished pork. These savvy marketers had found a way to bridge disparate worlds with a natural connection.

I don't think I'm a food snob, but I was lucky enough to spend about four years in Italy, and I had some amazing prosciutto there, so my expectations weren't all that high. Joanie rang me up, and I thanked her for the suggestion as I dodged my way through the crowd (that's what we call a gathering of more people than you have fingers in this part of Vermont). I opened up the sandwich when I got to my car, took a bite, and walked right back into the store to ask Joanie for the name and the location of the company. I couldn't believe it—the prosciutto put me right back in Italy.

As soon as I got to the office, I looked up the company's website, only to discover that the owners of La Quercia, Kathy and Herb Eckhouse, had put together a business founded on about every principle I valued in an agriculturally based enterprise: biodiversity conservation, sustainable agriculture, humane animal husbandry, cooperative growth strategies with farmers, commitment to their employees, waste reduction, and, of course, quality food. I immediately picked up the phone to learn more. The receptionist answered and said that Kathy was out of the office at the time and Herb was on his "Ham Independence" tour at venues all around the country. "His what?" I asked. The Fourth of July was only a few days away, but I hadn't linked ham to patriotism before (although I did link it to the South, having grown up in Smithfield). Then I looked up at my computer screen and saw the banner on their website: "In the prosciutto of happiness." These were my kind of people.

I connected with Kathy several days later, and she walked me through the evolution of their business, with a focus on their various efforts at biodiversity conservation. What was particularly unusual was their dual commitment to conserving both agricultural biodiversity and natural habitats. They came to agrobiodiversity by way of taste and experimentation. As Kathy explained it, their prosciutto products are 96 percent meat and 4 percent spices, so the quality and characteristics of the meat matter, and there are significant differences in the meat from different pig breeds.

There was once a tremendous genetic diversity in the pig breeds found in the United States, as each region and culture seemed to have its own preferences for what a pig did on the farm, how it grew, and what its ultimate destiny would be. Pigs were divided into two rough types: bacon and lard. Depending upon the primary focus of the pork products, farmers tended to lean (call that a counter-pun) one way or the other with their breeding. A lard pig was bred to gain quickly and to put on fat for rendering into lard, whereas the bacon pig was a slower grower that put on muscle as well as fat. The downfall of the diversity in pig genetics has a rough correlation with the agricultural intensification and specialization of grain-growing regions, along with the overall concentration of U.S. agricultural markets. Pig breeds have disappeared with alarming speed over the course of the last century, and most of the traditional pig breeds are considered highly endangered. The one positive trend in favor of these rare breeds is the emerging niche market for pork products.[1] La Quercia is at the forefront of this resurgence of taste, tradition, and rare-breed conservation—in a business model that works for them, their farmers, and their thirty employees.

Kathy explained that their approach to prosciutto was analogous to a vintner's focus on terroir, the unique combination of land and varietal, except that their pork products were the unique combination of breed and animal husbandry practices. As they tested out the differences in breeds, they began to link certain breeds to specific products. For example, for their bacon and prosciutto, they prefer the Tamworth, a long, lean pig that is an excellent forager, helping to create a leaner product through the development of its musculature and its diet . . . which brings us to animal husbandry.

The Eckhouses will use only pigs that have access to the outdoors, adequate space to move about and congregate, and deep bedding for rooting and nesting. Their farmers may not use synthetic growth stimulants, and they may not feed their pigs animal by-products. They never use meat from CAFOs (concentrated animal feeding operations). One challenge in their efforts to use rare breeds and to maintain such high animal husbandry standards has been finding farmers with sufficient numbers of pigs to provide La Quercia with an adequate supply. However, Kathy and Herb have seen this challenge as an incubation opportunity, a way to support farmers and contribute to livestock-breed conservation. For example, in one case, they have worked for six years with a farmer breeding Tamworths so that he could ultimately provide a sufficient number of pigs to La Quercia.

La Quercia has initiated a new, albeit far from novel, practice to the husbandry of some of their pigs. Europeans and their colonial descendants in the United States long depended upon feeding their pigs mast—the variety of nuts available in the fall—just prior to slaughter. These nuts were a boon to rapid weight gain for the pigs, and they also imparted a particular flavor and quality to the meat. In light of this traditional knowledge and its link to an age-old form of ecological management when done well, La Quercia helped create two husbandry models for some of their farmers. In Iowa, B&B Farms will be raising pigs on a diet based in large part on pasture and acorns. These acorns will be purchased from off the farm, with some scattered about for foraging and others ground into meal. Missouri farmer Russ Kremer, part of B&B Farms' Ozark cooperative of four farms, will be piloting the "woodland" pork: his pigs will get a substantial diet of acorns and hickory nuts through careful management of one-acre wooded paddocks, through which he rotates the pigs to prevent ecological damage.

In addition to the conservation of breeds, landscape, traditions, and local economies, Kathy and Herb have been quite judicious in reducing the energy consumed and waste generated by their operation. They recently installed high-efficiency (and superior-quality) lighting that has cut their energy consumption by 25 percent. They have selected the packaging products that have minimal environmental impact whenever possible, although any items contaminated with blood or meat are required to be sent to the landfill. Unfortunately, lamented Kathy, Herb's dream of a wind turbine may not be realized, as their property doesn't provide them with the required fall-line buffer in case of a blowover.

Kathy sees another example of their commitment to environmental stewardship right out her window. "I'm looking out at our replication of an Iowa prairie," she told me, "even though people in the know would recognize a few exotics here and there. Nonetheless, it's clearly an important little piece of critical habitat." She then related the story of the monarch butterfly migration that most of us never see. Few us have restored a habitat to the degree that monarchs actually land there en masse for a rest during their long journey from as far north as Canada to central Mexico, as they did at La Quercia. "They come here because of the biodiversity," she said. I can't help but think that the monarchs might have arrived knowing the company they might keep there.

Before we wrapped up our conversation, I had to ask Kathy about what it was like to be in Iowa, virtually the epicenter of the demise of traditional pig breeds and husbandry practices—and not in a place associated with Italian products.

"Well, when we started, people were really skeptical that we could do it. Iowa is a very small market for this kind of thing, even though there are nineteen to twenty million pigs here, more than in any other state. But because of our animal husbandry criteria, we select from only 0.5 percent of that pool!

And the Iowa Pork Board has decided that niche pork is good for them and their markets since it's good for the pork industry to be 'cool' in the eyes of the consumer, no matter how different the farming and processing methods."

Not all states seem quite that open to such novel ideas, as we'll see soon. But some groups in Iowa seem intent on cultivating something other than corn and soybeans—namely, new farmers with some refurbished old ideas.

Fading Genes on the Farm

What I love about local is that it doesn't allow you to stay in one spot for very long. I'm not talking about horizontal movement, going from place to place. I'm talking about vertical momentum—digging deeper. The more you dig where you stand, the more you discover. And just when you think you're done, you realize you're nowhere near the bottom of things.

As we unearth the myriad cultural issues surrounding local food systems, we soon run into questions of biological diversity. Cultural and biological diversity go hand in hand, and anytime we see a culture or a food tradition being eroded, there's a good chance that some form of biodiversity loss will follow in its wake. In fact, there's a term for it: genetic erosion.

Genetic erosion is the loss of a particular gene or combination of genes. As organisms of a specific type—a livestock breed, a vegetable or fruit variety, a wild plant or animal—diminish in number to the point of nearing extinction, the gene pool of the remaining organisms loses its diversity and therefore its resilience to disease, parasites, climatic shifts, and even certain husbandry methods. The loss of those individual genes and gene combinations is as great a threat as the loss of topsoil—perhaps even greater. To a degree, we can rebuild lost soils. However, once a gene or gene combination is gone, poof!—it's gone. Not even a wisp of smoke to signal its exit. Seldom is there even a whimper, much less an outcry.

Let's go back for a moment to the large-scale poultry processing facility. It's not a pleasant place for the workers inside, nor is it a good situation for the individual birds . . . or poultry breeds in general. Think about what the poultry industry wants in a chicken destined for the oven, the pot, the nugget: a fast-growing big-breasted easy-to-pluck broiler. Now consider consumers' preferences: a tender meaty bird with no colored pinfeathers to remind them of the bird's previous form . . . oh, and better yet, boneless. Cross producers' preference with consumers' desire, and you get what's known as an "industrial strain." Simply breed to the point of utter conformity and standardization, and you have a bird that reaches slaughter weight in forty to fifty days instead of the hundred days its early-twentieth-century forebears required. Doesn't matter whether it's a chicken, a turkey, a hog, or even a dairy cow; the recipe—for disaster—is more or less the same. That uniformity is the killer. All of that genetic diversity that used to outfox—sorry, outchicken—pestilence and disease has disappeared. Same thing with fruits, vegetables, grains, and wildlife—we're emptying Mother Nature's toolbox, just at the point that we're really testing our own limits as a species.

As we lose the genetic diversity that is so critical to our increasingly vulnerable food supply, we are simultaneously losing a storehouse of cultural diversity. With the disappearance of breeds, seeds, trees, and grains, we lose cultural links not just to foodways but also to more sustainable ways of producing and consuming food. Not only do heritage foods provide links to our varied cultural histories, but they also hold the genetic imaginings to problems that we don't even know we have yet—and to some dilemmas that are far too obvious.

I'll give perhaps not the best example I know but the example I know best. As my wife, Erin, and I became increasingly interested in the issue of genetic erosion, we decided to experiment with heritage-breed cattle. We wanted a triple-purpose breed (dairy, beef, and draft) that would do well solely on grass-based production (i.e., no grain), unlike many of the modern breeds that have been developed to grow quickly and convert high-protein grains into vast quantities of milk and meat. We also wanted a rugged breed that would maintain condition through the harsh Vermont winters with minimal shelter. Many of the cattle bred for modern dairies and feedlots simply couldn't meet our requirements for low-input, hardy livestock. We homed in on the American Milking Devon, the first breed brought to the English colonies in 1623. Renowned ever since for their thriftiness (ability to survive on poor forage), intelligence, and low-slung musculature, they gradually dwindled in number until there were fewer than five hundred breeding females in the United States in the late twentieth century. Interestingly, it was the diehard New England ox teamsters, the folks who wanted to maintain the tradition of working cattle, who kept the breed alive.

So why does it matter? Just a bunch of hay-burning, methane-belching lawn ornaments that belong in a museum—why bother? Well, in part because the stony hillsides of Vermont aren't appropriate for vegetable tillage on a large scale, given the likelihood of soil erosion. Such rock-laden slopes are typically best used as pasture—well-managed pasture, that is. There is also growing interest in discovering how milk from various breeds can create different types of artisanal dairy products. Cheesemakers are always interested in how the components of milk—proteins, fats, and solids—affect the quality and characteristics of different types of cheeses, and there can be tremendous differences in milk among various breeds.

Perhaps most importantly, however, we are experimenting with how to minimize fossil-fuel inputs into our farming operation through extended grazing—leaving pasture grasses uncut and ungrazed late in the summer and having the cattle graze the stockpiled areas as far into the winter as possible, without feeding them hay. Our record so far is January 9th. Our goal is to graze right into the next growing season, but that's primarily because I think the quest for resilience often requires a certain amount of audacity!

But audacity can come in several forms, and it can manifest itself in different ways in the conservation world. There's a fierce conservation battle going on in Michigan, and it involves some audacious "conservationists" on two sides of the same electric fence.

Mark Baker retired from the Air Force after twenty years of service and began his family farm, Bakers Green Acres in Marion, Michigan. In an effort to conserve rare swine genetics and develop an alternative market, Mark, his wife, Jill, and their six children raise Mangalitsa pigs and Russian wild boars on their farm. But in December 2010 the Michigan Department of Natural Resources (DNR) issued an Invasive Species Order (ISO) condemning their animals to seizure and destruction. The DNR considers the Russian boar to be an invasive species that threatens the Michigan landscape, potentially wreaking serious damage on farms and in woodlands if the population of wild Russian boars increases. But they have to be loose—running rampant. And that does not seem to be the case on a well-managed agricultural operation such as the Bakers' farm.

The Bakers' farm uses the Russian boar genetics in order to create a dark red meat and a much different fat composition from that of many other breeds. Is the meat really all that different? I can testify firsthand that it is quite distinct, as we raised two pigs that were half Russian boar. As soon as I began butchering the hogs on a cold winter day, I could see the brilliant red meat in the bright sunshine. Texture, taste, curing—everything was distinctive about it. I confess that I had no idea that I was on the verge of finding myself in hot water right alongside the hog! (Incidentally, I have never had a pig get outside of its electric fencing except once when a gate was left open. Like they say back home in North Carolina, "Pigs is smarter than dawgs, ya know!" They seldom challenge an electric fence if they are appropriately trained to it.)

The Bakers are raising not only Russian boars but also the magnificent Mangalitsa breed, a type of "woolly" pig developed in the 1830s in Hungary. Its tightknit curly hair resembles wool and bestows upon the pig a distinct appearance not found in any remaining modern breeds. (Actually, the same name is used for three similar breeds that differ somewhat in appearance: the Blonde, Swallow-Bellied, and Red. But let's not get too technical . . . that's the job of the bureaucrats.) Although the pigs would vehemently deny it, they were developed not for their cute and cuddly appearance but for their tremendous lard production and delicious meat.

Like many breeds, the Mangalitsa went into serious decline throughout the latter half of the twentieth century. But its meat products are once again taking the culinary world by storm, thanks in part to the recent redemption of lard but especially due to the extensive marbling and exquisite taste of its meat. In fact, I've had the meat while at the Brunnenburg Castle and Agricultural Museum in Italy, where I lived and worked for several years. The Mangalitsa pigs are featured as part of the museum there. The meat's flavor and texture are indeed extraordinary, but the pigs are even more a delight in their intact form—docile, affectionate, and intelligent.

The Bakers are crossbreeding some of their Mangalitsas and the Russian boars to create a unique type of pork that provides them with an important market niche in the exploding culinary world of "charcuterie," setting their product apart from the industrial strains of hogs being raised in large-scale confinement facilities.[2] According to Mark, "My chefs love it. They like the dark red meat, the woody flavor and the glistening fat."[3]

However, since the DNR has classified the Russian boar as an invasive, the agency intends to seize any of the Bakers' hogs that are suspected of having any Russian boar genetics. Furthermore, since these animals are not considered "livestock," the farmers implicated would not be paid for the value of the animals destroyed. The case remains in court, and the Bakers have been waging a national campaign to alert other niche pork producers of the seemingly capricious possibilities that can be unleashed by bureaucrats scrambling to create definitions as they act.

One fiery aspect of the controversy is that the DNR has had to create a list of characteristics that define a Russian boar phenotype (appearance), but according to the policy, an animal displaying only one of these characteristics can theoretically be declared an invasive. There is, therefore, a concern among some farmers that this attempt to define an invasive species may only cater to the current industrial strains of swine and potentially include some domesticated animals rare breeds or crosses. To further complicate matters, the state's use of the word "feral," meaning "wild" in common usage, as we know, is problematic when applied to pigs living in a secure environment on a farm, as is the case with Bakers Green Acres.

It seems only fair to say that there are probably good intentions among most people on both sides of this conservation dilemma, although you can't help but wonder if state senator Darwin Booher of Michigan was onto something when he published an editorial in the *Manistee News Advocate* newspaper that provided his opinion as to the reason for this bizarre chain of events:

> The small farmers I have talked to wonder why the DNR is singling out their pigs and joining forces with the Michigan Pork Producers Association on this issue. They believe the association wants all pigs to be raised in confinement facilities, and the best way to achieve that is to make it illegal to raise certain swine, especially those offering alternatives to the white pork raised in confinement.[4]

The conservation questions here are certainly muddied. What is the state of Michigan most interested in conserving? The integrity of local ecosystems? Corporate interests? Innovative options for the survival of local family farms?

It will take some rooting around to find out.

Notes on a Napkin

The conservation of local family farms has a direct link to the conservation of agricultural biodiversity. And such biodiversity is the backbone of community resilience: we have control and flexibility in our regional food supplies as long as we have in our possession the means by which to adapt to change, no matter whether that change is forced upon us by climate change, fuel shortages, or economic shocks. We are nimble and equipped only when we are in possession of the foundation of adaptation: the plant and animal genes of our past . . . and our future.

However, guarding such a treasury for future generations involves first rediscovering what genetic diversity related to our food supply actually surrounds us. As a culture, we are far removed from the foods and foodways of those who preceded us. In fact, many of us don't even live in the same place as our ancestors, so the likelihood of creating any kind of inventory seems all the more complicated. But one person figured out how to do it on a napkin.

Gary Paul Nabhan is a fascinating and eclectic conservation scientist and author, and he was trying to find a way to somehow capture the food traditions of the United States in a way that might compel people not only to care for the traditions of place-based foods but also to actively conserve them. So the thought came to him during a meal that he could recast the geography of the United States into a map based on food traditions. What better concept to jot on a napkin than that? His napkin-based brainstorm became one of the most intriguing conservation efforts in the history of our country, in my view. Why? Because one person's vision for conserving our local foods became a collaborative grassroots effort to hold onto disappearing treasures, treasures that are central not only to our future adaptations but also to celebrating the ecological and cultural niches in which we dwell.

Gary's sketch on a napkin became a nationwide brainstorming event. His reconfiguration of the United States' political map into regions based on food traditions provided the strategy for this novel effort, an initiative that he called Renewing America's Food Traditions, or RAFT (see map 1 in the color insert section of this book).[5] After gathering an alliance of collaborators including the Chefs Collaborative, Slow Food USA, the American Livestock Breeds Conservancy, and others, he set up meetings of working groups in each designated region—farmers, seed savers, fruit tree specialists, fishers, gatherers, and educators—with the task of creating lists of place-based foods unique to each region. After barnstorming to all of these brainstorming sessions across the country, Gary set to work with the regional experts to create a full inventory of traditional foods and foodways, carefully noting the level of threat for various items on the list. Each regional group functioned in its own way, ultimately creating a publication—titled, eponymously, *Renewing America's Food Traditions*—that shared not just the lists but also the stories and the recipes associated with various foods. Efforts are now under way to actively conserve many of the heirloom vegetables, heritage fruits, and rare breeds of livestock identified and promoted through the RAFT initiative. Gary, a fun-loving guy who recognizes the need to talk with your mouth full in such an endeavor, and the RAFT Alliance have gone on to sponsor regional food celebrations all across the country.

The scale and scope of the RAFT Alliance are unrivalled in the conservation of agricultural biodiversity. In many ways, Nabhan and his colleagues have made it clear what we risk losing. The question before us now is whether our culture will sacrifice diversity for sameness, local for multinational, and culture for consumerism. It means doing things differently, though—differently from how other regions are doing them and differently from the mainstream thinking. The brief vignettes that follow represent some of the pioneers and the possibilities. Consider them appetizers for a more meaningful and resilient repast . . . or was that re-past? Perhaps it is both.

Native Seeds

The Appalachian region boasts the highest extant food diversity in all of North America, and the history of the indigenous Cherokee people is one of developing vegetable varieties adapted to the mountainous region. Kevin Welch, a passionate seed saver of Cherokee descent, noted a serious decline of seed exchange among the Eastern Band of Cherokee Indians (EBCI) due to an increasing scarcity of these rare seeds, so he initiated a seed-saving and seed-exchange program through the EBCI Cooperative Extension. The consistent growth of the Center for Cherokee Plants' seedstock is based upon his request that recipients of seeds give back to the center 10 percent of the seeds from the plants they grow. In 2011, Welch received seeds for a traditional Cherokee white corn flour, and he is now distributing seeds so that Cherokee growers can help bring this traditional corn back into their tribe's food traditions.[6] The center also maintains a "memory bank" of oral histories shared by the Cherokee growers. As Welch notes, "Without a story behind a seed, a seed is just a seed."[7]

Fruits

Felix Gillet arrived in California from northern France in 1859 as part of the Gold Rush, but the rich storehouse he left behind was in the form of superb fruit and nut varieties well adapted to the West Coast. He established and ran one of the first nurseries in the West from 1871 to 1908, providing customers with cherries, plums, apples, figs, strawberries, asparagus, chestnuts, filberts, hundreds of grape varieties, and just about any perennial eaten in California today. He worked hard to ensure that he sold hardy stock, growing plants out and testing them in the poor soils of his upper-elevation land in Nevada City. He also bred plants to adapt to the harsh environment and sold only what thrived. Fortunately, California organic farmer Amigo Cantisano and several friends began to recognize the immense value of the legacy left by Gillet all through the region in the 1970s, so they began conserving the plants bred and sold by Gillet and also archiving his writings for a future website. The Felix Gillet Institute is in its early stages, but Amigo and amigos are working hard to ensure that Gillet's perennials are available to all—perennially.[8]

Food Forests

The concept of "community food forest" is cropping up all over the United States, although it's more of an annual concept than a perennial reality in most places thus far. Seattle lays claim to the most ambitious contiguous planting thus far, with a seven-acre area dedicated to the Beacon Food Forest, a permaculture planting of what ecologists would call "synusia," or levels, ranging from berry bushes to shrubs to fruit trees.[9] Uninhibited by acreage or design, the Boston Tree Party is hoping that "civic fruit" will take root as supporters of this conceptual art initiative plant pairs of heirloom apples around the city until a patchwork of historic plantings begins to weave its way throughout the city.[10] Meanwhile, San Francisco is working toward planting multiple urban orchards that will be maintained and gleaned through the city's housing authority, with the goal of distributing the fruit to low-income citizens.[11] The Community Food Bank of Southern Arizona is blending issues of history, culture, and food security in its efforts to plant a food forest in Marana, Arizona. The group is planting desert-adapted varieties of peaches, figs, apricots, pomegranates, and other fruits that were first planted by Father Kino, a Spanish priest who came through the region in the 1600s.[12] The Portland Fruit Tree Project cares for and harvests fruits across the city and then gets the fresh fruit to those most in need. As the group says, "Money doesn't grow on trees, but fruit does!"[13]

Annual Grains

Sometimes the past is the best way to the future, and Prairie Heritage Farm is one example of young new farmers using ancient and heritage grains to find their way to markets and a viable farm operation. Courtney and Jacob Cowgill of Conrad, Montana, were trying to find a way to return to their rural Montana roots as farmers. They decided to focus on heritage grains and seeds, organic vegetables, and heritage turkeys as their market niche, and they took the unusual step of adopting the CSA model for their grain operation. CSA members receive such unique items as Painted Mountain Indian corn, a heritage variety of spring wheat, Prairie Farro (a special emmer), spelt, Kamut, and white or crimson lentils. Like many other farmers in Montana, the Cowgills are working to build markets that are not tied to commodity grain markets by embracing varieties of crops that have long been appreciated for their resilience and their taste.[14]

Perennial Grains

Wes Jackson is famous for his admonition that our concerns are not the "problems in agriculture" but the "problem of agriculture." Such a statement could set you up for a serious research agenda, but Wes Jackson, founder of the Land Institute and also a Post Carbon Institute fellow, has no lack of ambition. As the grain basket of our country confronts massive soil erosion rates, climatic extremes, and serious downstream impacts, the Land Institute is focused on taking nature as our guide for grain production. Our use of annual grains

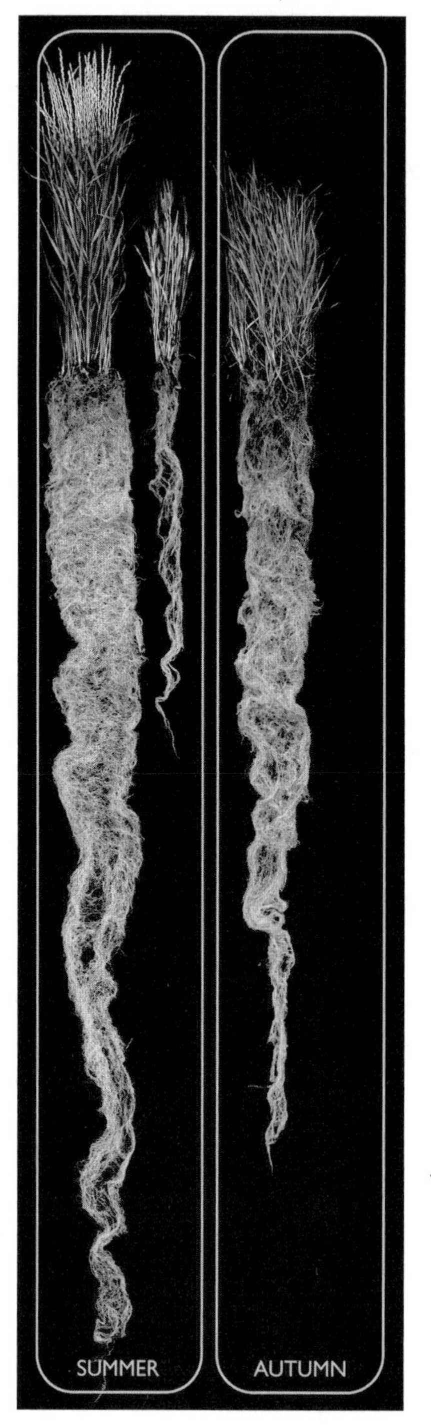

requires plowing and planting on an annual basis, disrupting numerous natural cycles and subjecting these annual crops to challenging growing conditions. Jackson's premise is that "natural systems agriculture" can guide us in mimicking the natural processes of the prairie in developing more sustainable grain production. The path, according to Jackson and his colleagues, is to develop perennial grains that do not require constant tillage and to find ways to plant them in polycultures (i.e., plantings of multiple species, often quite diverse in composition). Perennial polycultures can capitalize upon the benefits of plant diversity and less disturbed ecosystems, while also reducing fossil-fuel inputs and vulnerability to extreme weather events. The Land Institute's perennial grains breeding program has been recognized by numerous foundations and governmental agencies.[15] Genius happens at the edge of the mainstream, not in the middle.

Traditional Diets for a Healthier Future

O'odham corn: the fastest-growing corn in the world. Saguaro fruit: it makes the rarest syrup in the world.

FIGURE 8-1. Resilient Roots: Hybrid Perennial Wheat versus Conventional Annual Wheat. Image courtesy of the Land Institute.

Tepary beans: the most heat- and drought-tolerant beans in the world. Cholla buds: flower buds hand-harvested from a cactus, tasting similar to asparagus or artichokes, and with very high calcium levels, helping to control blood sugar.[16] All traditional foods for the Tohono O'odham people in the Sonoran Desert, the Native American community with the highest rate of diabetes in the United States. Nearly one-half of the adults suffer from adult-onset diabetes, so the path back to traditional foods makes a lot of sense here. The local Desert Rain Café features these traditional foods, working to make sure that each dish has at least one traditional food in it, thereby perpetuating not only the consumption of these traditional foods but also their cultivation and collection by the Tohono O'odham people. The café was founded by Tohono O'odham Community Action (TOCA) in order to heighten awareness of these foods and their link to healthy lifestyles, and proceeds from the reasonably priced café go back to support TOCA.[17]

It Takes a (Fishing) Village

As I drove with my family from the southern reaches of North Carolina back up to Vermont in the summer of 2012, we opted to follow the coast and be slowed by beauty instead of stalled traffic along the I-95 corridor. Our kids love a ferry ride, and so do I (it's great way to improve gas mileage), so we decided to make our way up the Outer Banks, the small sandy strip of islands along the easternmost flank of North Carolina. I'd traveled through there by car, bike, and boat several decades earlier, and I was curious to see the changes, particularly on Ocracoke Island. I had in mind that we might see some of the damage and repair from recent hurricanes, but it was actually the tempest in the modern fishing world that had set the stage for most of the damage and repair that we saw.

As in so many places around the world, commercial fishing fleets managed by large corporations were impacting not only harvests but also market opportunities for the small-scale commercial fishers from Ocracoke. If those economic pressures weren't enough, the community's one remaining fish house (sometimes called an "ice house") was put up for sale in the spring of 2006. That would have been the death knell for the thirty or so local commercial fishers, not to mention the island's three-hundred-year-old fishing culture.

A fish house provides a docking and unloading location for small-scale fishermen, as well as a place to restock the ice that is so critical to maintaining the freshness of the day's catch, and keeping a local fish house in business saves the substantial time and fuel that it would take to get to a distant processing center. Without the fish house, fishermen would have to make at least a four-hour round trip to sell their goods. As local fisherman Hardy Plyler said, "You advertise this town as a fishing village, well, if you don't have a fish house, that's a stretch."[18]

Community members came together and developed a plan to purchase the fish house, with the support of several foundations, including the Ocracoke Foundation, a 501(c)3 organization. The fishers formed a group called Ocracoke Working Watermen's Association (OWWA), which oversees daily operations of the fish house and develops long-range plans. All profits from the fish house, now called the Ocracoke Seafood Company, are either reinvested in the business or put toward mission-related expenses, as determined by OWWA. Ocracoke Seafood is now a for-profit subsidiary of the Ocracoke Foundation. In the end, the elimination of any middlemen also helped maximize profits for the commercial fishers.[19]

Of course, marketing was still critical for the OWWA, so they began to assess the values and methods that defined their type of fishing—in sharp contrast to the enormous commercial fleets competing with them. The phrases that they chose to describe their work and differentiate it from the competition included "tradition," "by hand (no mechanical harvest)," "fresh," "local," and "environmentally conscious."[20] Market distinction.

Today, OWWA members and community volunteers are immersed in a number of different projects. They have opened an exhibition of the OWWA, instituted research projects, and initiated several coastal restoration projects. Their products are sold at the fish house's storefront, as well as at local restaurants and bait and tackle shops. The work in continuing Ocracoke's fishing tradition will have to continue through the building and maintaining of markets. Captain Plyler notes that "consumers in [North Carolina] need to make a special effort to educate themselves about seafood—don't buy it from Ecuador or Thailand—concentrate on what is caught in North Carolina."[21]

FIGURE 8-2. Ocracoke Fresh. Photograph by Philip Ackerman-Leist.

It keeps the tourists coming back, the locals in their homes, and the fishers on the water. All because a community turned conversation into conservation—of food, traditions, the environment, and the local economy.

Chapter 9

Market Value

Maps matter. Why? Because she who names it wins it. (Remember—women are the driving force in this food systems landscape.) The game of the name is geography. One story always comes to mind first in this regard for me, as it involves a region in Italy where I lived and farmed for several years. Some call it Südtirol; others call it Alto-Adige. That's the point. It wasn't called either until someone decided to start playing the name game. Born in 1865, Ettore Tolomei was a geographer and ethnographer turned Italian nationalist who wanted to see Italy solidify its status as a powerful nation-state. Italy's coastal boundaries were relatively straightforward, but an ideal northern border was not so clear. Tolomei determined that the watershed for waters flowing south of the Alps near the Reschen and the Brenner Passes would be a natural northern border, despite the fact that very few Italians lived in this region. As part of the Austro-Hungarian empire, the area was almost all German-speaking, save for a minority of persons who spoke the ancient and mysterious language generally called Ladin.

To make a long and painful story short and painless, Tolomei began to give all of the villages and physiographic features of this region Italian names, sometimes using phonetic similarities, other times translating the meaning, and quite often just making them up. Eventually, he came up with an Italian name for the region, Alto-Adige, in reference to the high-elevation (*alto*) reaches of the Adige River that flows eventually to Venice. Actually, he translated the Napoleonic "Haut Adige" into Italian, essentially customizing a previously engineered geographical invention.[1] In the end, through this imaginative and rather devious exercise, he had created, on a map and increasingly in more and more people's imaginations, an Italian region. Through his cartography and his rather deceptive namings, Tolomei had set the stage for expanding the Italian reach into territory that was historically and culturally aligned with the Germanic aspects of the Austro-Hungarian empire. As the Fascists moved into the region following World War I, having gained power with the March on Rome in 1922, all German place-names were declared illegal, as were family names and the use of German in schools and publications. By the time it was all over, the Alto-Adige, known by the local German-speaking population as Südtirol (South Tirol), had become part of Italy.

So what do maps and names have to do with the economy and local food systems? Everything. Geographers talk about "appropriation" and "erasure" when discussing how colonial powers move into a territory and take over. They "appropriate" territory, first by naming it. At the same time, they work hard to "erase" the vestiges of the original culture in an effort to maintain control. Maps of the powers-that-be often show intent as well as reality. The tables can also be turned: the most strategic response for a community feeling marginalized or disempowered is to reimagine the realities, map out the possibilities, and redefine the territory.

I am not a conspiracy theorist. In fact, I tend to lose patience with conspiratorial spinnings from either end of the political spectrum. But I do believe that the realities we face in rebuilding local food systems are formidable, and there are citizens, policymakers, and corporations dead set against any push for relocalization. Ultimately, though, no single corporation is the concern. After all, when you look at corporations in the food systems world, they get eaten up by one another on a regular basis. Rather, I worry about the overarching culture of this consolidation-driven frenzy and how it consumes and breaks down everything that it ingests: local farmers, small-scale fishers, regional distributors, midscale processors, local agricultural suppliers, long-standing cooperatives, sensible ecological practices, local waste management companies, and sensible regulations for small businesses. The emphasis on efficiency, scale, uniformity, dependability, and price marginalizes anyone not willing to play the game, with the rules already well established and in favor of the visiting team. The result is standardization and homogenization—and the subsequent erasure of local economic diversity and community spirit.

Study the maps in the color insert and the black-and-white maps below for a few moments. What do they mean from the perspectives of appropriation and erasure? Who moved in and took over? Where are the areas of concentration of different agricultural and economic activities? Where are the gaps, and why do they exist? What are the current links between landscape, agricultural infrastructure, and economic activities? What might be the implications for all kinds of diversity within this framework: cultural, biological, and economic?

Begin with maps 3 to 6 in the color insert (starting after page 180), which depict the changes in improved farmland in the United States from 1850 to 1997. The story behind the westward migration of people and agriculture is, of course, the most tragic of the appropriation and erasure stories told in all of these maps. This migration displaced and decimated Native American populations, over the course of two centuries, in order to claim the farmland that would set the stage for the so-called greatness of America so grandly depicted in Armour's Food Source Map of 1922 (map 2). Increased transportation and communications infrastructure laid the tracks for an increasingly dispersed food system, with certain areas commanding reputations and markets for the goods that they could easily produce and ship nationwide with increasing complexity and efficiency, thanks to enhanced means of communication for orders and deliveries.

Maps 7 to 10 then depict our current era, showing the astonishing concentration of crops and livestock, with much of the country more or less squeezed out of the diverse national marketplace.

Now move back to the maps in this chapter. Figure 9-1 depicts the locations of Walmart stores across the United States from 1967 to 2007. (For an intriguing perspective on the rapid growth of Walmart stores across the country, look at FlowingData's time-lapse map of the progression at http://projects.flowingdata.com/walmart/.) Take note of one year in particular: 1988, the year that Walmart opened its first grocery store. In a span of just over two decades, Walmart has now become the nation's largest grocery retailer, boasting grocery sales of $140.6 billion,[2] more than double its closest competitor, Kroger.[3] Add to that the fact that Walmart is now the largest retailer of organic products in the country as well.[4]

Shift next to figure 9-2, depicting the distribution center locations of the nation's two largest distribution companies, Sysco and US Foods. When looking at their food distribution coverage, it's hard not to admire their

FIGURE 9-1. Location of Walmart Stores in the United States, 1967–2007. Source: http://www.econ.umn.edu/~holmes/data/WalMart/index.html.

expansive geographical penetration and accompanying ability to move vast amounts of food quickly over great distances. It's an extraordinary logistical feat. In contrast, local and regionally based food hubs (figure 9-3, and described later in this chapter) are just beginning to enter the scene, and the questions of market penetration and market differentiation loom large for these relatively nascent business entities.

Finally, go back to the color insert and examine map 11, which depicts the estimated value of direct sales to consumer, by county, taken from the 2007 Census of Agriculture. While those numbers have undoubtedly grown since that last census (as will be evident from the record-breaking local food sales numbers in 2010, detailed later in this chapter), the contrast between these direct-marketing sales and the overall volume in grocery sales by Walmart alone can look daunting, to say the least.

Do these maps indicate a grim scenario for those of us interested in rebuilding local food systems? After all, there's a difference between being bold in the face of reality and being naive in a worldview shaped by denial. The realities are intimidating, but it's better to be audacious than naive if you intend to be strategic and successful. In fact, it's best to be what some

FIGURE 9-2. Location of Sysco and US Foods Distribution Facilities. Source: http://www.sysco.com/about-sysco/our-locations.html and http://usfoodserviceworkers.org/content/know-your-company.

FIGURE 9-3. Local and Regional Food Hub Locations in the U.S. Source: http://ngfn.org/resources/food-hubs/food-hubs#us-food-hubs-map.

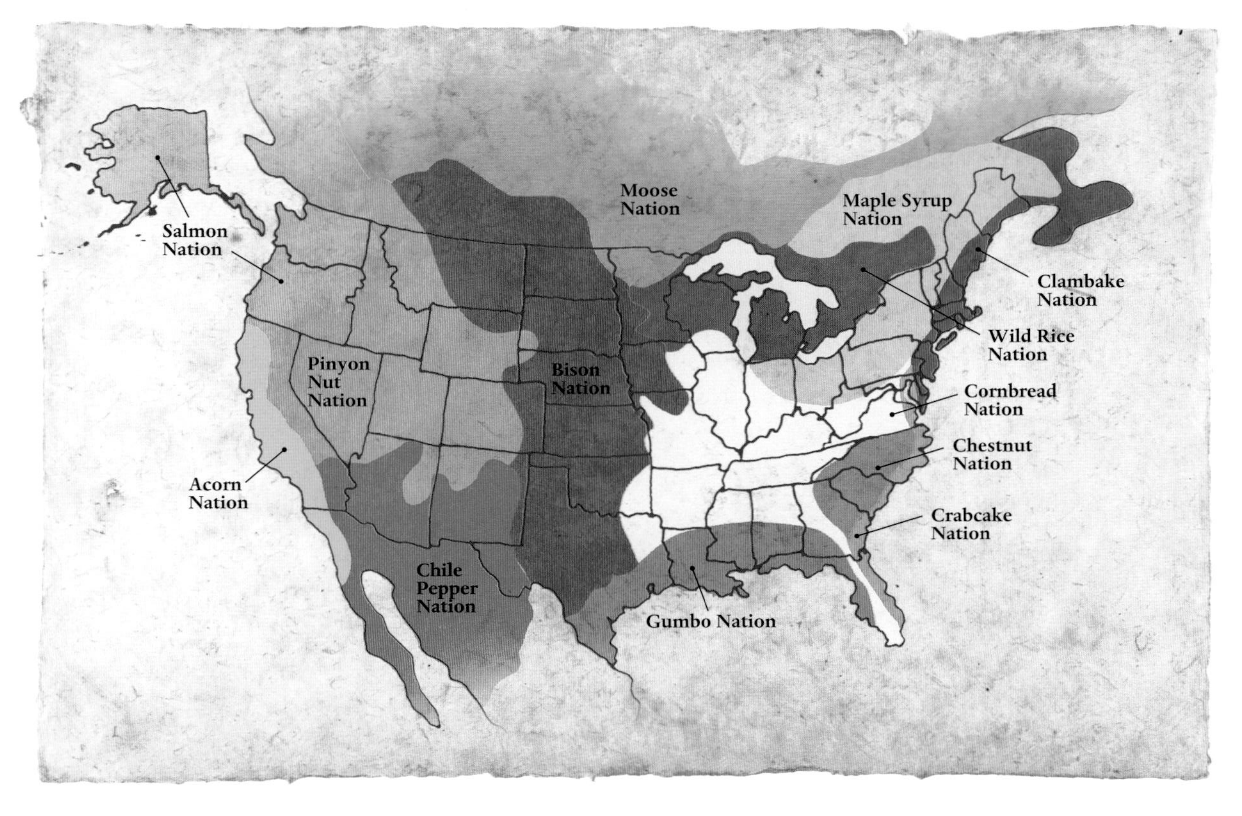

MAP 1. Renewing America's Food Traditions (RAFT) Map

Map courtesy of Gary Nabhan, ed., *Renewing America's Food Traditions: Saving and Savoring the Continent's Most Endangered Foods* (White River Junction, Chelsea Green, Vt. 2008).

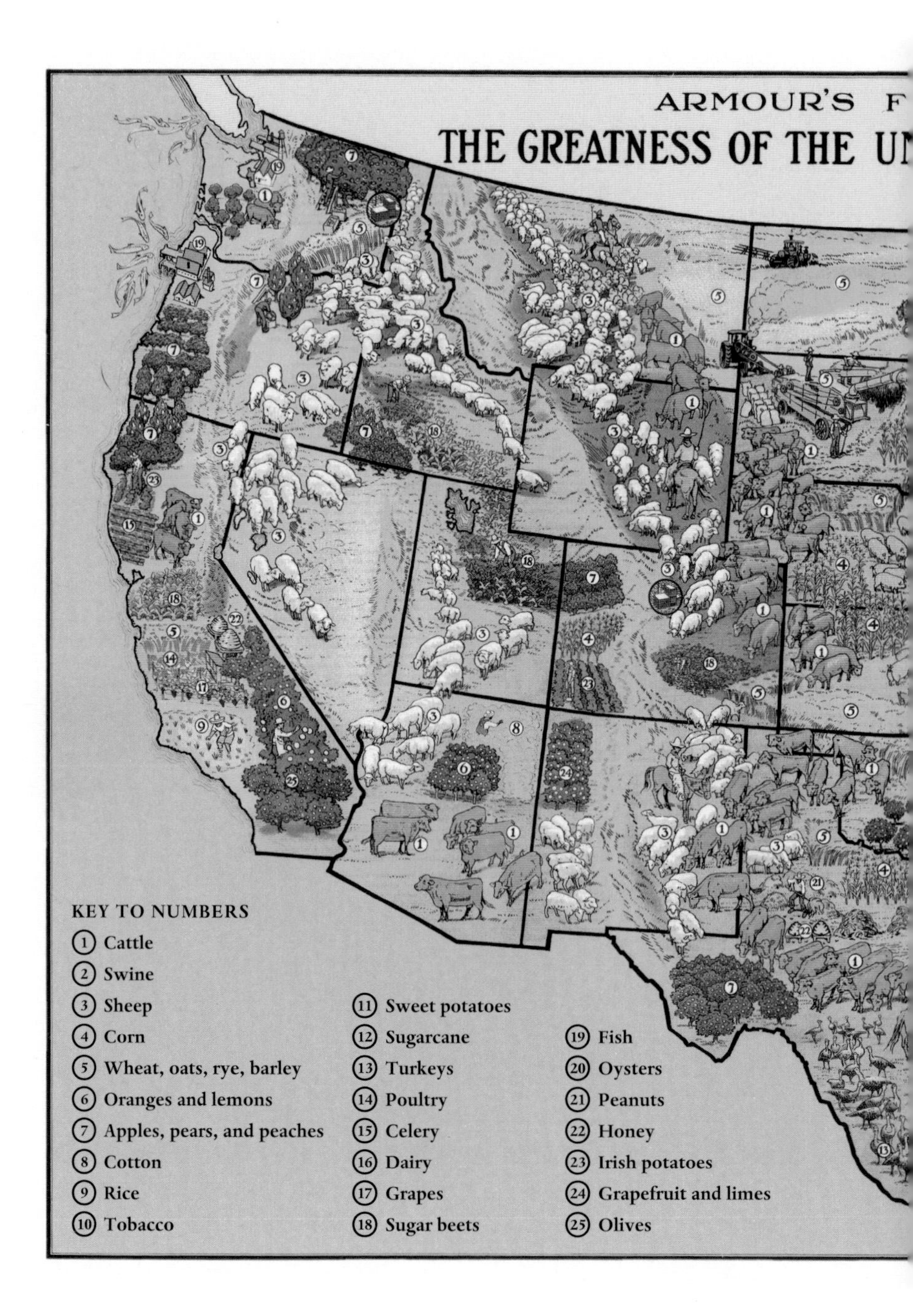

MAP 2. Armour's Food Source Map: "The Greatness of the United States Is Founded on Agriculture" (1922)

Map courtesy of the American Geographical Society Library, University of Wisconsin–Milwaukee Libraries.

SOURCE MAP
STATES IS FOUNDED ON AGRICULTURE

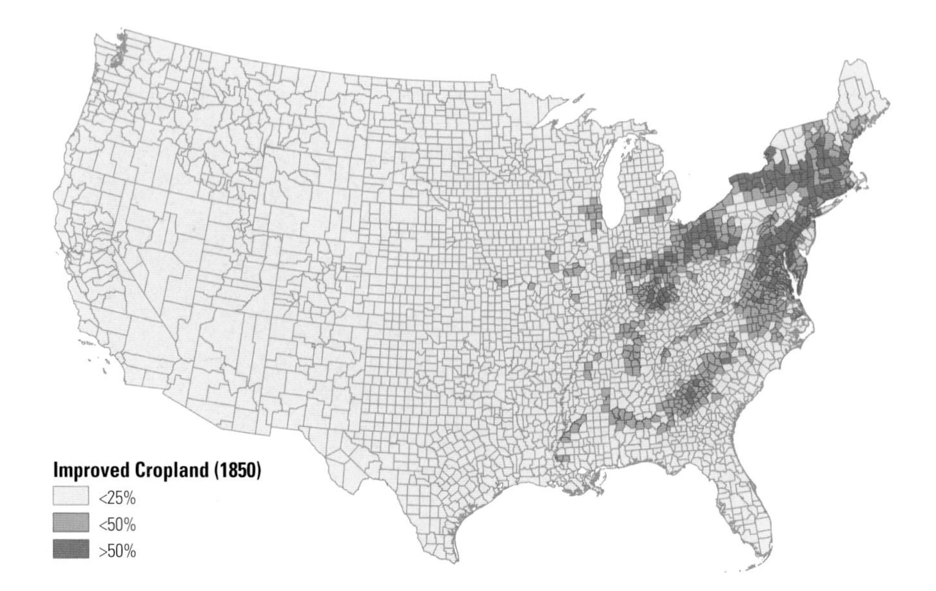

MAP 3. U.S. Improved Cropland, 1850
Source: http://landcover.usgs.gov/cropland/

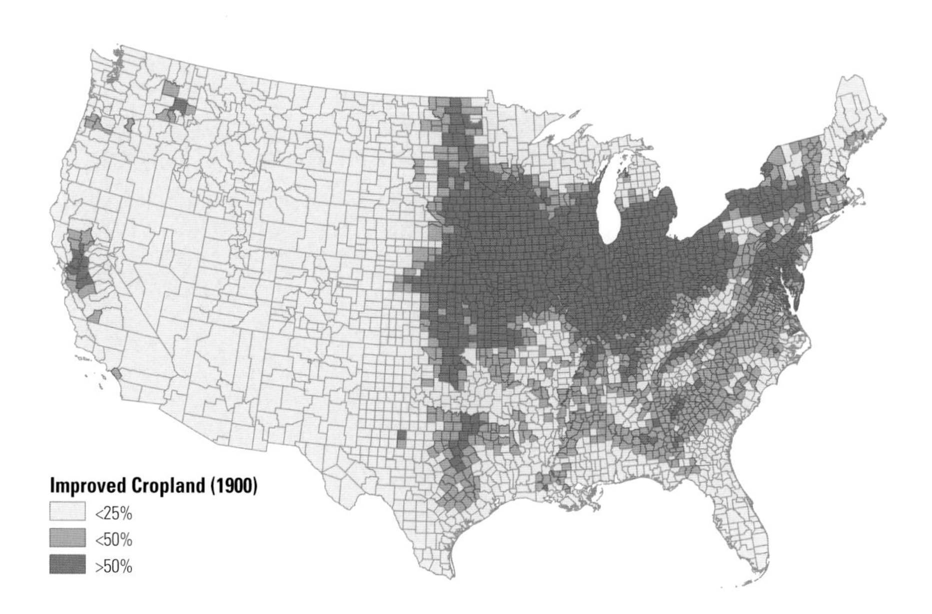

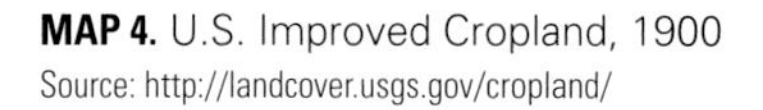

MAP 4. U.S. Improved Cropland, 1900
Source: http://landcover.usgs.gov/cropland/

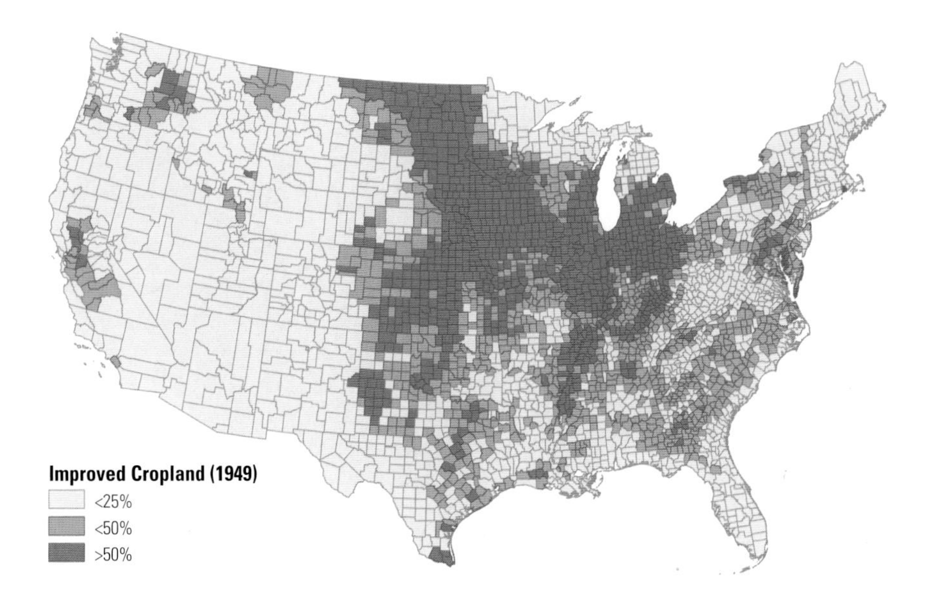

MAP 5. U.S. Improved Cropland, 1949
Source: http://landcover.usgs.gov/cropland/

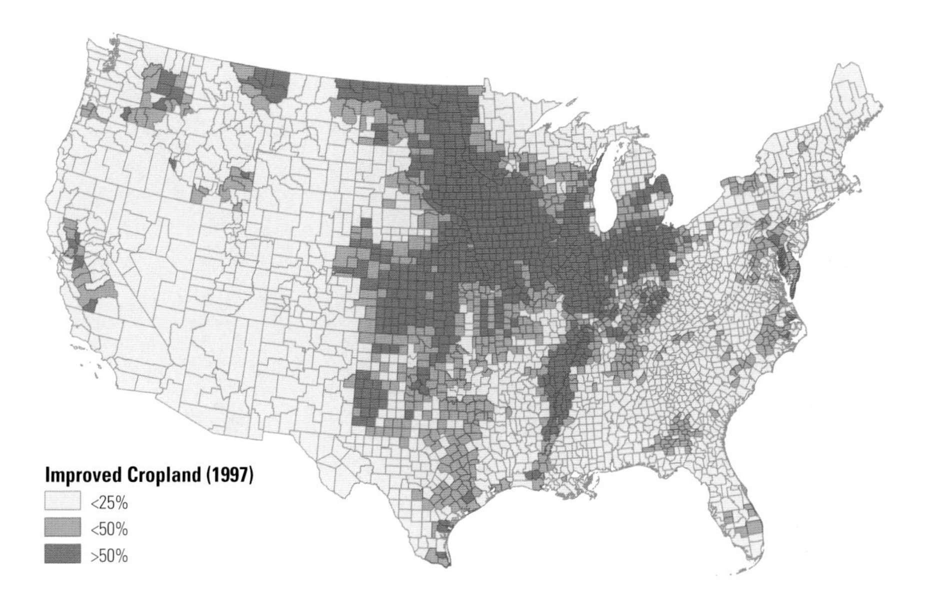

MAP 6. U.S. Improved Cropland, 1997
Source: http://landcover.usgs.gov/cropland/

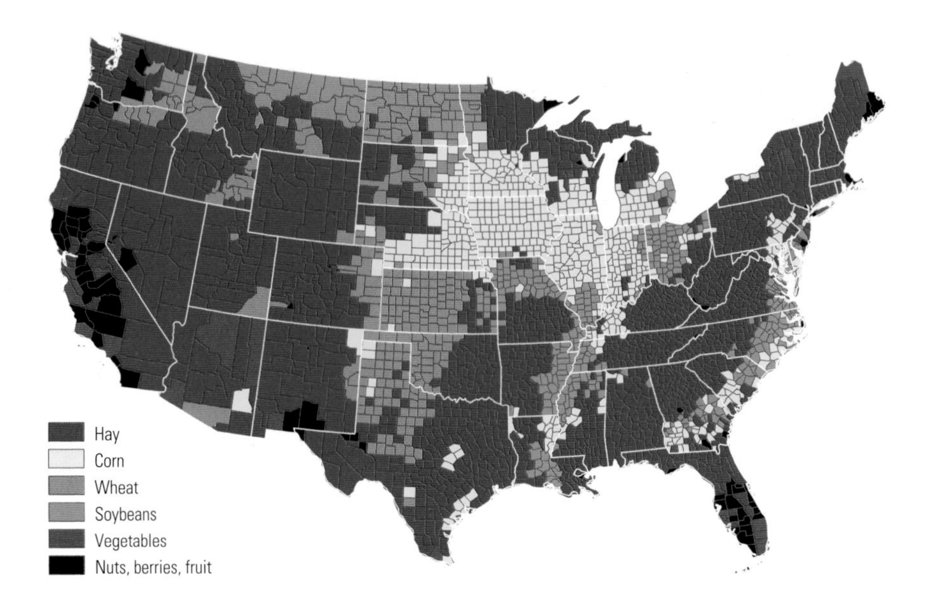

MAP 7. Dominant Crop by County (by percent of total acreage)
Source: 2007 USDA Census of Agriculture data

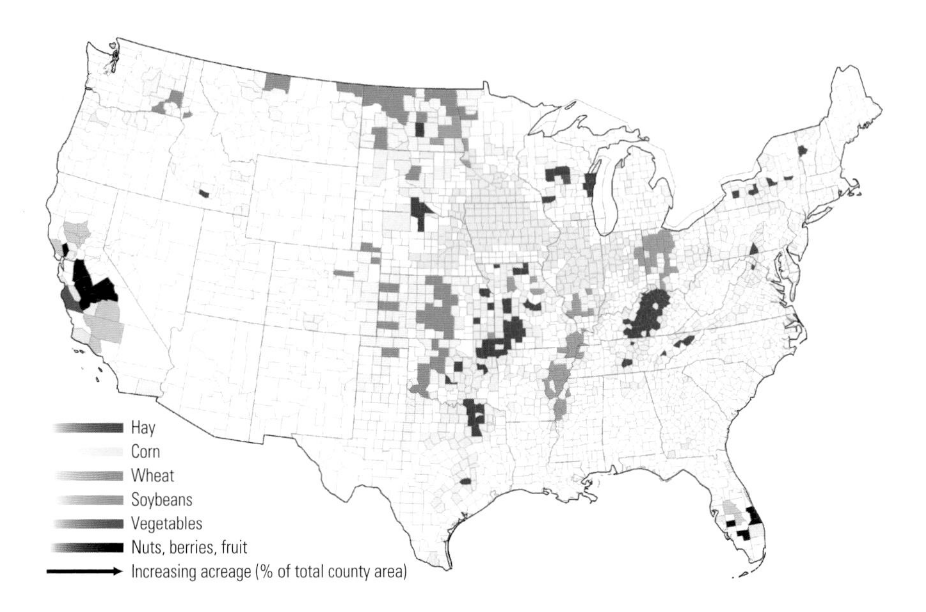

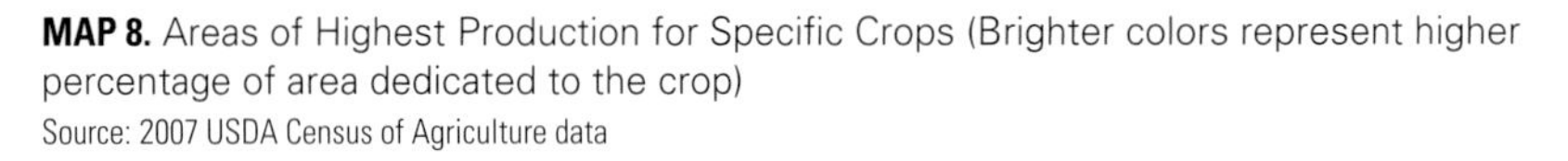

MAP 8. Areas of Highest Production for Specific Crops (Brighter colors represent higher percentage of area dedicated to the crop)
Source: 2007 USDA Census of Agriculture data

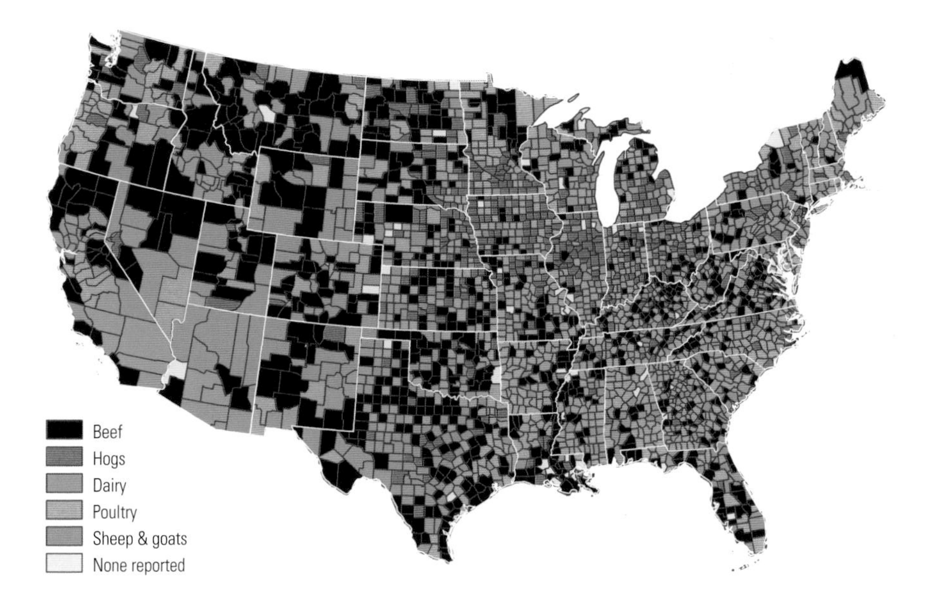

MAP 9. Dominant Livestock Species by County (by total count of animals)
Source: 2007 USDA Census of Agriculture data

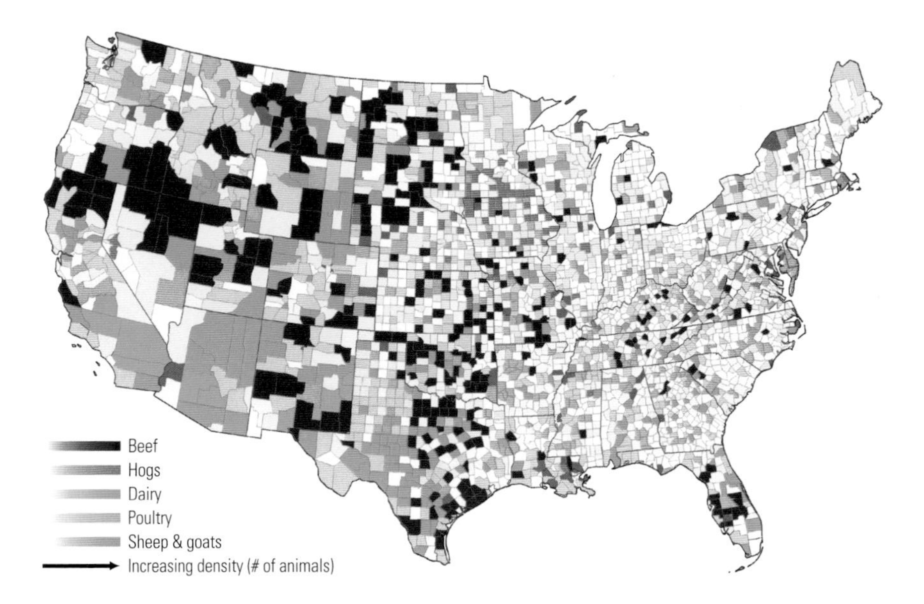

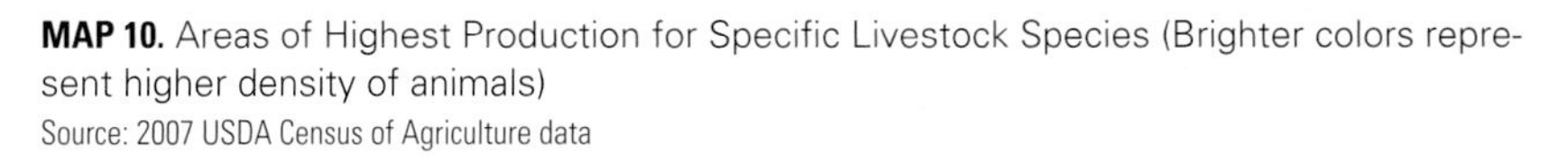

MAP 10. Areas of Highest Production for Specific Livestock Species (Brighter colors represent higher density of animals)
Source: 2007 USDA Census of Agriculture data

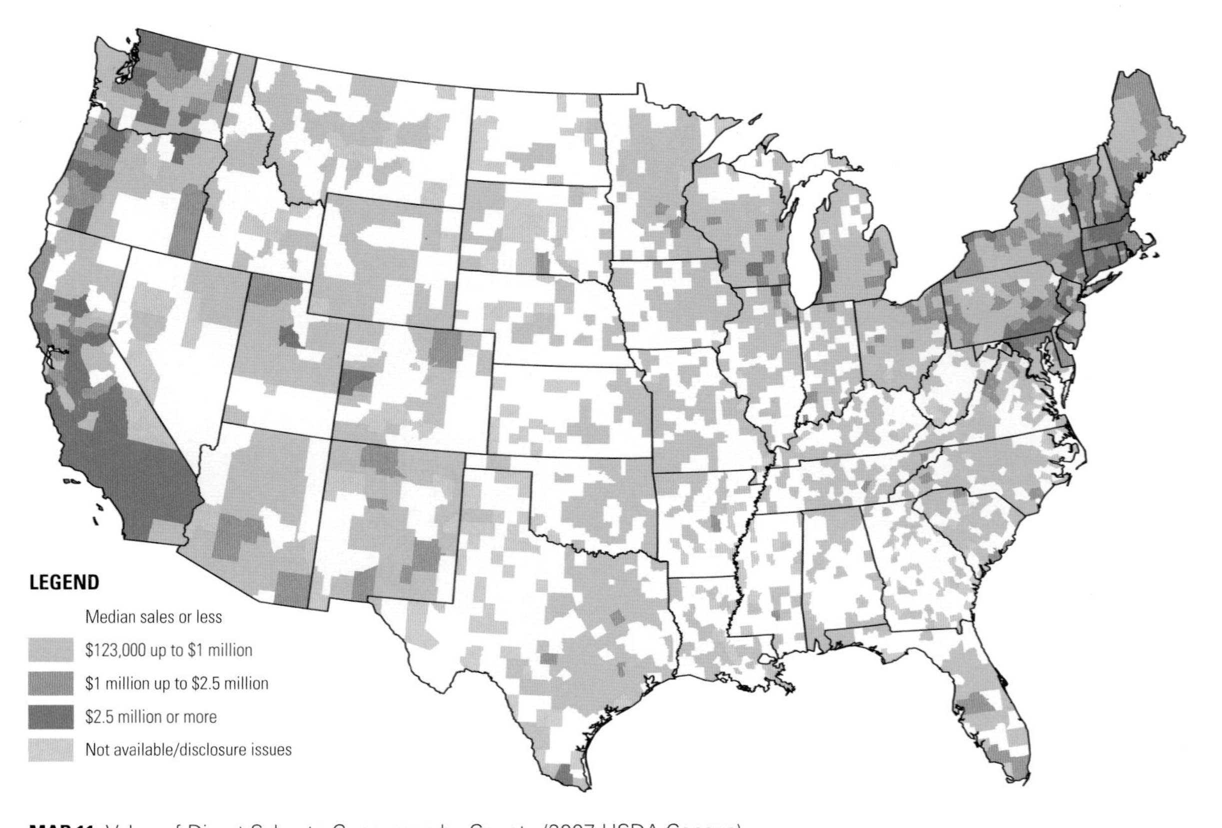

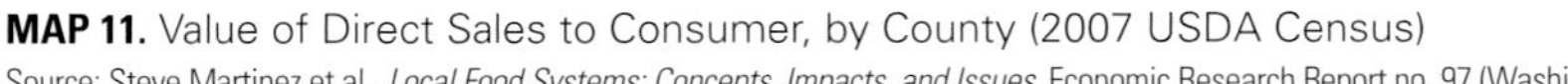

MAP 11. Value of Direct Sales to Consumer, by County (2007 USDA Census)
Source: Steve Martinez et al., *Local Food Systems: Concepts, Impacts, and Issues*, Economic Research Report no. 97 (Washington, D.C.: USDA Economic Research Service, May 2010)

would call "bodacious"—bold and audacious, all rolled into one. The initial strategy required in taking on the national trend of food industry consolidation and concentration involves naming not only what was lost but also what is desired. For some, this naming will come in the form of marketing initiatives; for others, it will arise through a loud and impatient declaration of food sovereignty.

While these maps are illustrative, none of them provides the whole story, nor is any one of them without bias and assumptions. But when viewed together, they start to convey a common story of concentration and consolidation. Consolidation has been a trend in our food system for years. It's the classic "big fish eats little fish" story, and it's been happening in every sector of the food system for decades, impacting farmers, fishers, and food processors. Concentration is a bit different, and even more problematic: there are only a handful of those big fish left, and the rest of us in the food system are essentially krill. But that doesn't mean we have to be spineless. What's a krill to do?

First of all, swim like hell. And remember that the mainstream trends of consolidation and concentration are not limited to the conventional markets. They are also occurring in the organic and natural foods sector. As figure 9-4 shows, there is only one specialized national distributor of organic and natural foods left in the United States. The existence of but a single national distributor of natural and organic food products is a serious problem, since it limits options to farmers and processors marketing their products on a national scale. While a single national distributor can provide some important efficiencies to farmers and other food-based entrepreneurs working in the natural and organic foods sector, competition and a diversity of options are always central to maintaining a healthy marketplace. Let's hope that there's not an even bigger fish out there that has a hankering for some organic small fry.

This dilemma brings out another important but oft-neglected point, however. The freewheeling use of "corporate" and "corporation" as finger-wagging curses of any food-related business that opts to become a corporation is problematic. Farms incorporate all the time in order to protect farmers from potential financial ruin. Relatively small businesses incorporate for the same reason. Some larger corporations are guided by executives and managers who are desperately trying to do the right thing

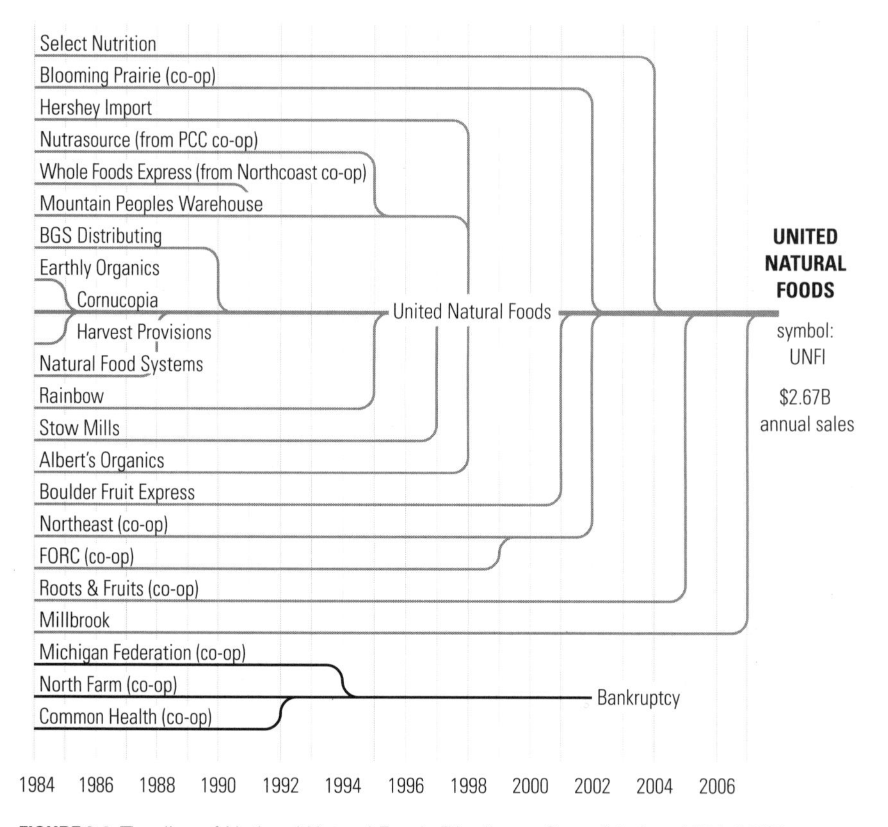

FIGURE 9-4. Timeline of National Natural Foods Distributor Consolidation, 1984–2007. Source: Philip Howard, "Visualizing Food System Consolidation and Concentration," *Southern Rural Sociology* 24, no. 2 (2009): 102.

but are operating within the tight and brutal constraints of the marketplace. Some of the most disenfranchised members of our communities are employees of various corporations working within the food system, and these employees are thankful to have the jobs that keep them afloat.

Therefore, for the purposes of trying to engage as many players as possible in the dialogues toward a better food future, there is some wisdom in reserving the tongue-lashing and finger-pointing specifically for the trends of consolidation and concentration. Avoiding wholesale corporate castigations allows us to see the picture more fully, while also lessening the polarization that turns dialogue into diatribe. Perhaps this point will become clearer as we think more about the variety of economic models cropping up in local food systems across the country. The economic models for remapping the way our food system functions have to be not just "farm tough" but also "farm savvy." To do so requires the right tools at the right price, along with some good old-fashioned determination.

When my wife and I began our farm in Vermont, we didn't have a tractor for many years. The lack of hydraulics meant relying on what we had (muscle power), and using it wisely. This strategy holds true for grassroots organizing and start-up businesses, too. I'll share the three indispensable strategic tools we used constantly in those days, because they're also appropriate for confronting the consolidation and concentration in our food systems:

- **The wedge:** *Market initiatives* are the wedges that find the cracks in the dominant food system and work their way in, gradually prying apart the strangleholds in the system.
- **The lever:** *Market distinction* is a long lever that puts as much distance as possible between the farmer and the immense weight of the consolidated industries he or she is trying to budge.
- **The fulcrum:** *Regional-scale infrastructure* then acts as a fulcrum for leveraging sales of local food.

In this and the following chapter, we'll explore the uses of the wedge, the lever, and the fulcrum as tools for reshaping our national food economy.

Farmers and Direct Sales

In November 2011, the local food systems bandwagon got a lot of attention in the media. Farmers, food entrepreneurs, and even politicians were falling all over themselves to get a perch on that rollicking piece of vintage farm equipment full of blaring brass and Holstein-patterned banners. The headlines said it all: "$4.8 billion dollars in 2008 local food sales!" When those figures were released by the USDA on November 14th, the farmers and activists who had been fighting the erasure of local agriculture over the past few decades celebrated with blog and brew. But the USDA's forecast for 2011 was all the more reason for Tweets and honks: 2011 looked to be an even bigger banner year, with $7 billion anticipated in local sales![5]

Those of us used to hearing about the hijacking of our local food efforts found a reason to celebrate the news with a cup of the day's first compromise—coffee or tea, that is (I diluted my globalized liquid travesty with a hefty dose of cream from our very local cow). So what did we learn about our farms from that USDA study about the growth of the local foods market?

- **The promising news:** For the first time, the data included not only direct sales from farmers to consumers but also "intermediated" sales (use of a middleman) to grocery stores and restaurants, providing a more comprehensive view of the current realities and the future potential. And the data were impressive: the sale of local food products hit an all-time high of $4.8 billion in 2008, with projections expecting a significant jump beyond that figure in the past few years, once the data are analyzed.
- **The sobering news:** These local food sales comprised only 1.9 percent of total gross farm sales in the United States.
- **The "celebrate your local fermentation specialist" news:** Direct farm sales had not been that high in nearly three decades, and there is a potential market share of 98.1 percent still out there.
- **The mediated news:** Most local food sales are not direct-to-consumer sales, like at a farmers' market. Fifty to 66 percent of local food sales are intermediated sales—that is, through a local store or restaurant.

It's worth looking further into how the sales data broke down between direct and intermediated marketing (figure 9-1). *Direct marketing* simply means that the farmer gets the product directly to the end user, with no broker or distributer involved. When a farmer opts for an *intermediated marketing* approach, he or she determines that there are advantages involved in having an intermediary involved. The advantages typically include a consideration of the scale, speed, ease, and/or consistency of the transaction. These advantages often come with costs, however. The sale price of the product in an intermediated market tends to be lower—sometimes significantly lower—than in a direct transaction with the consumer. Farmers also face additional fees for processing, distribution, and other services. Of course, the higher volumes also tend to require more specialization and infrastructural costs. Additional off-farm labor may be necessary, and increased volumes can propel farmers into a stricter regulatory environment that may stipulate additional or upgraded infrastructure, often at a significant cost to the farmers.

Intermediated sales far outweigh direct-to-consumer sales. Of the $4.8 billion in local food sales in 2008, farms that used solely direct-to-consumer markets captured only $877 million in sales, while intermediated markets hit approximately $2.7 billion in sales, and farmers utilizing a combination of direct and intermediated marketing captured an estimated $1.2 billion in sales.

Even more interesting, albeit a bit more complex, is the breakdown showing how farm size relates to marketing channels. Table 9-2, from a recent USDA study, shows the differences in how small-, medium-, and large-scale farms market their products. Small farms depend heavily upon direct marketing, but as farm size increases, so does the dependence upon intermediated markets.

The dependence on intermediated markets at the larger scale is primarily due to the volume of products and the need for efficiencies in marketing to a large demographic. Adding intermediation puts one or more steps in between the farmer and the customer. The only way for a farmer to continue commanding fair or premium prices in this context is to ensure quality, reliability, product safety, clearly marketed values, and perhaps appropriate certifications. With every step away from direct-to-consumer relationships, the farmer must increasingly ensure that the messaging and the products' allure are stronger and louder.

TABLE 9-1. Direct and Intermediate Sales of Local Foods in U.S. (2008)

	DIRECT-TO-CONSUMER MARKETS	BOTH DIRECT AND INTERMEDIATED MARKETS	INTERMEDIATED MARKETS
Sales	$877 million	$1.2 billion	$2.7 billion
Number of Farms	71,200	22,600	13,400
Outlets for Products	Farmers' markets, roadside stands, farm stores, CSAs	Mixture of markets	Grocers, restaurants, regional distributors

Source: Adapted from Sarah Low and Stephen Vogel, *Direct and Intermediated Marketing of Local Foods in the United States*, Economic Research Report no. 128 (Washington, D.C.: USDA Economic Research Service, November 2011)

TABLE 9-2. Local Food Marketing Channels Used, by Farm Size

SALES CHANNELS	FARM SIZE			
	Small (sales of less than $50,000)	Medium (sales of $50,000–$249,999)	Large (sales of $250,000 or more)	All
	NUMBER			
Local Food Sales Outlets Used	121,198	15,202	5,301	160,795
Average Number of Outlets Used per Farm	1.4	1.7	2.1	1.5
	PERCENT			
By Marketing Outlet	100.0	100.0	100.0	100.0
Direct-to-Consumer Outlets	**78.0**	**70.7**	**55.5**	**75.3**
Roadside Stands	34.1	24.9	23.7	31.8
Farmers' Markets	34.6	25.9	14.7	31.8
On-Farm Stores	8.3	17.4	15.7	10.4
CSAs	1.1	2.5	1.4	1.3
Intermediated Outlets	**22.0**	**29.3**	**45.0**	**24.7**
Grocers and Restaurants	17.2	26.0	23.7	19.2
Regional Distributors	4.8	3.4	21.4	5.5

Source: Sarah Low and Stephen Vogel, *Direct and Intermediate Marketing of Local Foods in the United States*, Economic Research Report no. 128 (Washington, D.C.: USDA Economic Research Service, November 2011)

Small and medium-scale farms tend to cater primarily to direct markets. They utilize that long lever of market distinction, putting as much distance as possible between themselves and that "other" 98.1 percent of farm sales (nonlocal, conventional, industrial, noncertified, etc.) in order to maximize the leverage. Ideally, this leverage attracts more customers who are willing to pay a premium price to the farmer. Face-to-face relationships have been the rule of thumb for this strategy.

Things are changing fast, however, and those farmstand nibbles are quickly being complemented by online bytes. The joy of birdsong on the weekly CSA farm visit is now in competition with tweets from the farm's database. Farmers are utilizing a vast array of online tools to draw in customers; indeed, these tools are one of the most promising wedges farmers can use to force their way into the cracks of the larger food system without investing in enormous amounts of new infrastructure.

Some farmers simply rely on their own websites to bring in additional customers, keep their clients informed of product availability, provide them with recipes, and engage them with farm and garden blogs. Others are going much farther, adopting online ordering systems so that local customers can put in customized orders for the farmers' market, their weekly CSA shares, and/or various home delivery programs. Market-savvy software and social media developers are cranking out a variety of online ordering systems for farmers. In some cases, farmers pay for a software package, but most of these systems are now operating on a transaction-fee basis or a monthly subscription fee.[6]

The Internet also opens up distant markets barely dreamed of a decade ago, and some farms are eager to widely market items unique to their particular region. These regional specialties are moving across the country and even internationally at an astounding rate: crawfish from Louisiana, maple syrup from Vermont, pecans from Georgia, Rocky Mountain oysters from . . . well, let's stop there. Local anesthetic, anyone?

Numbing, too, are the countless ways in which emerging technologies are linking local farmers to one another in an effort to build the overall local food market for everyone's benefit. Sometimes local farms link together through an online sales platform, pooling their resources and the variety of products offered, and increasing sales and the overall customer base. These "virtual" farmers' markets may still feature a location and specific time for customers to pick up their orders, but the convenience of a regularly scheduled delivery straight to customers within a given radius seems increasingly popular among farmers and customers. Farmers in these situations may even opt to share one or more delivery vehicles.

These types of arrangements can be particularly beneficial to small or new farmers who are low on infrastructure, start-up capital, product volume, and marketing networks. People producing value-added products (such as jams and baked goods) on a relatively small scale also find these virtual farmers' markets especially useful in their business's incubation stage. In all of these scenarios, the relationship between farmers and customers can be maintained, albeit in a different form and perhaps with less personal interaction.

Another increasingly popular use of technology and social media networks is in search engines that feature regional or national databases of farms, with contact information, product availability, ordering methods, maps, and images. Such websites allow customers to tease out the variables that are most important to them in establishing relationships with multiple farmers: location, prices, values, products, certifications, convenient pickup sites, and delivery options.

Although there is no question that these technological advances and opportunities provide farmers and local food advocates with an important wedge for cracking open new markets, farmers are still trying to assess the impacts on relationships with customers. There is some concern that the online convenience of these models may displace not only the depth of relationships between farmers and customers but also the hard-won customer participation in traditional farmers' markets and CSAs. Hopefully, a combination of approaches will help create a much broader market base for locally produced foods, with customers choosing models that best fit their individual lifestyles and values.

CHAPTER 10

Marketplace Values

In order to remap the American foodshed, we must do more than simply shift boundaries and demographics. We also have to change the legend on the maps. That is to say, we must change the ways in which we identify regions, markets, products, and populations. This remapping is an opportunity to imbue those maps with color. Rather than accepting the black-and-white views offered to us by corporate giants, our depictions of the foodshed require the tint of color in order to differentiate by way of values.

Differentiation based on values is critical to small-scale farmers but also to midscale and large-scale producers who sell primarily to intermediated markets, not only for gaining more customers but also for ensuring that they can be "price makers" instead of "price takers." The idea is to capture as much of the American food dollar as possible, and the national average of farmers capturing a mere 15.8 cents of that dollar is the trend to beat. The closer a local product edges into that enormous and concentrated arena of 98.1 percent nonlocal sales, the more important it becomes for the product to stand out in order to compete on a basis other than price. "Local" may carry the most weight, but being local in and of itself may not be enough to capture the optimal prices in those intermediated markets.

Once "local" enters the restaurant and grocery sector, those businesses become responsible for heightening the value of the product to their customers. In that scenario, *value* is immediately tied to *values*. The diner and the shopper willing to pay prices considered fair to the farmer justifiably expect to connect dollar value with appropriate farm practices. Such consumers are all the more intrigued when they feel the entire supply chain that got the food to their plate or their grocery basket is well grounded in its ethics and its commitment to the local and regional community. Hence the birth of what many now call "values-based supply chains," often abbreviated to "value chains." These supply chains are strategic collaborations and business relationships between farms, processors, distributors, and retailers that operate on the basis of explicitly conveyed values—shared values that create a collaborative business opportunity and, ideally, customer allegiance. The farms generally share agreed-upon management practices, whether it be grass-fed beef or organically produced fruits. The point of commonality becomes the lever that seeks to move aside the heft of the highly concentrated and often more "industrialized" competitors, while the cooperative arrangement, usually within a regional context, is the fulcrum that magnifies the potential of local.

Value chains are one of the most promising developments in rebuilding our local food infrastructures and in increasing local food consumption. That's not to say that these value chains are simple. In fact, it's probably best not to think of value chains as direct conduits from farm to plate. Rather, they tend to be a clustered network of the farms and businesses represented in figure 10-1. Just as farmers are likely to use several market outlets, food service and retail operations almost always utilize multiple distributors.

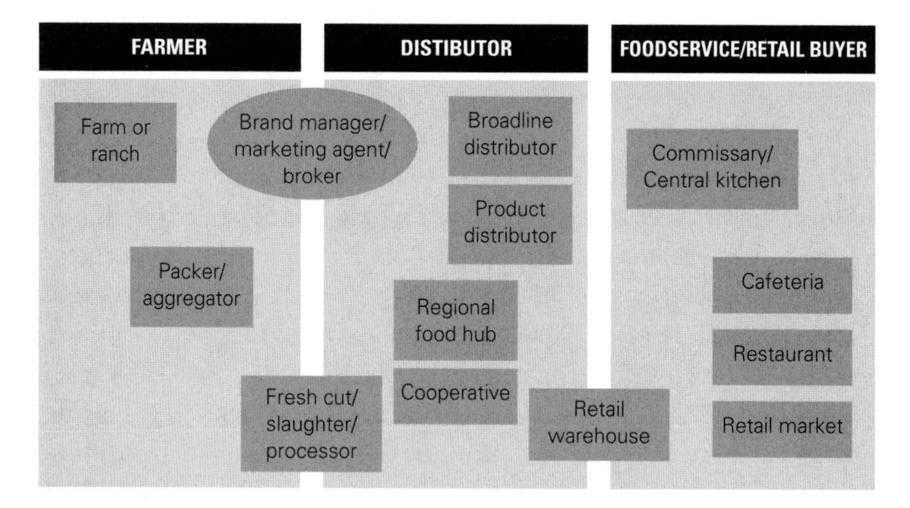

FIGURE 10-1. Components of a Values-Based Supply Chain, Including a Variety of Possible Linkages. Source: Gail Feenstra, David Visher, and Shermain Hardesty, *Developing Values-Based Distribution Networks to Enhance the Prosperity of Small and Medium Sized Producers: California Case Studies* (University of California–Davis: March 28, 2012).

In addition to stimulating demand, value chains can also increase farmers' profits. The USDA has begun to measure the difference between the farmer's share of the retail dollar for "food at home" and "away-from-home food." Farmers capture on average 24.3 cents on the dollar for food consumed at home, but only 4.7 cents for away-from-home food. You can be sure that the "value meals" at many fast-food restaurants are a raw deal for farmers—and an even worse deal for farmworkers.[1]

It may come as no surprise that these evolving models start to blur the lines between local and regional, often by necessity. It's a tough business world for local food purists, particularly in regions challenged by seasonal growing constraints. When you look at the maps throughout this book, it becomes evident that local will go only so far in many parts of the country, whereas there are countless opportunities for linking local and regional markets, thereby maintaining many of the values and relationships inherent in highly localized production.

The Leopold Center for Sustainable Agriculture at Iowa State University has been a leader in the development of value chain partnerships (VCPs). The center helped organize working groups from different agricultural sectors, including fruits and vegetables, niche pork producers, and grass-fed livestock, among others.[2] These working groups brought together individuals and businesses seeking a stronger stance in the marketplace, a position secured for their scale only through collaboration. The working groups assessed needs for each sector and then worked to develop plans for addressing those needs, such as supporting markets and infrastructure for niche pork operations—a challenge in a state dominated by large-scale pork producers. Basing any such collaborations on high standards of environmental and community stewardship earns the respect and the patronage of clients throughout a targeted region.

This linkage is the fulcrum described in chapter 9. When distributors of various types become part of a farmer's marketing and distribution efforts, the line between local and regional food systems frequently becomes porous. In most parts of the country, as local starts to push up against the fuzzy edges of regional, both the volume of products and the customer base increase. A local marketing approach provides a clear market lever, while regional marketing provides that much-needed fulcrum—the product volume and diversity desired by the food service and retail sectors. This increase in capacity helps move the fulcrum closer to the thing we're trying to budge: a highly concentrated national food system that has little use for diversity at the local and the regional levels. It's all part of making room for local opportunities in the bigger system. Nothing like a good lever to lift your spirits.

Scaling Up: The New Middle

It makes sense that the span between farm and fork is where a frenzy of innovation is occurring. After all, most of the work in rebuilding local food systems centers on increasing farm production to supply the local food market while building a stronger and more diverse consumer base. The span between farm and fork is perched on the fulcrum of what we could call "the New Middle."

The New Middle consists of aggregation, processing, and distribution, in a variety of possible forms and combinations, ranging from community-based nonprofit efforts to savvy entrepreneurial initiatives. Much of the large, conventional food-processing infrastructure is not scaled for smaller local volumes, nor does it necessarily allow for the product segregation and market differentiation so important to local food production and marketing. The same is true for much of the distribution sector, as the majority of that sector is national and international—as evidenced by figure 9-2 on page 180, which shows the range of distribution centers for Sysco and US Foods, the nation's two largest distributors.

The competitive edge for local foods these days is often not price, although there are certainly exceptions, particularly for foods prepared at home. Skyrocketing energy prices or an elimination of crop subsidies could change these economics, of course, and having efficient and cost-effective local/regional processing and distribution systems in place could help avert a food security disaster if any worst-case scenario comes to pass. Nonetheless, it's important to base revitalized local food systems on other existing marketing advantages for now.

I've used the phrase "rebuilding local food systems" frequently throughout this book. Nowhere is this rebuilding more dynamic and complex than in the

New Middle, and it is important to consider some of the different entrepreneurial models available here. An astounding menu of possibilities exists for communities and businesses to choose from in their forward-thinking business models in the New Middle. One important concept here could prove to be the "food hub": a central facility that provides capacity for some combination of aggregating (pooling), processing, and distributing food products.

No matter what specifically a food hub should or should not include (never fear, however: the feds are working on a definition right now!), it's time to get down to business and create whatever systems seem to work best for local communities, businesses, regions, and states. If it looks like a duck and walks like a duck, then it must be a food hub . . . or some other webbed creature.

Below is an à la carte menu, featuring a range of potential components that can be incorporated into a food hub. These items can be combined in most any manner to fill warehouse pallets and satisfy consumer palates.

The diversity of business models for the New Middle can be attributed in part to entrepreneurial creativity, but the cracks in the concentration of our food supply tend to manifest themselves in different ways across the country. The ecological realities, food traditions, workforce potential, and available infrastructure of a region can often help create these small but widening fractures in the system. Even the lack of certain infrastructure can prove a boon to these models by minimizing the threat of large-scale competition. Local

TABLE 10-1. À La Carte Menu for Cooking Up the New Middle

AGGREGATION	PROCESSING	DISTRIBUTION MODELS
Combining	Chopping and pureeing	Nonprofit
Washing	Canning	Farmers' cooperative
Cooling	Baking and confection	Farmer/consumer cooperative
Grading, sorting, and packing	Dehydrating	Specialized local distributor
Repacking	Freezing	Combined local and regional distributor
Storing	Labeling	Charitable food system
Marketing	Facility use for farmers	
	Facility use for food processors	
	Food business incubation	
	Workforce development	

Source: Adapted in part from *Building Successful Food Hubs: A Business Planning Guide for Aggregation and Processing Local Foods in Illinois* (FamilyFarmed.org, 2012)

food entrepreneurs find these cracks and create marketing wedges that allow them to work their way into the local and regional food systems by virtue of capitalizing on products and systems that don't fit into the infrastructure or business model of large processors, distributors, and food service providers.

Perhaps the best way to convey a sense of the possibilities is to describe briefly a few of the models from around the country. First, however, bear in mind that these models are adapting rapidly according to the new lessons learned each year. The volatility of the business and marketing aspects of local food systems can be attributed in part to the fact that agriculture tends to have its seasonal highs and lows throughout the year, fostering a tendency in the local food systems world to assess, refine, and even reinvent on an annual basis.

In the end, this constant morphing is a tribute to the adaptability and ultimate resilience of local food systems. Being lighter in infrastructure and less tied to hard-and-fast business models means that local food systems can be much more fleet-footed (even without the fleets) than their mega-counterparts in the big-ag world. They take vastly different approaches to the sometimes nerve-wracking work of carefully placing a fulcrum under a heavy load before exerting force on the lever.

Ecotrust FoodHub

Already one of the gold standards of innovation and collaboration in the New Middle is Ecotrust's FoodHub, an online nexus between farmers, ranchers, fishers, wholesale food buyers, and distributors in the Pacific Northwest (Alaska to northern California). As an incubator and capital vehicle for moving innovative social enterprises forward, Ecotrust established Food-Hub in 2009, representing Ecotrust's commitment to supporting "reliable prosperity" and serving as a model for other such initiatives around the country. FoodHub has developed an online model that is both streamlined and simple, but without sacrificing a complex social network for its subscribers. Not only can wholesale buyers, food processors, and distributors search for farm products online, but they can also let farmers and others know of bulk items they are searching for. FoodHub also provides a variety of ways for members to connect with one another for collaborative opportunities extending well beyond their online matchmaking service for food products. Website: http://www.food-hub.org

Oklahoma Food Cooperative

Founded in 2003, this state-based co-op is a fast-growing online ordering platform that allows farmers, gardeners, and ranchers to sell their goods directly to consumers. Both producers and customers are members of the co-op. Farmers sell at retail prices, not wholesale; they pay a nominal annual fee to join and then contribute 10 percent of all sales to the co-op. The co-op estimates that 80 percent of each dollar spent stays with its producers. Customers for the co-op tend to prefer natural, organic, and sustainably produced foods, and producers are encouraged to market their stories and educate others about rural Oklahoma. Volunteers sort the products and do the delivery for each monthly order—with the monthly schedule admittedly a challenge for producers of vegetables and fruits. Not intimidated by scale—"small," that is—the OFC is also encouraging urban farmers to participate as sellers, inspired in part by the SPIN (small-plot intensive) method of urban gardening promoted by Wally Satzewich and Gail Vandersteen in Saskatoon, Canada (http://www.spinfarming.com). Website: http://www.oklahomafood.coop

New North Florida Cooperative Farm-to-School Program

Begun in 1995, the New North Florida Cooperative was established as a cooperative of black farmers from Florida, Georgia, and Alabama. Since the inception of the cooperative, farmers have built a solid market by selling a steady stream of produce to public schools. Emphasizing high-quality produce and highly professional and courteous delivery service, the cooperative has focused on several crops that its growers can produce in sufficient quantities and with superior quality for delivery to the schools two or three days per week. As the orders increased in volume, several grants helped the cooperative acquire additional processing and distribution equipment. The cooperative works to negotiate prices that are fair to the farmer and appropriate for school budgets, as well as to educate students and the community about agriculture and nutrition. Website: http://bit.ly/nnfc-farm-to-school

Market Mobile

Either in spite of or because of its size, Rhode Island is a hotbed for local foods, with a long history of surf-and-turf cuisine. Farm Fresh Rhode

Island, a nonprofit founded in 2004, is dedicated to "growing a local food system that values the environment, health, and quality of life for Rhode Island farmers and eaters." The group's new Market Mobile program takes its mission across the border into Massachusetts, linking farmers, fishers, and value-added producers with wholesale buyers in restaurants, schools, and hospitals through an online ordering platform, product aggregation, a shared delivery truck, and a single invoice. Not only have the developers of Market Mobile been assiduous in tracking the dollars and details of their progress in building their local food economy, they are also gutsy in declaring their ambitions. Other local food initiatives would be wise to adapt Market Mobile's careful data analysis of its impacts. Displaying both recorded successes and ambitious measurable mileposts on the group's website helps generate excitement and inspire confidence (and perhaps funders). Linked to Market Mobile, Farm Fresh Rhode Island now also delivers Veggie Boxes to workplaces, community centers, and schools on a weekly basis. Move over Massachusetts—Rhode Island is expanding every day! Website: http://www.farmfresh.org/hub

Appalachian Center for Economic Networks (ACEnet)

When looking at models for any concept, it's a good idea to look at an organization's longevity and track record. ACEnet is a national model in helping to build community food systems because it has thrived for nearly three decades, pioneering economic development in a rural and economically challenged region, albeit one with a diverse agricultural heritage filled with food-based traditions. Located in Athens, Ohio, this nonprofit organization fosters innovation, sustainability, and networking within the economy of Appalachian Ohio. ACEnet's Food Manufacturing and Commercial Kitchen Facility helps train and incubate food entrepreneurs, serving over 150 businesses every year by providing production, storage, refrigeration, freezer, and distribution space. While a number of communities across the United States have created shared commercial kitchens, the failure rate is disappointing. ACEnet's advantage has been its strategic approach: not only does it provide excellent business training for food entrepreneurs, it also encourages networking among users of the facility. The comparison of business models and experiments provides clients with important lessons,

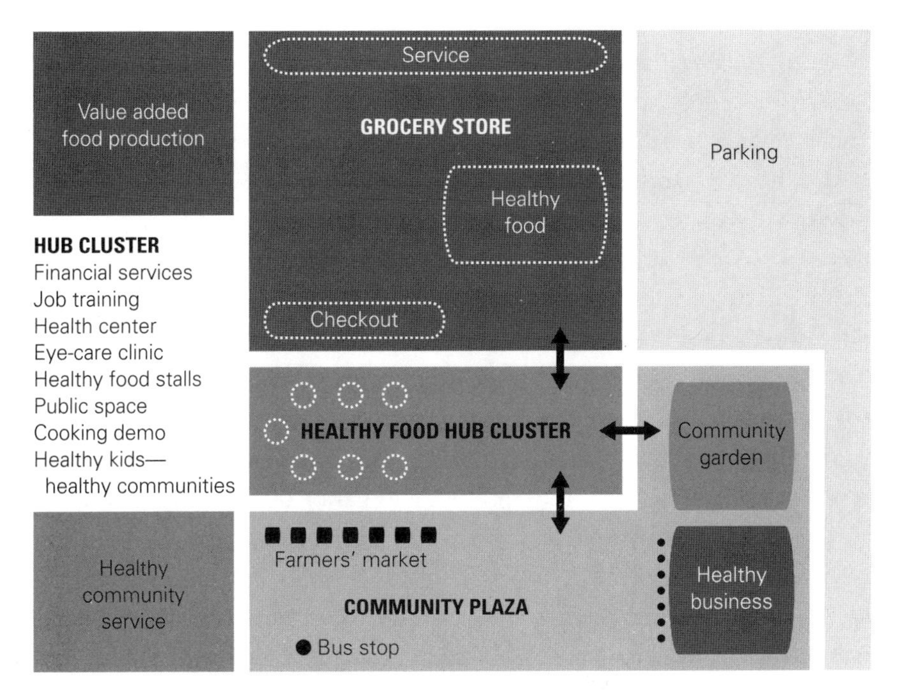

FIGURE 10-2. "Healthy Food Hub" Concept: One of Many Possibilities. Image used by permission of the Project for Public Spaces and Wholesome Wave.

and collaborative purchasing of needed supplies in bulk helps reduce costs to individual businesses. Cofounder Leslie Schaller sums it up: "The assets we have in our region are our traditions . . . our relationship-based culture."[3] Website: http://www.acenetworks.org

These vignettes of the New Middle offer a peek into the entrepreneurial dynamism emerging from all parts of the country, but it's all too easy in the excitement of it all to get caught up in software, social media, stainless steel, and grease-burning delivery vehicles, inadvertently bypassing the toughest issues surrounding food security and food justice. Fortunately, organizations such as the Project for Public Spaces, Growing Power, the National Good Food Network, Wholesome Wave, and the USDA are pushing the food hub concept a bit further, promoting the idea of what many are now calling "healthy food hubs." Healthy food hubs would bring together multiple organizations and businesses in a single physical site, providing a nucleus of entrepreneurial and health-oriented activity that would promote both healthy lifestyles and local economies. As with many food hub concepts, an anchor tenant is considered important; in this case, a grocery store serves as the anchor tenant. A concept map for such a site appears in figure 10-2.

Lest we get too carried away with the necessity of capital-intensive infrastructure or profit-centered activity, it is worth reminding ourselves that some of the best ideas for the New Middle are relatively simple and altruistic by design. Gary Oppenheimer, an early expert in the high-tech communications world, saw a gap in the food system and built a bridge to span it with an online matchmaking service for home gardeners and local food pantries. Oppenheimer set up a website, AmpleHarvest.org (http://www.ampleharvest.org), that provides a national database that helps gardeners and nearby food pantries find one another so that gardeners can help increase the inventory of fresh fruits, vegetables, herbs, and nuts available to individuals in need of assistance. With an estimated forty million Americans growing food at home, the potential for increasing the consumption of fresh, unprocessed foods via food pantries seems virtually endless. It's one more example of finding a crack in the system and using a carefully selected wedge to expand opportunities and innovation in the New Middle.

Leveraging Local Buy-In: Retail and the Food Service Sector

So much of the work necessary to rebuilding local food systems involves bringing people together and finding common ground. But there is the equivalent of a "consensus grenade" that one can toss into the conversation and watch good people scatter to all corners of the room: it's the Walmart conundrum. Can Walmart justifiably propose that it supports local food systems while it leads the charge for consolidation and concentration in the global food system? In 2008, Walmart announced its own "locally grown" program. At the time, Walmart claimed that it "purchases more than 70 percent of its produce from U.S.-based suppliers, making the company the biggest customer of American agriculture. This year (2008), Walmart expects to source about $400 million in locally grown produce from farmers across the United States."[4]

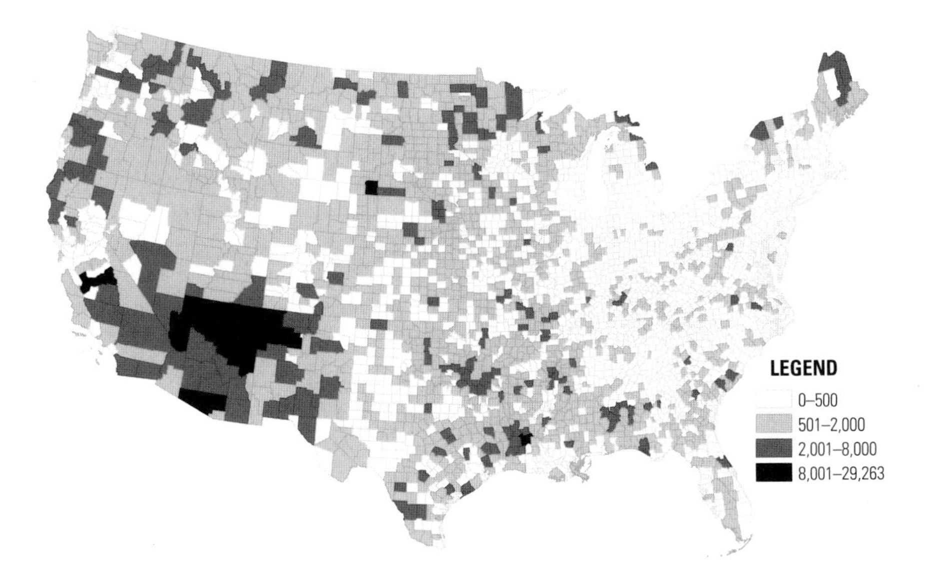

FIGURE 10-3. Number of Low-Income Residents Living More than 10 Miles from the Nearest Food Store, 2006. Source: USDA Economic Research Service Food Environment Atlas, accessed July 11, 2012, http://www.ers.usda.gov/data-products/food-environment-atlas.aspx.

A visit to the Walmart website will reveal a sampling of farms from which the company purchases products.[5] As the nation's largest grocery retailer, what Walmart does with its purchasing priorities obviously has a significant impact. However, if a primary driver for building local food systems is increasing farm profitability, should we trust such a multinational entity whose primary market differentiation is "always low prices"? On the other hand, is it fair to denigrate either legitimate or token efforts by Walmart to purchase and sell organic and local foods when it is already the only local source of groceries for some rural and urban communities? Figure 10-3 provides some insight into the challenge of distance to retailers in many rural areas across the country. The urban dimensions and the challenges of food deserts (areas with no local access to fresh foods) in these densely populated areas add yet more complexity to the considerations.

Walmart's mere presence often comes at the expense of shuttering smaller and locally owned businesses, as was documented in a Chicago study done by researchers at Loyola University in 2009.[6] A more recent examination of the potential impacts of Walmart store openings in food deserts in New

York City focused on the possible opening of a Walmart store in Harlem. The conclusion of the study, conducted by the office of Manhattan borough president Scott Stringer, was that Walmart's entry into the neighborhood would have significant negative impacts on existing fresh food retailers—including several fresh food retailers that had been supported by New York State public financing administered through the city's FRESH (Food Retail Expansion to Support Health) initiative.[7]

The questions become complex, and they differ according to the specific situation. If Walmart is already the only retailer of fresh foods in an area, should its purchase of more sustainable products be criticized as disingenuous marketing, despite the fact that such an effort brings these communities more organic and local choices? That said, if there are fresh food retailers currently in business in a location targeted as a new venue for Walmart, what are the questions that need to be asked about likely negative impacts for these existing businesses?

And then there are the questions that need to be asked from the farmers' and processors' perspectives. Shouldn't a farmer, a cooperative, or a nationally recognized sustainable business be free to decide whether selling to Walmart is a good business option or not? (A dairy farmer friend of mine, whose farm was actually depicted on the side of the trucks carrying Cabot cheese, has more than once reminded me and my students that Cabot—a farmer cooperative—was delighted to get a contract with Walmart, as it pushed the company's national sales and reputation significantly.) If local food purchases are less than 2 percent of all food sales in the country, does it not make sense to hit the mainstream markets such as Walmart and heighten the options for farmers and consumers?

Dissent on these issues is high. Not everyone you think would be skeptical of Walmart's ventures is necessarily of that mind-set. After receiving a $1 million grant from Walmart for his Growing Power initiatives, Will Allen stated, "Walmart is the world's largest distributor of food—there is no one better positioned to bring high-quality, locally grown food into urban food deserts and fast-food swamps."[8]

I am personally no fan of Walmart. In fact, I believe that Walmart functions primarily on the basis of profit and market opportunity. Nonetheless, I am not comfortable weighing in with a universal ethical caveat, simply because such caveats trump the spirit of local decision making. I will say, however, that one caveat may be more important here. In this case, *caveat venditor* ("seller beware") is probably a better caution for the realities inherent in this business transaction than *caveat emptor* ("buyer beware"). There's ample opportunity for anyone selling to Walmart to become part of the Walmart food chain, as in "gobbled up."

Assessing the payoffs to farmers in "buy local" campaigns always proves challenging. It is difficult to determine the economic benefits to farmers in these situations, and "local" can mask a variety of questionable practices on the part of the vendor and the farmer. Nonetheless, getting local foods into the mainstream and onto Main Street involves multiple strategies and some inevitable compromises. For some farming or food processing operations, not selling to Walmart or other large retailers may not seem wise, given the potential opportunity and the competition. Nonetheless, the stability and security of these markets is not something the farmer or processor can easily read, much less control. Locally and regionally owned opportunities can be somewhat easier to assess, if not trust, based on relationships, many of which stretch over years and even generations.

But there are contrasting retail models of all scales and types that we can all feel good about and even trust. Local Harvest, an independent grocery store in a resurgent St. Louis neighborhood, grew out of a highly successful farmers' market in the area. Local Harvest is committed to purchasing and marketing local foods, and its website features a tracking of the company's financial contributions to the local community, including an estimate that 85 cents of every dollar spent at Local Harvest stays within the community.[9]

In a similar vein, the Good Food Store in Missoula, Montana, touts itself as "One of Montana's Natural Wonders." In some ways it is, particularly when we consider its dramatic contrast to the predominance of commodity agriculture surrounding it. Started more than forty years ago as a natural foods co-op, the Good Food Store clings tightly to its mission of supporting a healthy community. The big-timbered storefront with its unique roofline draws Missoula residents in to find local, organic, and bulk foods, as well as a café that serves as a hub for a community that takes food justice issues as seriously as any city in the country—thanks in part to the Garden City Harvest program described earlier and the University of Montana's leadership.[10]

The People's Grocery in West Oakland, California, puts food justice issues at the core of its mission. For a decade, this "health and wealth organization" has worked to integrate issues of food justice, urban agriculture, neighborhood revitalization, and innovative marketing in West Oakland—without owning a grocery store. The founders of the People's Grocery determined after significant research that they needed to pursue other projects before actually creating a grocery store for West Oakland. So they developed Mobile Market, a converted postal truck that delivers boxes of fresh produce straight to the neighborhoods in need of fresh produce—a concept replicated in several other parts of the country (e.g., the Farm Fresh Rhode Island initiative mentioned earlier). Nutrition and cooking education are also critical outreach components of the organization, but with an important (and often overlooked) emphasis on culturally appropriate cuisine. Even without a storefront, the success of the People's Grocery's efforts have moved its people and its programs into the national spotlight as models for integrating cultural diversity, economic development, and food justice issues in a challenging urban environment. There's no one-size-fits-all philosophy at work here. Perhaps that's why the People's Grocery is now working to move into a new 15,000-square-foot home—they need more space for all of their big ideas![11]

Serving Up Local

Larger-scale local food purchasing opportunities come in other forms, too, and the relationship between farmer and chef can add a heightened element of trust and appreciation not always found in retail operations. The restaurant and institutional food service sectors pose tremendous opportunities for the rebuilding of food systems, albeit with some tough challenges. The white-tablecloth restaurant world has been a critical ally in putting the local food issue into the public spotlight by showing the public the culinary possibilities and introducing them to the local economy on the plate. This part of the restaurant world has also footed the bill for many of the early initiatives during the last two decades. What is less understood, however, is how food service operations are upping the ante among their peers—sometimes quietly, sometimes by throwing down a gauntlet.

One of our neighbors down in the bucolic Mettowee Valley of southwestern Vermont is Barbara Moore, owner of the Good Table, a food service provider and catering operation based in Westchester and Fairfield Counties of New York and Connecticut, just outside New York City.[12] Barbara grew up on a quintessential Vermont farm; her family had what has all but disappeared in Vermont—a midsize egg farm. She purchased her business from a colleague in 2002, and today the Good Table is a relatively small but much-respected business with approximately thirty employees and serving about 243,000 café meals a year (not including conference-room catering, executive dining, and special events catering).

Traditionally—if the last several decades serve as tradition—owners and chefs in this kind of business would simply place an order for the bulk of their needs to a major distributor once or twice a week by phone, fax, or computer. The primary considerations for which items to order would be "what" and "how much," not "where." Things are changing in this regard, and Barbara—inspired by her agricultural background—started running counter to the trend earlier than most. She has been exceptionally creative translating her rural farm knowledge into a business ethic in a periurban environment. She began by establishing a relationship with a farm bordering her family's land in Vermont and later transitioned to working with a larger organic farm in New York State, the Denison Farm.[13] From May until February, she receives a list of vegetables and some fruits that will be available the following week from the farm. Using this RSA (restaurant-supported agriculture) system, she orders as much produce from the farm as possible, planning her café and catering menus for the coming week around what is available.

Barbara says that there is never a question of how much to purchase from the Denison Farm because the produce is of such high quality. Not only do the products store well, but they are so fresh and carefully harvested that there is virtually no waste. In fact, leftovers from the kitchen prep work and the salad bars are often transformed into soup stocks. She also uses local meats and fresh seafood, which are often more expensive than their more conventional counterparts available through mainline distributors, but Barbara continues to search for additional sources at price points that work for her operation. She has discovered that serving ethnic dishes in which meats are used in smaller quantities can help offset the costs.

The Good Table also promotes the Denison Farm by offering workplace CSA shares to employees of the companies where its cafés are located. This increased volume of sales helps the farm and further promotes the Good Table's fresh messaging.

Barbara continues to seek ways to expand the company's purchasing of local and organic products, and she already has one account that pays a premium for organic meals. Expanding her sustainable purchasing effort presents some challenges. Mainline distributors sometimes feature local products, but on a limited basis. Specialty distributors offer the products, but often with premium markups that are out of line with the café budgets. Many of these distributors mark up local products by 35 percent, putting the products out of reach for economical café operations such as the Good Table.

The New Middle is coming into play in New York City and other regions across the United States, however. Barbara recently had the opportunity to begin working with FarmersWeb, a new start-up company serving as a matchmaker between farmers and food service operators.[14] FarmersWeb is designed specifically for wholesale buyers. Cofounder Jennifer Goggin explained to me that many food service operations give the excuse that local purchasing is too much work, too much hassle, and too time-consuming—particularly with separate orders and billing for each farm. FarmersWeb is designed to make it easier for buyers to make payments to multiple vendors and keep permanent records.

Farmers benefit from this system by maintaining their individual branding, since the products aren't aggregated. Furthermore, FarmersWeb has the buyer use a credit card payment—with a small fee going to FarmersWeb—and all payments are made to farmers within two weeks. This approach runs counter to the food service industry standard, which actively dissuades local purchasing by withholding payment to farmers for thirty to ninety days. Thus far, FarmersWeb operates on a basis of having the farms deliver their own products to the accounts, although they are discussing an alliance with a start-up delivery company in the area.

Total transparency is part of the mission for FarmersWeb, which is not always the case with the sourcing from some distributors. Thus far, FarmersWeb has been defining "local" as being within 250 miles, allowing for enough regional diversity to satisfy clients. However, the company says that it is considering updating its software so that clients can choose the radius from within which they would like to order.[15]

The Good Table and FarmersWeb are both operating in a world of giants, but they each carefully utilize the advantage of their small size in finding their niches within the bigger system. As Barbara notes, "There's no hierarchy of decision making that I have to deal with here, and we're very much in touch with our suppliers—we *know* the people we're purchasing from."

Change is happening within much larger businesses, too. And when things work out well, big changes in big corporations can have important ripple effects throughout the industry. Fedele Bauccio and Helene York are two key players in transforming the food service sector through their leadership at Bon Appétit Management Company (BAMCO) in Palo Alto, California.[16] Coming from a culinary background, Fedele is a visionary cofounder of the company, while Helene is the director of purchasing strategy, as well as a keen observer of food ethics and industry practices in her column for the *Atlantic Monthly* blog.[17] Bon Appétit provides food service to more than two hundred colleges and universities, businesses, and specialty venues across the United States, and the company also repeatedly sets the trend for sustainability in the food service industry. It does so despite its relatively small size in comparison to some of the food service giants (although it served over 135 million meals in 2011). Both Fedele and Helene provided me with a deeper understanding of how Bon Appétit has leveraged the food system and nudged its competitors to increase local and regional purchasing, as well as to address farmworker issues and animal welfare reform.

Fedele came to the local issue by way of a culinary search for freshness and flavor, but he then had to find a way to make it work economically. Like Barbara Moore, he found that buying local products produced less waste and thereby saved money that could then be reinvested in more local purchasing. "We work on the issue of waste," he says, "in the field and on the plate. Then we can buy the very best." Helene explains that the company's policy of making all dishes from scratch helps minimize waste as well. When you compare the cost of a fresh ingredient (such as chicken thighs) versus a processed and packaged product (such as a "chicken nugget"), the economic benefits of the fresh product win out, as long as you are willing to process the fresh product yourself. At Bon Appétit, chefs take the raw products and turn them into food: no packs, no cans, no bags—no business as usual. (There are a few exceptions: Bon Appétit chefs are permitted to buy plain canned tomatoes, which they then turn into sauces and soups.) Kitchen prep scraps are used for soup stocks. "If there's waste in our kitchens, it's really waste," Helene says. "If it gets tossed out, it's clearly inedible."

Helene describes the company's second most effective approach to reducing food waste: proper portion sizes. Proper portioning not only minimizes food waste but benefits the consumer's health, providing a visual cue of when to stop eating. Helene says, "We tend to feel satisfied when we finish a full plate, whatever size it is." So Bon Appétit chefs and other servers are trained to create portion sizes targeted to the small female, not the enormous football player. "When we take over a new account, after evaluating their equipment one of the first things we often have do is change out all of the serving spoons, scoops, and ladles. And if our clients are willing, we shift from 12-inch serving plates to 7-inch plates." Cutting down food waste brings down food costs as well as disposal costs. All of the savings can be reinvested into the quality and sourcing of the food. Putting the whole equation into context, Helene notes, "Food waste is a choice at every step from preparation to consumption."

Not that buying local, seasonal products came all that easily at first. Fedele notes that he and his colleagues had to help farmers get over some trust issues. Back in 1999, when Bon Appétit formally launched its Farm to Fork program, the notion of "farm to fork" was unheard of. Bon Appétit was leading the charge in the food service industry, and according to company lore, one farmer actually said, "You don't deserve my melons." Fedele and his chefs had to convince him otherwise. "We had to demonstrate that we were willing to support the communities where we work, live, and play," says Fedele. The company also had to confront a distribution system that did not always work in its favor. The mainline distributors had little interest in working with producers of smaller scales and outside of their national contracts. The costs, efficiencies, and scale of these operations didn't fit the distributors' business models very well—and there was a certain lack of imagination and confidence in the potential of the local and regional foods market. In many cases, Bon Appétit teams initially used their own delivery trucks (or in a few cases, actually bought trucks for farmer co-ops) to pick up local foods for them, heightening quality control and minimizing expense, since eliminating the middleman often helps provide cost savings. As the systems evolved, however, Bon Appétit began to find ways to utilize regional distributors that support a wide variety of farmers and provide a vast array of products through highly efficient distribution channels. In reflecting upon the company's waste reduction practices and its reexamination of distribution options, Fedele seems confident in the decisions: "When you get right down to it, I don't think our food costs are necessarily any higher than those of our competitors." In this case, linking sustainability to business practices not only paid its way but also reinforced a powerful mission.

And it's not a narrowly focused mission, as Bon Appétit has also led the way in standing up for animal welfare reform and sustainable livestock production. It was a leader in requiring its seafood to have been sustainably harvested (it partnered with Seafood Watch in 2002), rejecting the purchase of poultry raised with nontherapeutic antibiotics, and demanding cage-free eggs. Fedele observes that back in 2005 when the company made the switch, buying cage-free eggs was initially cost-prohibitive, but it was the right thing to do—and now it's virtually cost-neutral. The week I spoke with Fedele, he was being barraged with media interview requests resulting from Bon Appétit's announcement of the most comprehensive animal welfare requirements in the food service industry. "I'm not against large farms," he offered, somewhat conciliatorily. "But I am against the way they farm."

Nor will he abide farmworker mistreatment. When Fedele got word of the enslavement of migrant farmworkers in the tomato fields of Immokalee, Florida, he traveled there himself with one of his bilingual chefs and a *Washington Post* reporter to investigate. They had to crawl under a barbed wire fence to enter the fields, and they spent two days with the largely Spanish-speaking workers. After earning their trust, they joined them in the evening, crammed into the miserable housing to listen to their stories of abuse. The third day, Fedele traveled around to the growers in the area, alerting them to the fact that he would be teaming up with the Coalition of Immokalee Workers to create a code of conduct that he expected them to sign, or else he would boycott their tomatoes. Sure enough, for four months that winter, Bon Appétit accounts east of the Mississippi had to serve only grape tomatoes, not "slicers," because it could only get one producer into distribution—which made for some interesting burger toppings. With Bon Appétit on board, the Coalition of Immokalee Workers was able to exert pressure on Whole Foods, Burger King, McDonald's, and Chipotle to join these efforts. In the end, the collective pressures did indeed result in improved conditions for the Immokalee workers, while also letting them know that others do care.[18]

I don't know about you, but I want more CEOs who will not only go the extra mile for local but will also crawl under barbed wire to search out injustice in our food system and then work to address it.

Part Three

NEW DIRECTIONS

The following chapters present a variety of models that are challenging and changing the food system in innovative ways . . . even if some of what they are doing might sometimes appear to be ordinary. It is important to note that the individuals, initiatives, organizations, and approaches mentioned here are part of a fast-paced change in the movement to reform our national food system. I think that this change is due to both the dynamism of these efforts and the marketplace, but it also comes back to seasonality. If you work with food in any fashion, there tend to be points in the year when you pause, reflect, reassess, adapt accordingly, and move forward. Certain models provided here will have changed by the time you read this book—but it is my hope that many of those changes bode well for the future of resilient community-based food systems!

CHAPTER 11

Bringing It All Back Home

Finding the entry points for rebuilding local food systems can seem a daunting task at times, so this chapter is intended to help the reader think about the impacts of choices at home. It is not a how-to guide for locavores or homesteaders, but it does offer a few local carrots for thinkers and doers alike. The sections that follow consider how the things we do and the food choices we make as individuals and families play a vital role in increasing the sustainability and resilience of our local food systems. These entry points for change at the home scale are not meant to be exhaustive or exhausting. Rather, the intent is to inspire and to encourage more adventurous thinking.

Wildcrafting

You might consider wildcrafting, also known as gathering or foraging, to be an odd starting place for thinking about rebuilding local food systems at the personal level, but it's the natural place to start, so to speak. I teach an upper-level college course called "Sustainable Farming Systems" every year, and I begin the course each time with a "wildcrafting walk" around the edges of the college farm. My friends Nova Kim and Les Hook, extraordinary wildcrafters who started a unique wild-harvest CSA, lead the tour for the students.[1] As we walk the edges of the farm, Les and Nova do a fine job poking holes into the whole concept of agriculture, making it clear that, while a lot of work goes into cultivating all of those plants in the gardens, there's also plenty to eat on the periphery.

The lesson is important for two reasons. First, it's important to rethink what we consider to be appropriate and edible as we discover the astounding variety of foods in lawns, sidewalk cracks, and local forests. A classic example of our cultural biases and odd oversights might be lamb's-quarter, a plant considered an aggravating weed by gardeners today but originally brought to the United States by Europeans as an edible green high in iron. Lamb's-quarter is in the beet family, and it's only through some odd fluke of history that we've forgotten its taste and value.

Second, food security begins by knowing what is available and how to access it. Sometimes food is literally underfoot, or at least within the distance of a healthy walk. Of course, it's also important to have knowledgeable mentors for any wildcrafting activities in order to avoid harvesting any toxic plants or mushrooms, and anyone engaging in such harvesting should consider both the ethics and the ecological ramifications of collecting. In fact, it's a good idea to step into the world of wildcrafting together with other people immersing themselves in these culinary adventures. Groups of interested folks can gather for teaching sessions on identifying, harvesting, and preparing wild edibles. I've watched the thrill of students finding and cooking Jerusalem artichokes, Japanese knotweed, and stinging nettles—all discovered on just a walk around the block. There's nothing like watching people savor the taste of the plant they had always cursed for being an invasive or nuisance species!

Les and Nova take their work seriously, and they see it as a form of food security. In fact, they are publishing a free downloadable field guide to wild edibles geared specifically to the food insecure. In case you think such an approach to food security is sheer whimsy, consider the fact that when Nova and Les take my students out for a "wild edible walk," they invariably find at least a dozen wild edibles between the college library and the heart of the college farm—a distance of less than two hundred feet, including manicured lawn.

Urban and suburban areas are surprising treasure troves for wildcrafting. Groups like Urban Edibles in Portland, Oregon, maintain maps and databases of wild edibles in their area.[2] Fallen Fruit in Los Angeles maps fruits located in or hanging over public places in L.A., provoking public discussions about land ownership and the shared urban landscape.[3] The Salt Lake City Fruit Gleaning Project opts for yet another angle, asking fruit and nut tree owners to register their trees online if they would like gleaners to come and harvest the bounty at the appropriate time. The proceeds go to the charitable food system, and the city saves dollars and disposal costs by keeping the streets clean of debris.[4]

All of these efforts help educate us about the startling biodiversity of edibles all around us in every type of landscape, as well as the threats to that biodiversity. Furthermore, long-term involvement in wildcrafting activities can lead to important observations about climate change and our food supply. Nova and Les have been working to document the early and erratic emergence and readiness of the wild edibles that they collect around Vermont. The signs of climate change are becoming increasingly obvious . . . but only if you're paying attention. And that's a wildcrafter's primary task.

The Kitchen

Even the more ancient food skills usually bring us back to the modern kitchen. And what happens in our kitchens determines what happens in our landscapes. We live in an interesting time, with a curious juxtaposition of celebrity-chef shows and low-income cooking classes both extolling the virtues of fresh, local foods. The point of intersection is the kitchen.

We can't lose sight of the importance of the kitchen. Hours spent in the kitchen and our time at the table are both critical investments in relocalizing food systems. Food is where the comfort of tradition and the allure of exploration are in creative tension, nourishing our bodies and sustaining our connections to our communities and the natural world.

Coincidentally, just as I wrote that last paragraph in my upstairs office, a smell resembling that of garlic or onions wafted up the stairway, accompanied by hushed but obviously excited children's voices in the kitchen. Not hearing my wife's voice in the mix, I wandered down to check on things, only to discover all three of my children, aged nine, seven, and two, hard at work making wild leek pesto completely of their own volition and under their own supervision. Asa and Ethan, nine and seven years old, respectively, had watched Tomer, a visiting student from Israel, invent a wonderful wild leek pesto the week before. Inspired, the boys had taken the dog and walked a quarter mile through the woods earlier in the day to a patch of wild leeks, harvested them sparingly with a trowel, and proudly brought them home in a cloth bag.

In a wonderfully effective effort to surprise the adults, they pulled out the manual food processor, the electric food processor, and kids' safety scissors for two-year-old Addy and went to work just before the dinner hour. Tomer came in to check on the status of dinner, only to find his pesto nearly replicated. He added a little more olive oil and a few more walnuts to get the consistency correct for pasta, while I sliced and lightly heated some *speck* (a smoked and aged Tirolean bacon that needs no refrigeration) that I'd made from our pigs the year before. Cultures, technologies, and generations mixed and mingled, and the dinner was perfect, save for the fact that Addy kept raiding everyone else's plates for their *speck*. It was as serendipitous a moment as I could ever imagine, and it only reinforced my thinking on the importance of the kitchen in the orbit of our daily lives.

Granted, any call to the kitchen, especially by a male who presently is not pulling his weight in shared culinary contributions, should induce some head-scratching and perhaps finger-poking. American culture faces a seemingly eternal conundrum in trying to loosen the fetters to household work while simultaneously glorifying "home cooking." The solutions and the inevitable compromises have to be tailored to the individual and the family. I am advocating not for particular decisions, but simply for conscious and conscientious decision making that links the kitchen to nature, nurture, and nutrition.

Efficiency is clearly a concern, not just for those who labor in the kitchen but also for our overall energy diet, as discussed in chapter 4. Remember that home refrigeration and preparation comprise about 31 percent of the energy consumed between farm and plate in the United States. Energy-efficient appliances are, therefore, an important part of building your own food resilience, not to mention securing your own financial future. Storage and preparation choices make a big difference. Energy savings can be applied to your grocery bill, and the cumulative effect of wise energy choices can, in the long run, help a community reduce its overall energy needs and potentially minimize not only energy investments but also the risks of power-grid interruptions. Consider, for example, the increased energy demands that occur during a heat wave: refrigeration demands increase, cooking generates heat, and cooling and further refrigeration demands both rise as a result.

Investing in new appliances can be a serious challenge for many families, but several states have created excellent incentive programs for the purchase of energy-efficient appliances. Oregon, Michigan, and New York have experimented with incentive programs geared especially to low-income residents, since low-income families are particularly hurt by fluctuations in energy prices. No household should have to choose between food and energy, particularly when the two are so tightly interrelated and there is often less flexibility in an energy bill than in a food bill.

Waste reduction in the kitchen comes in other forms, too. The first link between growing local and reducing waste is the potential for minimal spoilage and waste with fresh products. Fresh vegetables and fruits provide the most nutritional value when used right away, but they also can last longer if handled properly. Many of the scraps can also be used in soup stocks, as can the bones, skin, and organs of whole poultry or even pork and beef bought in bulk quantities. Buying these products at a farmers' market or from some other local source also can mean reduced packaging (less transport and storage typically means less need for packaging). In most cases, the closer a food is to its point of origin, the less potential there is for waste.

Obviously, not everything gets eaten, but a "soil to soil" philosophy means that we can largely toss the concept of waste right out the window. Things grown in the soil need to be returned to the soil (and that doesn't mean being buried in a landfill). Effective composting and anaerobic digestion programs depend not only upon municipalities' effective decision making but also upon residents' cooperation in any organics recovery programs. Recognizing that 36 percent of its landfill waste stream was in the form of food scraps, San Francisco took a bold step in 2009 and became the first city in the country to establish a mandatory composting program for its residents.[5] All residents are required to subscribe to a waste hauling service, and those companies help outfit residents with the necessary equipment and knowledge to divert their organics into proper containers for collection. The goal of the program is zero waste, meaning that discarded organics will be utilized as a resource rather than put into a landfill or an incinerator. The city is already at a diversion rate of nearly 80 percent, and steadily closing in on zero. San Francisco is a model not only for the three-bin systems that the city has put into place for recyclables, compostables, and trash, but also for how it has carefully managed the gradual implementation of its efforts in diverting organics from the waste stream.

It's a good reminder that every drop in the bucket—or compost pail—is another step toward transforming local food passion into a local food *system*.

"You Can't Grow That Here!"

What you do in your kitchen is very much your own affair. But what you do with your lawn—well, that seems to inspire intrigue and even disdain in many parts of the country, particularly the suburbs. If you start a garden and hear something like, "You can't grow that here!" then you'll know that you're probably on to a good idea. There are three possible interpretations to such an exclamation:

1. It's against the law.
2. It's not what we do here.
3. It's not possible.

But doing anything new necessarily involves doing some things differently. One of the most benign and uncontroversial of all human activities has suddenly taken on radical significance. The act of gardening has become subversive, and therefore all the more fun and inspiring. Gardening drives a wedge into the corporate stranglehold on our food systems and begins to pry apart the possibilities of increased self-reliance. Even if it weren't so controversial in some places, it would still be worth doing! So how can we further such laudable "misguided" activities?

1. It's against the law. Zoning regulations and homeowner association covenants are meant to perpetuate the status quo, and sometimes that's a good thing . . . if the status quo makes sense. But does it make sense not to allow people to maintain gardens? Cases abound across the United States of individuals running into conflicts with the protectors of the public good, officials entrusted with ensuring the uniformity of inedible lawnscapes. Flowers (except dandelions, which happen to be edible) are fine, but not vegetables. Perhaps they didn't realize that the eggplant was first cultivated in British and American gardens primarily for its beautiful purple flowers and fruits?

It approaches the point of absurdity: In the summer of 2011, a Michigan mother who replanted her lawn with vegetables (after the city dug it up to replace a municipal sewer line) was threatened with ninety-three days in jail! Fortunately, sanity prevailed in that case and charges were finally dropped, but only after a media frenzy and one family's undeserved turmoil (read the whole story at http://oakparkhatesveggies.wordpress.com). The tide does seem to be turning somewhat, evidenced by large cities such as San Francisco and Seattle recasting their zoning regulations to accommodate edible gardening efforts. In similar fashion, New York City revised its zoning regulations in 2010 to accommodate urban beekeeping, but only after a lot of bureaucratic buzz about nothing.[6] For advice on how to negotiate zoning issues as well as suggestions for helping to develop zoning that is more accommodating to gardening in urban and suburban environments, check out *Seeding the City: Land Use Policies to Promote Urban Agriculture* (available as a free download from ChangeLab Solutions at http://changelabsolutions.org). The website UrbanAgLaw (http://www.urbanaglaw.org) also offers a vast array of resources related to legal issues surrounding urban agriculture initiatives.

2. It's not what we do here. Sometimes gardening in these environments is less about the law and more about perception and the status quo. If you need to shake things up in the neighborhood, just invite artist Fritz Haeg to drop in for a little artistic event. Haeg is what you get when you cross-pollinate guerilla theater with guerilla gardening. In 2005, Haeg began his long-standing "Edible Estates" art installation series, in which he selected public and privately held spaces—even ordinary lawns in very typical neighborhoods—and installed beautiful and provocative edible landscapes that challenge the paradigm of the American lawn. Give him an inch, and he'll take your yard . . . and turn it into a much more rational landscape.

In a similar vein but a broader scale, Transition US, a nonprofit dedicated to building communities prepared to face the challenges of climate change and peak oil, picked up on what was called the "350 Garden Challenge," an initiative started by Daily Acts in California. Originally, Daily Acts conceived of the project as an effort to begin 350 gardens in Sonoma County in one weekend. In reality, though, their efforts spawned 628 gardens, including some community gardens. Transition US expanded the concept and challenged its registered "transition towns" to follow suit.[7] The Transition US challenge encouraged individuals, organizations, and communities to improve their communities by developing edible landscapes, establishing composting projects, and creating rainwater catchment concepts—and participants in the challenge engaged in over four thousand such actions across the country in response.[8]

There are, of course, less dramatic approaches to returning some common sense to the American yard. California's "Healthy Eating Active Living Cities Campaign" pushes for changes in zoning regulations and community planning in order to accommodate private gardens, community gardens, and farmers' markets, with notable successes throughout the state.[9] The University of Maryland Extension is promoting an intriguing "Grow It Eat It" campaign geared to establishing a mere one million home gardens in the state, with an interactive map that lists the number of gardens recorded in each county. As that campaign enters the 2012 growing season, it has a total of almost ten thousand gardens so far.[10] Let's hope that advocates have proactively disarmed the zoning administrators and homeowner associations—or at least replaced their citation books with seed packets.

3. It's not possible. Some of us wither when we hear those words; others stand taller and more defiant. We may not be looking *forward* to the intensifying challenges of climate change, peak oil, and continued economic struggles, but we do have to look *ahead* to the magnification of those intertwined challenges. Preparation for any combination of those scenarios means finding ways to grow things with increasing constraints. In the northern tiers of the country, we are looking toward sensible season extension structures and methods that can protect plants during the coldest months, while southern stretches of the United States have to prepare for increased drought and high temperatures. All of us face the likelihood of having less oil and money to throw at our agricultural shortcomings.

How did we get to this point of laughable absurdity about what we can or can't do with a lawn, a vacant lot, or a rooftop? Journalist, gardener, and community builder Scott Carlson provides some interesting perspective here. Journalists typically do their muckraking in print; Scott does his in the garden. But he did discover that even this kind of muckraking can get things stirred up. Hampered by a large silver maple in his backyard, Scott decided to put some raised beds in his Baltimore front yard to grow some vegetables for the 2007 season. Simple enough, until the local homeowner association got involved and sent him a letter informing him that they preferred "a traditional landscape." As it turns out, they were only able to point out their preferences, not the actual association rules. In this case, the association could dictate architectural details but not landscaping details. It just so happened that Scott was writing a how-to article on this approach to urban gardening right at that time, so the event provided a perfect backdrop for the story.

The bountiful yields of Scott's garden gave at least some of his neighbors more to chew on, as he shared excess vegetables with them. A few others simply refused to talk to him, but the idea took root and expanded. He and his friend Joe put raised beds in the yard of an elderly neighbor who was unable to garden, and they shared the results from both gardens. Ironically, the 2008 recession hit soon after the first groundbreaking, and several of his neighbors were laid off. Gardening became a necessity, to the point that almost all of his neighbors now have gardens.

Since then, Scott has been thinking about the history of suburbia and the culture of the American lawn quite a bit. "Suburban development transforms open land that people enjoyed and converts it to something that is completely devoid of the life and vigor it had when it was farmed. The thing really missing from suburbia is meaning," he muses. "But when you create a piece of land with which you have a relationship—not just something you look at—it becomes something completely different. You've restored meaning to the place."

The conversion of farmland to lawn-land also suggests a break in generations-old traditions of passing on fundamental sustenance skills and knowledge. The 2008 economic downturn is certainly not the last of such scenarios, and it almost certainly will not be the worst economic situation we face in our lifetimes. As Scott puts it, "This is the tradition of the suburbs—deskilling." It's simply not a tradition we can afford.

Rooted Rebellion

It's ironic that a culture based on an astounding array of immigrant cultures and a coveted tradition of individualism should be the dominant homogenizing force in global food systems. Sameness knows no bounds in the modern world.

Enter the American home gardener, backyard livestock aficionado, urban homesteader, suburban permaculturalist, and rooftop beekeeper. These keepers of traditions play a vital role in maintaining agricultural biodiversity, although it's a role that warrants much greater prominence and understanding. Heirloom seeds, heritage fruits, and rare breeds of livestock tend to be ideal fits in smaller-scale, home-based scenarios, partly because that's where they originated. Many of the plants were grown for taste, not uniformity; for storage, not shipping; for disease resistance, not size. The livestock were bred for versatility, not massive production; for local forage, not imported grains; for long lives, not short nightmares.

Most of these plants and animals simply don't function within industrialized agricultural operations, and there can even be some challenges to using them on small and midsize farms. Home-scale uses, therefore, are critical to conserving these vital seeds and breeds and the traditions surrounding them. Sometimes the entry into this fascinating world begins with a taste of something unique, but more often than not it originates with stories. In fact, Gary Paul Nabhan describes these efforts as "restoring and re-storying" our landscape.

It is in this storytelling approach that community comes into play. Raising heritage turkeys in the backyard or Cherokee Purple tomatoes on a balcony isn't merely an attempt to have one big feast or a private bounty, nor is it just about conserving the stories and the recipes for success. Rather, a primary benefit is the actual act of *storytelling* that these actions can prompt, rekindling community and cultural connections through food. We love to talk about food, and sharing food dissolves boundaries. All kinds of unexpected relationships pop up.

I often take my students on a tour of my academic dean's home in the middle of the village of Poultney, Vermont, where we teach. Tom Mauhs-Pugh and his wife, Carrie, have a wonderful front-yard vegetable garden directly adjacent to the sidewalk, and they produce an amazing amount of food out of that small and beautiful garden plot—right into December, when they typically bring in a final harvest on Christmas Day. Tom tells the stories of how, while he and his wife are out working in the garden, many passersby stop on their walks to talk with them about the garden or other things. Besides the usual sidewalk banter, many of the conversations turn to gardening ideas and culinary possibilities. It was the first such front-yard experiment in the village, and others are beginning to appear. Relegating horticultural endeavors to the backyard may mean that you won't lose a few blueberries or cherry tomatoes that are just too tempting to some passersby. However, bringing it all front and center may well foster some undiscovered friendships.

Fortunately, garden clubs and tours around the United States are beginning to bring edible plants back into the picture, and there is a growing focus on edible landscaping in the garden tours that are cropping up in many towns and cities. There are also organizations like Ecology Action in Willits, California, that provide homeowners with resources for installing gardens that utilize permaculture and biointensive garden techniques. Not only do they provide a garden-starting resource called the RenewAll Garden Kit—including seeds—but they also help inspire and educate new gardeners through organized tours (information available at http://www.renewallgardenproject.net/GardenKit.html). The sharing of aesthetics and growing techniques on organized tours can inspire a whole host of people in the summer, while seed swaps in the winter can generate new alliances and experiments for the next season. Once again, the potential of shared maps and social media on the Internet can also pique curiosity and diminish timidity, as gardeners can easily share their locations, stories, and photos and make plans to share seeds and cuttings—and sometimes meals or tastings.

The fun need not exclude livestock lovers, as evidenced by the growing trend of chicken coop tours in cities across the country. Backyard poultry are on the rise, providing homeowners with meat, eggs, and superb fertilizer, not to mention pest control (chickens, ducks, and geese can help minimize weeds, insects, and slugs). Pigs, goats, rabbits, and even the compact homestead cow are also enjoying a renaissance on various "smallholdings."

In the end, it's all quite serious work, no matter how much fun it is. Whether we are saving seed and breeding livestock or purchasing new seed and stock each year, it is the participation in these subversive acts within a highly homogenized world that matters. Ultimately, the size of our individual contributions matters much less than the scale of our multiplied efforts.

Seasonal Stretches

When I moved to Vermont in the mid-1990s—a place where most winters have several days where temperatures drop to −20 degrees Fahrenheit or colder—greenhouses were being used primarily in northern areas for starting spring transplants. Some home gardeners and market growers were using cold frames, row covers, and low hoophouses for stretching the growing season on both ends, but large-scale season extension was still in its infancy. Then something happened. Growers like author and season extension pioneer Eliot Coleman started experimenting with larger structures and in-ground planting techniques (see fig. 11-1), oftentimes borrowing from and transforming European methods and crops. Iceberg lettuce began to lose its appeal (as if it ever had much), and microgreens became popular, including cold-hardy varieties that could be grown in surprisingly cold temperatures. The marketplace responded rapidly, and farmer innovations surged. In addition to tinkering with new season extension possibilities, more people began digging into older traditions such as root crops and root cellars. As a result of all this activity, growers and chefs found more and more ways to collaborate throughout longer periods of the year. The idea that local markets were viable only in the warmer months began to melt.

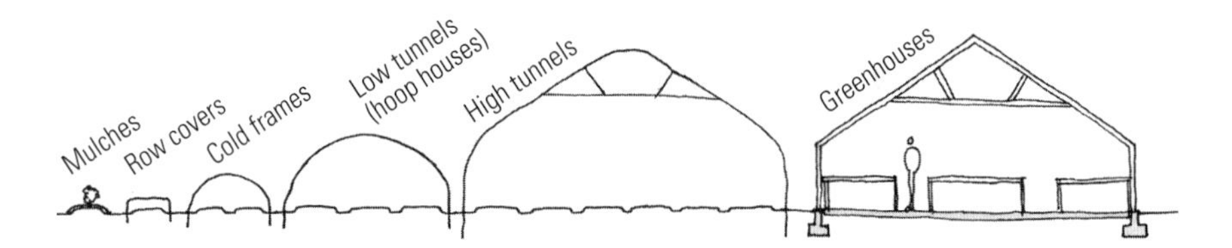

FIGURE 11-1. Season Extension from the Ground Up. Image courtesy of Philip Ackerman-Leist and Lucas Brown, *Season Extension: Agriculture, Eco-Design, and Renewable Energy*, a presentation at SolarFest, Tinmouth, Vt., July 2010.

Season extension techniques and technologies have had a profound impact on local food systems, and not just in northern states. In any climate with freezing winter temperatures, season extension can help maximize production and minimize the threat of crop losses on the shoulder seasons of winter. Extending the growing season also heightens the diversity of products that local farms can offer, an important advance for building alliances with customers who have come to expect a continuous exotic cornucopia from a globalized food system.

Season extension structures range from elementary to sophisticated, and the price tag runs from zilch to a fistful of zeros. A quick overview of the basic structures should help introduce the possibilities, for both home and market gardeners.

- *Row covers*: A row cover is a lightweight fabric that can be placed over plants in order to provide extra protection from freezing temperatures. Some plants can tolerate having the row cover simply tossed over the row, directly on top of the plants. However, in order to prevent damage to the plants, as well as to buffer the plants with a bit of insulating airspace, wire or plastic hoops are often placed over each row, so that the row cover looks like a giant caterpillar taking over the garden.
- *Cold frames*: As soon as glass became affordable, cold frames became popular in chillier climates. A cold frame is essentially a rectangular box with a low front, a higher back, and a translucent covering (glass, plastic, an old window, etc.), which you set over plants growing in the ground. The frame is aligned to give the sloped covering a southern exposure in order to collect heat during the day and offer protection at night. Greens, root crops, and seedlings can all be grown in cold frames. An insulating cover of some sort can be thrown over a cold frame on particularly chilly nights for extra protection against freezing. Some gardeners use cold frames as their first greenhouses or as places to harden off tender seedlings that need to transition from an indoors location to being planted out in the garden.
- *Low tunnels*: Definitions will vary, but for clarity's sake, a low tunnel can be thought of as a small hoophouse built at or just above head

height. A number of crops can be planted directly into the ground in a low tunnel, and it can even be used in combination with row covers to maximize insulating capacity.

- *High tunnels*: A semipermanent greenhouse, a high tunnel tends to maximize working and growing space with relatively high ceilings and interior infrastructure (usually metal tubing) that can accommodate efficient trellising systems for vining plants such as tomatoes and cucumbers. End walls are often rigid, and outside perimeters tend to be insulated belowground. A high tunnel can be used for seedling production or in-ground planting—or both, with a transition from seedlings to in-ground planting in late spring. It may be heated or unheated, and it can be in hoop form or feature vertical sidewalls with a Gothic arch (the latter is my favorite, due to its strength, ergonomic efficiencies, and maximum growing capacity).
- *Greenhouses*: Greenhouses are relatively permanent structures, either freestanding or attached to buildings. Greenhouses can serve both horticultural and passive solar collection functions (sometimes to a fault, so be aware that design matters) if attached to the south face of a house, or they can serve large-scale commercial functions. Greenhouses are used more for plants in containers than in-ground plantings, but create a rule and a farmer will find a way to break it! In colder climates, some sort of backup heat is usually part of the greenhouse system.

The beauty of season extension structures lies in their versatility. Designs can be tweaked relatively easily to meet different needs, and two methods (such as row covers and high tunnels) can be combined. Some of them are also mobile via rail systems—as Eliot Coleman has long demonstrated in his books on winter growing—while others can be taken apart and moved to other sites with relative ease. They can also be used in combination with renewable energy systems, and some creative farmers have found inventive ways to use them in combination with heat captured from decomposing compost. At our college farm, we have two high tunnels—one with root-zone heating from four solar thermal (solar hot water) panels and one without any such root-zone heating. There are clear differences in production between the two high tunnels, although we need several more years of data to determine whether the additional root-zone heating infrastructure provides farmers with a sufficient return on investment to make it cost-effective in a relatively short period of time.

Season extension structures can also help control moisture levels, wetting of foliage, and soil splash, minimizing diseases in fruits and vegetables. Growers are using high tunnels for strawberries, raspberries, blackberries, sweet cherries, and even peaches, praising the ability to control climatic conditions and also minimize predation by birds and some insects. With such structures, pick-your-own operations can offer customers rainy-day entertainment, and on-farm education possibilities are greatly enhanced.

Of course, it is important to consider the wonders of season extension from energy and life-cycle analysis perspectives, since the structures are often petroleum-based. However, it is also worth bearing in mind that what seems questionable in a relatively stable economic climate may be understood as incredible foresight in extremely turbulent times. While ecological perfection may elude us, the pioneering efforts of gardeners and farmers working on season extension technologies will undoubtedly serve us well.

There is one important final point to make here, and it is not my own. Rather, it comes from one of the most savvy and benevolent farmers I know, Greg Cox of Boardman Hill Farm in West Rutland, Vermont. In the midst of one of the best lectures I have ever heard inside a barn—with his pigs grunting with contentment in their dry sawdust bedding and a Jersey steer curiously nosing Greg's neck from the stall behind him—Greg observed that winter markets are one of the best ways to build local agricultural economies. "With winter markets, you don't have to get new customers to come close to doubling your income," he noted. "You just keep the same customers coming to you for the other half of the year. And then, once other people realize that year-round shopping is possible at your farm store or the farmers' market, they start buying local, too. It's a win-win that builds the local economy faster than we could have imagined a decade ago."

Perhaps we should raise a glass to "the other greenhouse effect." But we'd best not get too carried away . . .

Energy Drinks

Alas, approximately one-fifth of the energy that goes into our diets is linked to beverages, in large part due to the energy costs associated with the distribution of these heavy and often refrigerated goods.[11] Of course, the barely tapped resource in the situation is obviously tap water. Nonetheless, I don't think it's likely that our society will make any dramatic transitions back to drinking nothing more than plain water. In many ways, our beverage consumption comprises a set of rituals that frame our days and nights; they tantalize the palate while drawing us together in conversation or seducing us into meditative silence. We may not do much cooking as a culture; we probably do even less making of our own beverages. Perhaps that's why the beverage industry is one of the most consolidated and concentrated parts of our food system. But that's the glass-half-empty version of the story.

Here's the glass-half-full pitch: There are ways to minimize the negative impacts of our beverage choices—on our bodies and the environment. It turns out that the most effective ways of confronting the whole beverage quandary start at home—and the home-based solutions tend to yield healthier and tastier options than beverages purchased elsewhere. In addition, these home-brewed concoctions often have direct links to local food systems.

Home is where the creative juices start flowing. It's simple to make your own juices and syrups from local fruits and vegetables, whether bought from a local farmer, grown in your own yard, or collected from around the neighborhood. One of the best gifts my wife ever got for me was a soda maker with a refillable (*not* disposable) CO_2 cartridge. Mostly we use it to make our own effervescent spring water—"spicy water," our two-year-old daughter calls it. But we also make our own bubbly sodas from elderberry flowers, black currants, maple syrup, and most recently (thanks to our wildcrafting friends Les and Nova) wild ginger.

Homemade wine and beer can also substantially minimize beverage transport, refrigeration, and container use. Such ventures can also link us more tightly to the local resources available, whether in the form of grapes, grains, hops, or small fruits. (Braver souls can distill that idea even further, opting to crank up Grandpa's copper kettle to make Grandma's favorite elixirs . . . and if it's not palatable, you can just slip it into your gas tank!) Developing a palate for locally available beverages is not just a way to observe the natural world and restore ritual to the seasons; it is also an opportunity to connect with neighbors and new acquaintances embarked on similar pursuits. The results may bring you together at the table . . . or underneath it.

The bottle itself has ramifications worth considering. One of the most energy-intensive aspects of beverage consumption in the United States is the way in which we package and transport beverages and ultimately dispose of the containers. Not only do beverage containers take up significant space in landfills—one ton of plastic bottles takes up approximately 7.4 cubic yards of space—but their manufacture requires an enormous amount of energy.[12] In the *Post Carbon Reader* (a 2010 publication of the Post Carbon Institute), contributors Bill Sheehan and Helen Spiegelman recount how soda and beer used to be sold in returnable bottles that went back to the ubiquitous local bottling plants. Then the interstate highway system made bottling and distribution at the regional level simpler, while also leading to the much-regretted "no deposit, no return" practice, resulting in roadside litter and the disappearance of many locally produced beverages.[13] One more instance of a throwaway future.

Landfill disposal and incineration of beverage containers are both highly problematic for a variety of reasons, whereas recycling offers numerous benefits. Recycling an aluminum container saves 95 percent of the energy required to create a new container out of the same material; recycling plastic saves 70 percent; and recycling glass saves 30 percent.[14]

Although the analysis is complex and the results vary in some ways, life-cycle analysis studies indicate that refillable beverage containers generally make more ecological sense than one-way and recycled beverage containers. This area of research is important to local food systems because refillable beverage systems function most efficiently with minimal transportation distances. Countries that mandate refillable beverage systems report very high rates of landfill diversion, and refilling can even bring beverage prices down.[15] Creative local businesses that utilize refilling practices can appeal to the consumer based on the uniqueness of their products, the lower sticker cost, and the commitment to environmental values.

For example, on a recent trip to Montana I was amazed at the number of growlers that I encountered—nothing to do with grizzlies, though. People had microbrewery growlers of all shapes, sizes, and designs, and they definitely embraced the concept of refilling. It was clear that Montana takes its beer seriously, and it's a clear win for farmers with the state's "Brewed from the Ground Up" campaign. Montana brewers currently use more than 5.5 million pounds of grain in their breweries, about half of which comes from Montana.[16] These local buying efforts help farmers decrease their reliance on commodity markets and give them reason to create agricultural products that stand out in the marketplace. Being able to capture a higher price based on quality and distinctiveness often spurs farmers to do an even better job at farm management, since the price difference is dependent upon the farmers' products, values, and stewardship. In addition to advancing the local ingredient push, the Montana Office of Tourism helped establish the Montana Brewery Trail, an agritourism initiative built on the region's agricultural heritage.[17] Such efforts have multiple spin-offs for a variety of businesses in the tourism sector.

Ciders are another beverage with growing sales and agritourism potential, thanks to the local food frenzy and the focus on heritage foods in recent years. More knowledge of "cider craft" and its rich historical traditions should translate into more market opportunities for farmers, as well as promising agritourism possibilities for rural communities that can benefit from tourist dollars and a resurgent agricultural community.

Due to disease pressures, certain grains such as wheat and hops have always been challenging to grow in the cooler and wetter regions of the northern United States, so farmers there had traditionally relied on cider as the fermented beverage of choice, since apples thrive in cooler conditions with reasonable amounts of precipitation. With the homogenization of the apple industry (Red Delicious, everyone?), the gradual shift in the American palate from bitters to sweets, and the regulations of the Prohibition Era, however, hard cider waned and cider varieties of apples were replaced with commercially recognized "eating apple" varieties in orchards across the country. While seeing a resurgence of interest, hard ciders are still finding their way into consumer consciousness, in part because a shift in tastes is involved. Americans have to retrain their palates—or at least their expectations—in order to appreciate the bitter, dry, sour, and tangy aspects of traditional cider. Cider-tasting rooms are one more way to bring both locals and tourists directly to commercial orchards—and not just during prime harvest season. Orchardists looking to maintain a flow of customers through much of the year can capitalize upon cider tastings and sales, in addition to other products such as jams, apple butter, and stored apples. Here again, we see that value-added products are central to working our way out of commodity markets over which a farmer has no influence. Direct sales tend to be best for the farmer's bottom line, but there are times when it is also good for a farmer to sell to a retailer.

Interestingly, the more local cidermakers enter the market, the better, at least in terms of shelf space. One of the challenges facing cidermakers is that customers are drawn to a product line when there is ample shelf space dedicated to it. If there are more ciders and increased shelf space in retail sales venues, then consumers are more likely to consider purchasing them, particularly if the ciders speak to the traditions and history of the area.

Having a density of artisan producers in a given area tends to yield more collaborative opportunities than competitive challenges. The Hudson Valley offers an excellent example of this concept. New York is the second largest apple producer among all states in the United States, and the Hudson Valley has long been referred to as the "Apple Belt," so central has the orchard tradition been to the region's identity. Nonetheless, the number of acres in apples fell by 14 percent from 2002 to 2007, and the region lost 25 percent of its apple orchards during this same period.[18]

In an effort to halt further collapse of the apple industry in the region, the Glynwood Institute for Sustainable Food and Farming coordinated a three-pronged strategy dubbed "the Apple Project." It helped establish a "cider trail" running through the Hudson River Valley watershed and a Cider Week harvest celebration. The institute also founded the Apple Exchange, a transatlantic exchange for farmers in the Hudson Valley and the innovative Le Perche region of Normandy in France, with a goal of helping farmers in both regions solidify entrepreneurial endeavors steeped in the rich traditions of the past. In the words of the Apple Project's tag line, they're "saving orchards with cider."[19]

Perhaps you could claim that this focus on beverages is too much about refined tastes, but I think it is really more about refining tastes—and opening up the door of agricultural possibilities. One of the most exciting things about the momentum behind local foods is that it is forcing us to expand our palates and our sense of possibility. And it compels us to be more imaginative, even if by means of constriction. To illustrate, think about Vermont. What beverages come to mind? Probably a few microbrews, since Vermont has the most microbreweries per capita of any state in the country. Maybe hard cider. Perhaps Green Mountain Coffee, even. But not wine.

For the better part of a decade, I've watched two alums from our college embark on an intriguing business model that dares to be different. And so far, it's working quite well. Upon graduation, the appropriately last-named Andy Farmer was looking for a niche to get into farming. As he tells the tale, he knew that he didn't have the resources to compete with existing vegetable and dairy farmers in the region, so he felt like he had to find something no one else in the area was doing—at least not in the way that he thought it could be done. He wanted to start a nursery selling cold-tolerant grapevines to growers in the northern United States. So he went to Quebec—true wine lovers *north* of Vermont—to learn from the growers there, and then to the Upper Midwest in pursuit of a legend named Elmer Swenson. Swenson was a pioneering farmer who, from about the 1940s onward, had a passion for breeding grape varieties that would survive on his farm in the fierce Wisconsin winters, where temperatures could drop below −30 degrees Fahrenheit. The trick was to find hardy varieties that would produce enticing fruit and quality wines. Swenson's foresight set the stage for innovators like Andy and his wife, India.

After years of probing fruit specialists and vintners, Andy and India founded Northeastern Vine Supply, Inc., and they now sell thousands of vines every year. They not only found a niche, but they also developed a much-respected expertise. They are helping growers establish vineyards in places many people imagined would never have any grape growers, much less wine cellars. There is a trick in all of it, and this is the key point: the wines ain't gonna taste just like Merlot or Chardonnay. We can have good-quality wines in northern climes, but they will be different. They're *supposed* to be different—that's the whole idea of local!

I suspect that there's as much room for innovation in the beverages sector as in any other part of the food system, in part because the beverage sector was so easily and rapidly concentrated. Truly locally sourced beverages disappeared rapidly in the latter decades of the twentieth century. Not that "local" generally made soda pop or cheap wine healthy or necessarily desirable. But on a hot, dry day in a high mountain desert area of Oregon not long ago, I was delighted to be offered a complimentary local Hot-lips-brand soda from a galvanized tub filled with ice and brown bottles. Already enamored with the astounding abundance of berries in the Pacific Northwest, I chose the boysenberry soda.[20] "If you haven't had one before, just be ready for the pulp in the bottom," the vendor warned. I held the dark bottle up to the light. The pulp filled the lower portion of the bottle. I looked at the label. No corn syrup. No artificial flavors. I'd heard that Hot Lips purchased all locally grown fruits for its soda business—a side business for its pizza restaurants. I celebrated by buying a bottle of soda for my kids, for the first time ever.

Some of the beverage "innovations" aren't all that new, however: home delivery of milk in glass bottles, for example. Dairies around the country are reinstituting (and, in just a few cases, maintaining) the virtually lost practice of milk delivery into a cooler just outside the customer's door. The milk is delivered fresh in glass bottles—a boon for people like myself who believe that milk in glass containers tastes better than milk from cartons and plastic jugs. The glass bottles from the previous delivery are picked up, cleaned, and reused. Longmont Dairy in Colorado estimates that its glass bottle delivery service has avoided the manufacture and disposal of twenty-eight million plastic jugs.[21] Crescent Ridge Dairy in Massachusetts never stopped with its glass-container milk delivery since beginning the service in 1932. Not stuck in the past, however, it now delivers not only other grocery supplies but also a weekly CSA share from Ward's Berry Farm.[22]

Got glass?

No Small Matter

Got milk . . . jug? A bunch of them? Got a balcony? Container gardening, home canning, prudent shopping, leasing an urban garden plot, patronizing farms you want to invest in—there are multiple ways virtually anyone can get involved in the local food scene on a personal level. If you have the inclination, grab a few seeds and plant them somewhere—in a window box, on a terrace, even in an empty space somewhere in the neighborhood, guerilla-gardening style. Many of us got excited by the "square foot gardening" approach that was so popular a few decades ago. Now we have downsized our demands and scaled up our expectations with "square inch gardening," a topic R. J. Ruppenthal covers well in his book, *Fresh Food from Small Spaces*.

Shopping as a thoughtful citizen as well as a savvy consumer also matters. Such an axiom need not be considered elitist. In one of my wife's favorite blogs, on the Cook for Good website (http://www.cookforgood.com), Linda Watson of Raleigh, North Carolina, demonstrates how to cook healthily and inexpensively for less than $5/day/person. Linda also wrote an excellent book titled *Wildly Affordable Organic*, documenting how a family can pursue a fresh and healthy vegetarian or vegan diet while remaining within the financial constraints of several different food-assistance programs. She documents her own expenses each year and prices each meal in two ways: the "green plan" utilizes organic foods, while the "thrifty plan" opts for cost, health, and freshness but without the focus on organic foods. For example, she notes that a person following her "green plan" (organic) could save $2.08 per day, or a total of $760 per year, in comparison with the North Carolina food stamp allowance. She notes that a person could save $6,500 per year in comparison to using the USDA SNAP benefits with her "green plan"—or $9,000 per year using her "thrifty plan" (nonorganic).[23]

In the end, where you live may not always matter as much as you might think when it comes to access to fresh food (while for some it matters more than people like myself can even imagine). I'm reminded of a conversation with a former student and now close friend, Kris Jacoby-Stevenson, in the local co-op one day. Kris seriously surpasses my family in food preservation and tight local ethics—her kids rarely get bananas, and she is a canning and freezing machine. Nonetheless, she was comparing her own food preservation efforts on her farm to those of her sister, who lives in Detroit. "I walk into her apartment in the middle of Detroit, and she has food preserved in every way imaginable stashed in every corner!" Kris exclaimed. It just goes to show, when you're in the right mind-set, what grows around, comes around!

Chapter 12

Collaborative Possibilities

Large-scale change seldom requires unity, but it does require collaboration. Rebuilding local food systems may begin at home, but in many ways our individual efforts are just the starting point of the work. This chapter provides some methods and models for the transformation of local, community-based food systems. Finding pathways to more sustainable agricultural systems is complex enough, but creating resilient community-based food systems is perhaps even more challenging, since the social aspects become an integrated aspect of the broader ecosystem considerations. While sustainable agriculture involves the careful *conservation of resources*, the building of a resilient community-based food system involves *utilizing as many resources as possible:* farmers, entrepreneurs, social justice advocates, and technologies old and new.

The ingredients of success in rebuilding local food systems may be no great secret, but the recipes can be rather intricate. Cooking from scratch is fun; starting from scratch is less desirable. It pays to know what to add, when to add it, approximately how much to toss in, and how long to expect to cook a concoction to avoid getting a half-baked idea that no one wants to sample, much less invest in.

Most any recipe is an intentional conspiracy of available flavors, whether local in origin or sourced from afar, whether keeping with convention or fingering the cutting edge. Some of the recipes for re-envisioning and reconstructing local food systems that follow are time-tested, while others are fermenting wildly, with still uncertain results. In the end, the history of cuisine mirrors the future of local food systems. Both involve time-honored traditions, unexpected combinations, and a sense of who is coming to the table . . . as well as an empty seat or two for unanticipated guests.

Mapping the Foodshed

The concept of a "foodshed" might seem relatively novel, as might our contemporary concern for feeding urban populations in the face of increasing ecological and economic challenges. Yet it was back in 1929 that Walter Hedden, chief of the Bureau of Commerce for the Port of New York Authority, proposed the term in his book *How Great Cities Are Fed*. Hedden compared a foodshed to a watershed, noting that "the barriers which deflect raindrops into one river basin rather than into another are natural land elevations, while the barriers which guide and control movements of foodstuffs are more often economic than physical."[1]

The concept reappeared in 1991, revived by Arthur Getz in an article titled "Urban Foodsheds" in *Permaculture Activist* magazine. Impressed by how urban populations had begun to embrace the importance of protecting the watersheds that fed their cities (often from well outside their urban boundaries), Getz sought to revitalize the concept of foodsheds. Many others since have worked to refine the concept in hopes that foodshed awareness would lead to both stewardship of foodshed resources and investment in maintaining surrounding farms, markets, and infrastructure. The increasing sense of vulnerability in urban settings regarding food supplies in the case of some major disruption has helped to further interest in defining and protecting foodshed resources, while the locavore frenzy has sped up layperson applications of foodshed mapping projects.[2]

Imaginative yet important, the concept of foodsheds allows us the opportunity to explore and to tinker. While it's easy enough to choose a centerpoint for a foodshed, establishing boundaries—no matter how porous—proves to be a challenge. Foodshed edges may be based on distance, but landscape features, current agricultural productivity, future agricultural potential, demographics, markets, and political boundaries also fit into the calculations. The intellectual exercise of delineating your own foodshed is important for building awareness of the possibilities and the limitations, but the collaborative experience of having a diverse group of people discussing how to define a particular foodshed can be much more illuminating and instructive. If community self-reliance and resilience are the drivers of such an exercise, then a comprehensive database of local food resources becomes all the more critical. And you can't gather data until you establish some basic parameters like boundaries, no matter how porous or playful.

Foodshed maps are appearing in three different forms these days, and each has its merits. Some regions and states are developing complex geographic information system (GIS) maps that show layers of different kinds of information related to foodsheds. Cornell University, for example, has developed an interesting map of New York State that allows you to view a viable foodshed for each of the state's cities, represented by the surrounding area's ability to produce a balanced diet that can provide its citizens with the needed human nutrition equivalents (HNE). According to this mapping project, given optimal agricultural production and significantly lower consumption of some animal-based protein (30 percent of the current average intake of meat and eggs), most upstate New York cities and towns could potentially feed themselves using primarily the resources found in their local foodsheds. As you might expect, the study came to the unsurprising conclusion that New York City, in contrast, cannot function on a local-food-system basis.[3]

While the data provided by GIS maps is excellent for assessing a community's food system resources, the technology is not accessible to everyone, and there are other approaches to building a database of resources. Google Maps allows a more participatory approach to creating a map and a database, with contributors adding the locations and information about local farms, distributors, restaurants, and related businesses. Foodshed projects using Google Maps typically approach the foodshed concept from a city, regional, or state level, although a university, public school, or hospital may also choose to define its own foodshed using this relatively simple technology.

The complexity of a foodshed analysis can be interpreted imaginatively but skillfully into maps that speak to a variety of audiences. I have yet to see a more compelling collection of foodshed maps than those produced by the Dreaming New Mexico project. Cartoonlike in appearance, these maps nonetheless convey a number of ways to think seriously about New Mexico foodsheds at all different levels. Community planning involves ample doses of imagination and inspiration, and I'm willing to bet that New Mexico is mapping out new realities that leap off the map and seep into the future. Even those of us who don't like to ask for directions (I hear that's about half of us?) may nonetheless find ourselves relying on maps.[4]

Food Policy Councils

You saw the word "policy" and almost jumped to the next section. Wait, please! Instead of skipping this section, consider this topic a page-turner. There's some drama here, or at least a good dose of irony. We have councils and commissions at local and state levels that deal with the essential human needs of housing, water, air quality, and land conservation, but somehow food security issues usually get left out of the equation. Where's the sanctioned or self-appointed body ready to represent the interests of building a robust and resilient regional food system? Food policy councils, now in place in a number of communities across the country, are an attempt to redress this common (and somewhat bizarre) oversight. Think of these groups as the big tent for food system issues at the local or regional scale.

"Food democracy" at the grassroots level—this is it! Food policy councils have been around since the 1970s, but they are currently more popular than ever. At their most basic, food policy councils are ideally composed of representatives from every sector of the food system, as well as food security advocates, educators, health professionals, natural resource conservationists, and others. Food policy councils generally focus on creative solutions to maintaining vibrant food systems, with an eye on food access and security, nutrition, social justice, economic development, and environmental degradation. In other words, food policy councils have both a reactive and a proactive role: identify the challenges and propose innovation.

Food policy councils can be formed at the local, county, state, or regional level. They can be appointed by a government body or simply convened and designated as an independent body that will serve the needs of the community without governmental oversight, although governmental interaction and influence may be primary goals. The authors of the highly informative 2009 report *Food Policy Councils: Lessons Learned* by the nonprofit group Food First categorize the typical functions of such councils as follows:

- To provide a forum for bringing forward food issues
- To enhance coordination among different parts of the food system
- To review and guide policy

- To create or support important programs and services geared to local communities[5]

It is worth noting that these councils tend to vary not only in form but also in longevity. While some have endured a decade or longer, others have had shorter life spans. Structure, focus, and resources are therefore important considerations in forming a food policy council, and any group thinking about beginning such an initiative should review Food First's report, mentioned above, along with similar studies that analyze best practices of these organizations and how they sustain themselves in a dynamic political and economic environment.

Implementing a food policy council at the municipal, regional, or state level can be challenging. I recently spoke to a colleague working on developing a food policy council in a highly diverse metropolitan area suffering from extreme issues tied to poverty, obesity, and a long history of racial biases. The development of that food policy council at the city level was under way, but at a snail's pace. When I asked what the problem seemed to be, the response was that the members—even though they were elected and represented a variety of perspectives from within the regional food system—said that they did not feel as if they sufficiently understood the components or the functioning of a food system. Another challenge was that the metropolitan area was highly fractured due to the intricacy of the internal political boundaries and the ensuing difficulty in finding authority or consensus. The final difficulty—and the one that may make or break the ultimate success and longevity of the project—is the lack of an official home for the council. "It needs to be in the mayor's office," my colleague emphasized. "It has to have resources, power, and a way of getting the word out."

While that assessment of one ongoing effort may be a bit sobering, the good news is that the food policy council emerging at the state level in this area is really beginning to gain momentum. As we concluded our phone conversation, my colleague left me with a final irony. Despite all of the efforts that her region had pursued through formal channels, the state-level group came together informally, and not by means of nominations, vetting, and voting. "The group is built on passion as much as anything," she observed. "We're all people working to make a change, and we recognize the value in collaborating to achieve what is so desperately needed."

Given their grassroots and place-based nature, no sampling of food policy councils will tell the full story, but a few examples can help stir the imagination and perhaps inspire similar efforts elsewhere. As one of the veterans in the bunch, the City of Hartford Advisory Commission on Food Policy in Connecticut was formed in 1991 to integrate food security efforts coming through government programs and NGO groups. Mark Winne, one of the most important and experienced thinkers in the food policy council realm, was at the helm of this project through his role as executive director of the Hartford Food System, a nonprofit group working on food and hunger issues in the area. The commission, referred to as "the Department of Food," is engaged in work such as ensuring that farmers' markets are present in all parts of the city, banning trans fats in the city's food service establishments, lobbying for adequate public transportation to grocery stores from all areas of the city, and providing city residents with the results of a biannual grocery-store survey indicating best average prices for consumers.[6]

The Oakland Food Policy Council in California is a newer entity, officially formed in 2007, and the twenty-one-member council and its supporters have helped propel the city forward as a national leader in local food policy work.[7] The council's emphasis on promoting food justice has helped make Oakland a model for engaging a highly diverse citizenry in transforming the local system. The council has recently attracted attention for its policy work to open up possibilities for urban agriculture. It's been working hard to promote both agricultural production and the sale of fresh agricultural products in designated zones. One intriguing result has been the proposed zoning definitions of three types of urban agriculture: residential, civic, and commercial.[8] These zoning delineations help clarify expectations for urban farmers and their neighbors, while also allowing municipalities to plan constructively for the future. It is precisely this kind of creative problem solving that should move things forward—all the more reason to see what others are doing out there!

Cleveland, Ohio, is yet another city to watch, and the food policy work going on there and in surrounding Cuyahoga County is a model to mimic and borrow from. Interestingly, this work is occurring through a coalition, not a council, since it is not administered through a specific governmental organization. Faced with the irony of urban decline in the midst of a county that is simultaneously losing farmland to suburban sprawl at an unprecedented rate, the Cleveland–Cuyahoga County Food Policy Coalition has five working groups focused on the following areas: community food assessment, food waste recovery, health and nutrition, land use and planning, and local purchasing.[9]

Since the coalition's inception in 2007, the participants have been careful to articulate their mission and goals and to document their results. This coalition's work is somewhat unusual in its clear recognition of the fact that coordinated work must be done at several scales: city, county, and regional. One clear advantage of this approach is the way in which the needs and resources of rural and urban communities are examined together, with collaborative strategies continually unfolding.

The Northeast Ohio Local Food Assessment and Plan is one outcome of this kind of systems-based thinking. The goal of this major initiative is to develop an economic development strategy for the entire region that is built upon local food production, processing, and distribution.[10] Michael Shuman (author of *Local Dollars, Local Sense*, another book in the Community Resilience Guide series to which this book belongs), Leslie Schaller, and Brad Masi researched and authored a study that proposes a target of 25 percent local food consumption in the region, accompanied by proposals for achieving this ambitious goal. According to the study, such a move would create approximately 27,000 new jobs, nearly $1 billion in new wages, and potentially $4 billion in new economic outputs.[11]

Getting to such a point would take considerable strategy and coordination, and the study recommends the creation of a Northeast Ohio Food Authority (NFA). More comprehensive in scope than the Cleveland–Cuyahoga County Food Policy Coalition, the NFA could also be created as an entity other than a nonprofit, thereby liberating it to participate in lobbying efforts, business operations, and financing scenarios. The NFA would utilize bonds—ideally tax-exempt bonds purchased by local residents—to finance local food businesses in the region, businesses determined to have the greatest potential for wide-reaching positive economic impacts. The NFA would have the ability to coordinate efforts within the fifteen-county region, and it could provide loans for local food businesses while also leveraging other means of support for these local enterprises.[12]

By now, several themes should be emerging regarding the relationship between local food advocacy and policy. The first is that resource assessment (i.e., mapping the foodshed and the resources in it) is the initial stage in moving a community forward, whether the focus is on a town or a region; only then can the necessary creativity come take root in a way that safely bridges reality with potential. The second theme is that "business as usual" has created a highly centralized, unhealthy food system and is not the ticket home. Food policy councils and the ideas that spin out from them can comprise various components of the community-owned vehicle: the brakes, the steering wheel, and the engine.

Sometimes, these councils can go beyond the regional level. Some states have formed their own food policy councils, and for an example of how a state can utilize the concept to counter the riptide of mainstream economics, you need go no further than Hawaii. Hawaii currently produces only about 8 to 10 percent of the food that its citizens consume—a rather odd reality considering its climate and natural biodiversity. However, this reality is not such a surprise considering that the last two centuries of Hawaiian agriculture have been essentially a history of raising a handful of crops for export: pineapples, sugarcane, coffee, ginger, and other tropical items.[13] If you think you feel vulnerable in the continental forty-eight, imagine what a toll such isolation from 90 percent of your food supply might have on your psyche in Hawaii!

The Hawaii Food Policy Council is in its early stages of advocacy, but it is pushing for resource assessment through various bills related to increasing self-reliance in the Hawaiian food system. According to the council, the state does not seem to have adequate data on food production in Hawaii, so one of the initial pushes is to provide funding for a resource assessment that can support the development of appropriate future strategies. It will be fascinating to watch how this unique state works toward increased resilience in its food system. There is a rich agricultural heritage for Hawaiians to tap into, but there are also challenging economic forces at play. While islands create complications for self-reliance, they also provide the opportunity to carefully quantify imports and exports. Most any disadvantage has a flip side, and the agility of local food systems can capitalize upon such advantages.

Enough policy talk . . . *E 'ai kākou!* Let's eat!

Urban Ag: Getting in the Zone

Unfortunately, the travails of farming an island aren't limited just to places surrounded by water. Farms are becoming islands in a sea of development across the U.S. mainland. It's a truism among planners as to why this is the case: prime agricultural soils are generally found in locations best suited to building and commerce. Cities and villages were originally sited in low-lying locations with rich alluvial soils and an ample supply of water for crops, livestock, milling, and transport. As populations and infrastructure continue to spread outward from these centers of development, farms become surrounded not just by buildings and roads but also by market pressures. They become islands prone to total submersion in the tide of sprawl.

At least, that's how I would have framed the issue of land access and conservation had I written this text a decade ago. Now, however, it's a more complex and, arguably, richer situation. Or perhaps we as a culture just needed to get beyond equating "agriculture" with "rural" and start expanding our visions of agricultural possibilities. It is no longer just the large expanse of fertile fields on the outskirts of our cities and towns that need to be reserved for agricultural use. It's also the vacant lots, underutilized city lots, sidewalk edges, and rooftops in and around our cities that warrant our "undivided" attention. We've divided, built, paved, and limited access to future agricultural land with no obvious forethought—consensus by default.

And yet it all feels more hopeful than grim right now. The future of urban agriculture has the feel of a modern Green Rush. Prospectors are running to vacant lots with pickaxes and wheelbarrows, tweeting their successes to the masses. The tougher the terrain, the more promising the prospects, it seems. Staking a claim, however, is rife with conflict. That's where zoning, tenure, and ownership enter into the equation.

Zoning ordinances for agricultural production and commercial sales can both advance and stymie the growth of local food systems in urban areas, and it's an arena where reason and strong voices are often needed. Zoning in and of itself is not necessarily a detriment to urban agricultural endeavors; in fact, thoughtful zoning ordinances can lay the groundwork for successful urban agriculture initiatives in a serious and strategic manner. It can also minimize the potential for friction between neighbors. If not well managed, certain agricultural endeavors can generate odors, attract unwanted pests, introduce noise issues, and even create unsightly messes. In order to ensure a set of agreed-upon ground rules and expectations, some cities like Seattle and Madison have established farm management plans for urban farming operations of certain scales and types.[14] While these requirements add a layer of bureaucracy, they can also help minimize conflict and thereby further the cause. Good farmers plan methodically anyway, and transparency in these plans can help build confidence and create strong allies.

On the other hand, zoning can be an enormous impediment to the forward momentum of urban agriculture. Whereas some cities have virtually no zoning regulations regarding urban farming, other cities have zoning ordinances that make it virtually impossible. Nonetheless, models of accommodating and even proactive ordinances are out there. Recent publications such as *Urban Agriculture: A Sixteen City Survey of Urban Agriculture Practices across the Country* and *Seeding the City: Land Use Policies to Promote Urban Agriculture* are excellent resources available for free online.[15] The American Planning Association's 2011 report *Urban Agriculture: Growing Healthy, Sustainable Places* offers not only zoning and planning considerations but also a superb history of agriculture in American cities.[16] And UrbanAgLaw (http://www.urbanaglaw.com) is well worth mentioning again as a website that is continually updating its resources in an effort to help move these ideas into reality, without undue legal snafus.

In an economy in which competition for economic development is stiffer than in any recent period, no city wants to be behind the times. After all, cities are constantly vying for tourists, new residents, business opportunities, and government funding. Savvy advocates can try leading city leaders to these concepts, or, if all else fails, shaming them right out of office.

Sometimes the politicians really get it. Such seems to be the case with Atlanta, where the mayor's Office of Sustainability offered a $25,000 prize to the winning design for a 0.8-acre urban garden right across from city hall. (Yes, Walmart is offering the prize—it obviously doesn't feel threatened yet by the urban agriculture trend.) Dubbed the Trinity Avenue Farm Design Competition, this strategy has at the very least provided some prominence to the potential of urban agriculture in Atlanta and elsewhere.[17] Beacons are important, as long as they are not simple, isolated tokens. Fortunately, Atlanta is working to overhaul its zoning ordinances in order to actively encourage increased agricultural activity within the city, even setting forth a goal of making local food available to 75 percent of its residents by 2020.[18] "Hotlanta" seems to be getting hotter.

Richmond, Virginia, is another place where the local government has apparently been paying attention. The city's Green Richmond Initiative is providing public land for both community and commercial gardening initiatives. The city maintains an online map that identifies existing operations of both types and potential sites for such use. The map also links to an online application with which one can apply for a garden site at any of the given locations.[19]

Zoning ordinances that will encourage urban agriculture endeavors aren't likely to arise in a bureaucratic vacuum. Local citizens interested in seeing these projects move forward need to weigh in not just with well-grounded opinions but also with information gleaned from other parts of the country. One of the first tasks for a community is to define the different types of agricultural activity, since definitions are the backbone of any zoning regulations. Variations abound, although the categories that emerge from most planning processes seem to orbit around definitions of home gardens, community gardens, and commercial farms. Planning efforts must also take into consideration zoning ordinances regulating commercial sales. For example, areas zoned as residential may prohibit sales activities, including both "on-farm" and farmers' market sales.

Once again, GIS mapping technologies can play an important role in the planning process, in part because they can reveal some demographic trends that might otherwise be missed, but also due to their capacity to project future trends through engaging visual presentations. One of the best is a recent Columbia University report titled *The Potential for Urban Agriculture in New York City: Growing Capacity, Food Security, and Green Infrastructure.*[20] The maps in this study are fascinating, particularly in chapter 10, "Case Study Neighborhoods." These maps analyze production and marketing potential by highlighting public and private roof space, public and private vacant land, community gardens, unused open space, and farmers' markets. This kind of data allows for enlightened public input into the planning process, and it provides potential entrepreneurs with the opportunity to make wise business decisions for their farm-based businesses.

In addition to zoning ordinances, municipalities can adopt other policies that will pave the way—oops, wrong metaphor—*remove some of the barriers* to increasing urban agriculture. For example, local governments can impose penalties on development that creates new impervious surfaces. Agricultural operations, in contrast, capitalize upon the soil's absorptive capacity. As a result, citizens, planners, and politicians start to see the added benefits of gardening and farming operations over other possible uses and start to encourage such endeavors.

Other policies that can support these green-thumb ventures are also related to water. Farming and gardening tend to require a fair amount of water, so reducing water fees for agricultural use can be a boon to urban agricultural producers, as can incentives to create rooftop rainwater catchment systems. Finally, providing landowners with tax breaks for keeping land in agricultural use instead of other potentially more lucrative and infrastructure-heavy options can forestall development and provide a neighborhood with a diverse and educational landscape.[21]

Tax incentives aren't always sufficient to convince landowners not to develop their land, however. Long-term leases to gardeners and farmers can be effective, on both private and public lands, but there are times when having a clear title to a piece of property can make more sense. As cities face abandoned properties and delinquent taxes, they can actually open up possibilities for urban agriculture through a land bank. The land bank is typically a government entity that secures a property on which back taxes are owed, and it then puts a clear title on the property for legal transfer to another individual or organization. The price for such properties is often minimal, and this strategy offers ways to get properties back on the tax rolls and under proper management. The increase in the number of foreclosures in recent years, combined with the surge of interest in food security and urban agriculture, has inspired creative thinking among policymakers in some beleaguered communities.

This kind of strategy has been employed in Cleveland and Columbus, Ohio, as well as parts of Michigan, and it seems to be particularly helpful for securing land for innovative agricultural and green-space projects.[22] As well as having one of the oldest land banks in the United States, Cleveland can also lay claim to having one of the most progressive approaches to utilizing its land bank program. Not only are the eight thousand properties in the land bank listed in a database and linked to GIS parcel maps for the public to review, but the city also collaborated with Kent State University and several community organizations to create an outstanding guide for determining the best use of vacant lots, of which Cleveland has about twenty thousand.[23] *Re-imagining Cleveland: Ideas to Action Resource Book* is available online, and it features an assessment tool for reviewing a property, options for vacant lots, available resources, and vignettes and budgets of successful projects in the area. Sprinkled with engaging graphics and a full spectrum of possibilities—ranging from vineyards to pocket parks to sideyard expansions to rain gardens to market gardens—this publication is sophisticated in its approach but simple and inspiring to read.[24]

Land Access and Conservation

So far, we've been considering ways that urban communities can increase access to land and the conservation of open land. As we look at the next two approaches—land trusts and conservation easements—the strategies for urban and rural environments begin to converge. Land trusts and conservation easements are geared not only toward getting land into the hands of those who will farm it but also to ensuring that the land remains in agricultural production through a series of owners and into perpetuity.

Land trusts are formed at community, regional, state, and national levels. As nonprofit entities, these organizations provide critical functions in ensuring the future of farming, whether through outright purchase of properties or through the purchase of conservation easements. These organizations are often founded at the local or state level simply because that's where the desire and need for specific farmland conservation efforts are ordinarily best understood. Furthermore, the conservation of farmland requires trust and cooperation between the land trust and potential sellers and conservation-minded individuals, so relationships and connections matter.

Across the country, many community garden projects are threatened with the unexpected loss of the land that they have worked so hard to transform. Cities can change their mind about properties overnight—and indeed, all it might take is one evening vote of unsympathetic or unwitting officials. Private landowners can choose to capitalize on the good work done by a community garden project in improving a property's value and even its surrounding neighborhood and decide to sell it or develop it without consultation or input. In Philadelphia, recognizing these challenges, individuals representing the Pennsylvania Horticultural Society, the Penn State Urban Gardening Program, local businesses, and community gardens created an organization called the Neighborhood Gardens Association / A Philadelphia Land Trust (NGA) back in 1986. NGA worked carefully with various community garden projects to determine needs and strategies. As a result, this unique land trust now holds title to twenty-four properties in the Philadelphia area, ranging in size from individual house lots to a larger 3.8-acre area. In essence, the NGA holds the titles on behalf of the citizens of Philadelphia. In addition, the NGA helps a variety of local community garden initiatives in managing legal, taxation, and insurance issues "so that gardeners can do what they do best—garden."[25]

Translate this kind of work into a rural setting, and we land deep in the heart of Texas—and *big* hearts are at work here, albeit not without some fancy local rope work. Not all bragging rights are the best claim to fame. Unfortunately, Texas is unrivalled among states in its loss of farmland: in the first decade of the new millennium, it lost nearly 2.1 million acres to various forms of development. Maybe there's still a lot of land to spare in a place the size of Texas, but keep in mind that it's the best agricultural land that tends to get developed first.

And remember that quip about trust and local connections being such an important part of land conservation? Well, when American Farmland

Trust (AFT) officials first came to town—or the statehouse, at least—they weren't warmly received by Texans, not even by a lot of farmers and ranchers. Unsuccessful in their early efforts to promote land conservation in Texas, AFT got smart and hired a local Texas rancher by the name of Blair Fitzsimons to move the initiative forward, and she carefully herded the land trust concept through the Texas legislature. The current tagline for the Texas Agricultural Land Trust is no accident: "Created by landowners for landowners, TALT's mission is to protect private lands, thus conserving Texas' heritage of wide open spaces."[26]

One of the most important tools in the land conservation toolbox is the concept of a conservation easement. The easement typically moves with the title when ownership of the land is transferred, guaranteeing its enforcement either into perpetuity or until an established date. An agricultural conservation easement can take various forms, but in essence it is a legally binding covenant that protects the agricultural viability of a parcel of land. Easements can be purchased or donated, and they can be "positive" or "negative." A positive easement conveys certain rights to the holder of the title, such as the right to water or to a particular activity. A negative easement restricts certain activities on the property—perhaps gravel extraction or the construction of nonagricultural buildings.

Another important tool utilized in the conservation of agricultural land also can help farmers stay in business or get into business. The purchase of the development rights of a piece of agricultural property is intended to ensure that a piece of property is not converted to nonagricultural use. The value of the development rights is determined by establishing the difference between a property's "highest and best use" (a phrase that makes a fairly mellow person like myself absolutely livid, since it means the land's potential for *development*) and its agricultural value. The difference between these two appraised values is considered the appropriate purchase cost of the development rights. A farmer or rancher may sell the development rights of his or her land in order to ensure that it will not be developed and will, hopefully, remain in agricultural use. A landowner may also opt to donate the development rights. Obviously, the sale of the development rights can relieve a landowner of significant financial pressure, and there are often enticing tax incentives for landowners who sell or donate their development rights. A farmer or rancher who benefits from the sale of these development rights may choose to use the proceeds to retire after selling the property to a new landowner or to make improvements on the existing property.

After such a transaction has occurred, the subsequent landowner—often a new farmer—is usually able to purchase the property at "agricultural value," making it much more affordable while also keeping the land in what many of us would argue is its "highest and best use." (Touché! Take that, you heartless speculator!)

The ability of a new farmer to purchase a farm at agricultural value is one of the most important leverage devices in the conservation toolbox. Not only does it conserve land for agricultural use, but it is often the only way new farmers can afford to purchase a farm. As my grandfather's generation often commented, "The only thing they can't make these days is new land." If a truism can become truer, that one certainly has. If the markets are good in the Midwest, then land prices are out of reach for beginning farmers. Land in prime production areas such as California is never cheap. In high-density locations where there's no shortage of consumers, land is in high demand and out of reach of the ordinary working professional, much less a farmer.

As someone involved in sustainable agriculture education, I've seen several of our college's alumni begin their farms only because the Vermont Land Trust purchased the development rights to the land, allowing the farmers to purchase those farms at their estimated agricultural value. And once a land trust organization invests in a new farmer, it tends to protect its investment by making sure that new farmer is supported by the local community and a variety of state and federal programs. Two of those alums, James and Sarah Elworthy, established the first organic dairy in close proximity to the college, but not without a lot of hard work.

James often jokes that dairy farmers don't have money; they just have cows and independence. However, James and Sarah didn't even have much in the way of cows when they graduated in 2003—they had only four cows and a ten-year-old pickup truck. They spent the next few years building equity and were finally ready to purchase a farm in 2008, but it took two years to find the right farm and work out the financial and legal details. "Conservation was something we believed in when we were looking for a farm," said James. "We saw how hard it was to find ag land in an area that was saturated with land for housing development. It was hard to find farms that would stay farms after they were sold."[27]

The conservation of farms and gardens of all types doesn't matter much, however, if our communities aren't encouraging agriculture as a worthy educational enterprise or farming as a respectable profession. That brings us to the next arena in which citizens can support the long-term viability of local food systems—educational community gardens!

Community Gardens

The word "garden" does not immediately conjure up thoughts of complexity, politics, and long-term viability. However, beginning any kind of community garden is often more complicated than you might think. Sometimes the start-up phase is fraught with the most intense challenges, while some projects face more obstacles in providing long-term vision and support for the project. Few would argue against the multiple values of community gardens, since they obviously provide people of all ages with a direct connection to nature, nurture, and nutrition. However, not everyone is ready and willing to initiate such a project.

If it seems daunting, just think of Rosalind Brooks in Las Vegas. Yes, that city in Nevada that brings to mind an oasis of excess (although actually its name means "the meadows" or "the fertile lowlands" in Spanish!). When I reach Rosalind one summer day on her cell phone, she is hard at work at Vegas Roots Community Garden, a magnificent five-acre garden she founded in March 2010. As with any good conversation between strangers, I turn to the weather. I have to know how hot it is out there in the garden at 9 a.m. "One hundred and ten degrees," she notes, totally nonplussed.

With the full support of her husband, Rosalind had quit her job as a schoolteacher to begin Las Vegas's first community garden, on her own time and money. No grants, no fiscal angels, no financial support from the city ("They're busted," she observes, with only a hint of frustration). "When this land was offered to me to set up a community garden, I couldn't say no," she tells me. "God just led me to it. I never thought I'd be a farmer—in fact I never *wanted* to be a farmer! But here I am with five acres of desert dirt."

"Desert dirt" is a serious understatement. Working full time, Rosalind has pulled together a band of people who also never thought they would be cultivating the desert soils, and certainly not with such spectacular results. With the use of drip irrigation and careful selection of plant varieties well adapted to the desert climate, Rosalind has transformed an empty lot into a beacon of sustainability and food-centered community in Las Vegas.

Rosalind has long been into health and nutrition. "Being a teacher was a natural segue into doing this work," she says. "Now I'm just focusing my teaching on how to be green and healthy." And her educational impact is spreading far beyond a few gardening and nutrition lessons. Rosalind is actually helping change the way Vegas citizens see themselves and their community. "It's the first community garden in Las Vegas, so part of the work is just getting people used to the idea that 'Wow, I can grow my own food!' People come out and get a hold of it and then want to bring their children."

The enthusiasm spreads further as the site begins to create a sense of community. As Rosalind tells it, "No matter who you are, you want to talk to people. I mean, I used to go to parks, but nobody meets anybody in a park. But you go to visit or work with other people in a garden, and you actually *talk*!" The garden also starts to give the community a new vision. Rosalind believes that the project is clearly "a catalyst for changing the Vegas mentality," so she has been working hard to help others trying to find ways to make the city not only a more sustainable place to live but also a place with more community spirit. She is looking for ways to introduce more gardens in downtown Las Vegas in order to inspire citizens to rethink the future of the city. She is also the head of Nevada's farm-to-school program, an initiative to get more Nevada produce into the public schools.

It's not all easy, though. Besides the need to develop an income stream to support her commitment to the project, Rosalind's greatest challenge is the purchasing of soil to maintain the garden. "I can purchase soil—good soil," she says, "but it's our biggest annual expense, by far." The long-term viability of the project rests in large part on the ability to build and maintain soil fertility, and Rosalind is thinking about how to overcome this obstacle. She has a small compost pile, but given the large scale of the garden, adding a compost operation that could support it is a big project for which she has no current funds. But Rosalind will make it happen. She didn't get this far for no reason!

Resources for community gardens obviously vary from one place to another, but it generally makes sense to check with the master gardener program in your state early on, as well as reviewing the National Gardening Association website (http://www.garden.org/) for ideas and information. There are national organizations dedicated to supporting the start-up of these enterprises, including the American Community Gardening Association (http://www.communitygarden.org), which has an excellent online list of resources. It is also well worthwhile to search the Internet for other community garden projects in your region in order to grasp the parallels and the possibilities. Actually visiting several of those sites can inspire and create important connections. There's fertile ground in these collaborations!

School Gardens

What's a school without a garden? Not what it could be—or should be. However, school gardens are not always simple to start, and they are often even more challenging to maintain. But so are a library, and a gym, and a soccer field, and a music program. For some children, a school garden may be the only opportunity they ever have to see the seed-to-plate process or to taste a carrot straight out of the ground. So what does it take to start a school garden?

I recently spoke with Ramsey Cronk, who works for the Orfalea Foundation in Santa Barbara, California, as the garden installation coordinator. He has worked on the installation and improvement of approximately thirty-five school gardens, and he was invited to consult with Michelle Obama on school garden initiatives across the United States, so he has some sophisticated perspectives. Once a landscaper, Ramsey is now as much a community facilitator as he is a garden designer. School gardens are, by their nature, community-oriented projects, and it's important to have as many people invested in the project as possible.

"We'll have a hundred to a thousand show up to put in a garden in a few hours," he says. "It's a huge community workday, and the goal is to make it everybody's garden. The kids love to have an investment in the garden, and so does the community—that's one of the most important pieces." Ramsey goes on to make a distinction: "These gardens can be community gardens, not just school gardens. Then the people who come to participate in the garden are inspired and go back home to put in a garden." This kind of garden requires everyone to consider the process of installing and maintaining it, as well as what it will actually look like.

Ramsey adds that you have to know "who's coming to the table and what are they going to bring to the table. Everybody has to be invested. It's about facilitation and organizing." The schoolkids help design the garden, and they work with the community to install it. The end result is not the garden, in Ramsey's view. It's the long-term educational value of what the garden brings to the school. And he has witnessed one of the most important results: "If the kids grow it, they're more likely to eat it."

The Orfalea Foundation supports Ramsey's work, and he is quite grateful for how effective that organization has been in getting so many gardens in Santa Barbara County in prime growing condition. However, he worries about the future of school gardens, and he believes that they need to be part of our public policy so that every school can support the garden projects with targeted curricula that meet state standards, as well as appropriate staffing. "These gardens shouldn't be dependent upon outside funding and the PTA," he says. "We need to take JFK's idea of physical education and apply it to school gardens—that's been Alice Waters's perspective, and I share it. If it becomes public policy, then it will be a transformation of the way kids learn—and it can't be just standards-based, or else there's no real flexibility."

Fortunately, a growing number of people and organizations are thinking in similar ways, and the collaborative sharing is powerful. As mentioned earlier, the National Farm to School Network has ties to this work, and the Farm-Based Education Association is another excellent resource, particularly when the garden is fortunate enough to be linked to a farm operation of some sort.[28] The national FoodCorps is providing heft and vigor to the concept with its superb work in connecting AmeriCorps volunteers to schools implementing or enhancing their gardening efforts. The National Gardening Association has an innovative "Adopt a School Garden" program,[29] and certain groups that have been at this work for a long time, such as Life Lab in Santa Cruz, California, offer significant help in linking school gardens to standards-based curricula.[30]

The excitement about the burgeoning efforts across the country is shared by more and more citizens—even by people who don't have much, if any, garden experience—and Ramsey seems to channel that enthusiasm. His passion is obvious throughout the conversation, but he is perfectly clear about his primary concern in the move to build student gardens all over the United States: "What about having a gym teacher who is also a garden educator? Why can't we link the concepts of healthy food and activity—movement? But there's no public funding." That lament is delivered in a tone of not surrender but, rather, determination. There's a lot of work to do to create a nation of healthy eaters. And then the question remains, who is going to grow our food?

CHAPTER 13

Farmland Security

Forty to fifty million farmers. That's the minimum number of additional farmers the United States will need in order to deindustrialize agriculture in the face of declining fossil-fuel supplies. Such was the conclusion put forward by author and Post Carbon Institute senior fellow Richard Heinberg in a provocative address to the E. F. Schumacher Society (now the New Economics Institute) in Massachusetts in 2006.[1] Heinberg also suggested that we may well experience a "re-ruralization" of the United States in coming years, with farmers of various scales repopulating the rural landscape, oftentimes balancing farming with other professions.

Regardless of whether we concur with Heinberg's projections, most of us would agree that the current status of the farming population in the United States leaves our food system highly vulnerable to disruptions in our economy, energy supply, and transportation systems—not to mention a disturbing lack of farming expertise in the remainder of our population. It all depends upon how we count, but the 2010 census officially lists the number of U.S. farmers at 2,204,792. That's about 1 percent of the population, with the average farmer pushing sixty years of age.

Communities interested in resilience and relocalization have to move beyond farmland conservation by reframing their educational systems to encourage and prepare the farmers of the future. Leaders in education must ensure that farming is viewed as the esteemed profession and rigorous intellectual enterprise it is. Our culture made a grave error decades ago in shoving farming and home economics into a half-lit corner of our K-through-12 educational system, as if there were no future there at all. For several generations, we used education as the vehicle to get people off the farm. Now we have to make it clear that community food systems and community educational systems are inextricably linked.

The cultivation of aspiring farmers must begin much earlier in life. Given the small percentage of farmers in our society, basic exposure to gardening and farming—not to mention in-depth knowledge—has to come through school and extracurricular activities at a young age. I am always surprised at what rudimentary culinary and agricultural skills our farm staff often have to teach to our students at the college level—skills that young children traditionally learned more by virtue of discovery than instruction. We seem to have traded the mastery of the mundane for virtual expertise. Rekindling the interest of our youth in food and farming is not to be equated with turning on yet another mobile device and expecting it to get them where they need to go. Rather, it's about getting our children's heads out of the iCloud and getting their hands in the dirt.

Our educational system has long eschewed the farm (and the kitchen) not only for upward mobility but also for lateral mobility. In other words, it doesn't pay to stay. That goes for communities across the demographic spectrum stretching from rural to urban. Part of education should involve showing young people the possibilities that could arise if they decide to remain in their home communities. As activist and social entrepreneur Majora Carter describes it, "You don't have to move out of your neighborhood to live in a better one."[2]

Both time with farmers and gardeners and time spent in the field are critical for building a new generation of farmers, and it's thrilling to see the intensity of activity focused on gardening and nutrition in the public schools, all the way from kindergarten to high school. Parents and community members need to push for the incorporation of agriculture into schools by way of specific experiential activities, related coursework, and internship opportunities. In fact, some public and private schools have already been building their curricula around agriculture, creating well-rounded students grounded in experiential education. For example, Massachusetts has three such agricultural vocational schools, despite its small size as a state. On a different scale, thanks to the enormity of agriculture in its state economy, California provides a number of secondary-school opportunities for aspiring farmers. California has more than three hundred schools with agricultural education programs that integrate traditional courses of study with "supervised agricultural experiences" and FFA (Future Farmers of America) participation. I've had several students who have come from these specialized schools, and they consistently display a well-rounded educational perspective, complemented by a rich appreciation of agricultural theory and practice. They come to college ready to test ideas and to test themselves.

Urban communities can also support this kind of curriculum, as exemplified by the Chicago High School for Agricultural Sciences and the W. B. Saul High School in Philadelphia, both magnet public schools that boast large campuses with a variety of hands-on learning opportunities. Rural private schools such as Scattergood Friends School in Iowa and the Putney School in Vermont present students with an immersive farm-life experience in which students participate in daily chores and see the fruits of their labors appear in their dining hall on a regular basis. The key for all of these schools is to capitalize upon the resources in their immediate area and to help students discover the array of possibilities not readily conveyed by textbooks and traditional classrooms.

Students yearning for educational opportunities that provide hands-on learning in an outdoor setting are an important potential labor pool for the development of local food systems. Forcing young people who don't grow up on a farm to postpone their agricultural education and employment until after high school means that they are often not learning the basic lessons of farming early enough. Not only are they discovering the subject matter too late in life, but they are also missing an opportunity to absorb the discipline and physical skills associated with this type of work at a young age. Furthermore, building relationships with farmers and gardeners in their hometowns enhances the chance that these young people will stay there or return later to ply their craft. Part of rebuilding local food systems and working toward community resilience is giving young people a reason to cherish their roots and consider farming in or near their hometowns.

Formal education is, of course, only part of the answer. It does provide aspiring farmers with a breadth and depth of understanding not always discovered solely through experience, and it does so in a concentrated and efficient fashion. However, informal mentors and formalized apprenticeships are probably the most vital part of bringing new farmers into the fold and ensuring their success. There simply is no substitute for direct on-farm experience. Formal education approaches agricultural education from the outside in, whereas an apprenticeship-type situation allows learning from the inside out. Those two contrasting perspectives are both vital, and it's to the advantage of the aspiring farmer to have both experiences, when possible.

Formal academic programs related to sustainable agriculture are cropping up at colleges and universities all over the United States. As interest surged in these programs in the first decade of this century, there was a need for students and faculty from across the country to compare programs, share perspectives, and develop best practices that fit their curricula, their college gardens and farms, and their local communities and ecosystems. Graduate students Albie Miles, Damian Parr, and several others from the University of California (UC) system began to conspire with faculty members to offer the first national conference on sustainable agriculture education in 2006. This student-initiated conference led to the development of the Sustainable Agriculture Education Association, a key organization for advancing college and university programs in this arena that now has members from a variety of institutions, including land-grant universities, liberal arts colleges, community colleges, and others.

The UC system also produced what is probably the most renowned sustainable agriculture apprenticeship program at any college or university in the country. UC–Santa Cruz's six-month apprenticeship program in ecological horticulture immerses students in the day-to-day practices of ecological farming on two contrasting sites, a three-acre organic garden and a larger twenty-five-acre vegetable farm managed with a tractor. Work on these two sites provides apprentices with direct links between presentations, readings, and farmwork. Students manage different aspects of the farm on a rotational basis so that they become intimately familiar not only with how things are done but also with why they are done a particular way. This demanding, well-crafted program also emphasizes cooking and eating together, a celebration and a deeper inquiry into the food system realities beyond the field's edge. Established more than fifty years ago, this program has long been a beacon for other programs in sustainable education, and the more than 1,400 graduates of the program have created a "diaspora" (remember—that's a scattering of seeds!) of farmers and experiential educational models.

Despite my role as an educator, I believe that living in the shadow of an experienced farmer is the penultimate education for a farmer in the making. Sometimes this experience will be filled with lessons of what to do; other times, it may be rife with lessons of what *not* to do. Either way, farming with a mentor provides an aspiring farmer not only with lessons of what works for a veteran farmer but also of what fits his or her own personality, abilities, and interests. For example, it's better to learn early on that artisan cheesemaking can be profitable but you don't like constantly cleaning equipment, floors, and walls, rather than unwittingly following a formal cheesemaking training into a world of discontent and frustration. Or perhaps it is just a matter of learning that you like milking goats more than cows. (After all, goats have two fewer teats—and you can pick up a goat when you need to. Just remember that goats are smarter than cows, which can sometimes make life complicated!)

Many regional, state, and national organizations provide databases and even matchmaking services for linking aspiring farmers with experienced farmers. Groups like the Appropriate Technology Transfer for Rural Areas (ATTRA, or the National Sustainable Agriculture Information Service) and World Wide Opportunities on Organic Farms (WWOOF) provide excellent national listings of various farm apprenticeships, although regional apprenticeship programs sometimes provide more detail and can serve as go-betweens for farmers and apprentices.[3] If these matchmaking services don't exist in a particular region, it's an opportunity for someone to step in and take the lead. Aspiring farmers don't always know where to turn, and, in some cases, they are too timid to do any cold-calling. Farmers, on the other hand, don't always have the time or the networks to find these potential apprentices. Some brokering of the possibilities is important because the "word on the street" doesn't always get out on the rural byways or filter through the density of urban populations.

Examples of these regional matchmaking programs abound, and they are increasingly successfully thanks to the Internet. One example is Growing Growers, a collaborative effort of the Kansas State Research and Extension, University of Missouri Research and Extension, Lincoln University Cooperative Extension, Kansas City Food Circle, Cultivate KC, and Kansas Rural Center. These groups all saw the need to help the supply of local food products equal the growing demand in the Kansas City area. Part of this work involves linking apprentices and farmers, so they established a website and a two-year apprenticeship program. It offers a broad perspective on farming operations the first year and a second year with a more intense focus on management skills. Growers are required to have been in operation for at least two years before participating in the program in order to provide a well-grounded knowledge base to share with the apprentices.[4]

One of my favorite apprenticeship programs has gone national, although each region seems to imbue the program with its own style and distinctiveness. With its cool acronym and cumbersome full name, CRAFT (Collaborative Regional Alliance for Farmer Training) was created in 1994 by a savvy group of farmers in the Hudson Valley of New York and the Berkshires of Massachusetts to bring together farmers and apprentices throughout the region. Like Growing Growers, CRAFT farmers are experienced and eager to share their knowledge. The apprentices apply for acceptance into the program based on their skills and motivations. Throughout the growing season, the apprentices visit many of the participating farms for workshops and get-togethers, seeing how different farms are managed as ecological systems and as businesses. Not only do they learn from the farmers, but they also form a strong network and support group among themselves. Since its inception, the CRAFT model has expanded into the Midwest, Kentucky, Vermont, North Carolina, eastern Massachusetts, and Ontario. Replication of the model with local adaptations is a sign of success!

As apprentices develop their knowledge base and prepare to take the plunge into farming independently, they face a challenging transition. As discussed earlier, land access can be a challenge, but there are creative leasing and purchasing opportunities. More difficult is acquiring capital, which is typically not readily available to beginning farmers. Think for a moment of the challenges. Aspiring farmers have typically worked at relatively low wages, they are in a relatively dangerous profession that requires adequate health insurance, and they are too often saddled with significant student debt. They also tend to need time to hone their craft and develop a sound but flexible business plan based on what they learn in the first few years. And they have to gradually build a customer base, in an environment with much more experienced farmers.

One solution to buffering the challenges of this difficult transition stage is the "incubator farm" concept. Incubator farms have become somewhat more common over the past few years, but one of the very first was the Intervale Center in Burlington, Vermont (locally known simply as "the Intervale"). Officially opened as an incubator in 1990 in a floodplain area used for decades as a municipal dumping site, the Intervale was the vision of entrepreneur Will Raap, founder of Gardener's Supply Company. In 1985, Raap began envisioning the possibilities of an incubator adjacent to his new retail store and shipping facility in this fertile but long-neglected site. He began building a "green industrial park" that incorporated his retail operation, the city's first large-scale composting operation, an adjacent biomass electrical generation plant, one of the nation's first CSA farms, and a slew of farms in various stages of incubation. Borrowed from Europe, the idea of this green industrial park is that one business utilizes the resources of another—including wastes—so that together they are complementary and more environmentally responsible.

The evolution of the Intervale is a model for the strategic and collaborative development of a local food system, and its incubator farm program is at the heart of the Intervale's success. In a nutshell, an aspiring farmer presents a business plan to the Intervale, and if accepted, he or she receives training, land, and access to shared infrastructure. Beginning farmers are subsidized during the critical early stages of their business development, and more experienced farmers provide them with guidance in managing their land, products, and finances. Tractors, greenhouse space, irrigation equipment, washing facilities, and storage areas are all available to the farmers in the program. Currently, there are farms of all different scales and types operating on approximately 135 acres through the Intervale's program.

Not only does this type of incubator operation provide new farmers with mentors and access to markets, but the entire operation has enabled Burlington and the surrounding area to build a vibrant local food system. The increase in the number and diversity of farmers in the area has been accompanied by a growth in market opportunities and available products. Other new businesses have also sprung up as a result, and Fletcher Allen Health Care (the regional hospital), whose cafeteria purchases from area famers, has pushed the hospital to the point of national renown for its fantastic food—people actually choose to go there to eat, as if it were a restaurant!

Another of the Intervale's projects, the Intervale Food Hub, is expanding market opportunities throughout the city. Begun in 2008, the Intervale Food Hub aggregates products from a variety of farms and then distributes them citywide as CSA shares. Part of the distribution strategy involves dropping off CSA shares at customers' workplaces for added convenience and increased exposure for the CSA model. This food hub has been hugely successful. It is projected to produce $1.1 million in gross annual sales by 2016, with approximately $700,000 of that sum returned to the farmers. And from 2008 to 2010, more than thirty farmers participated in the program, farmers on average experienced an 80 percent growth in sales, and the prices they received were slightly higher than standard wholesale prices.[5]

Egg on the Incubators

Incubator farms are taking hold across the United States in ways that fit their particular regions, local demographics, and available resources. There is nothing wrong with starting out relatively small, as is the case for Clemson University. Clemson is providing land and resources for a brand-new incubator project sponsored by Lowcountry Local First in South Carolina, taking up to six farmers on an initial ten-acre plot, four acres of which are certified organic. This pilot project will allow all parties to test out the incubator farm concept without significant overhead costs.[6]

Incubator farms are excellent concepts, but they do pose fiscal and logistical challenges, and they are best developed in stages. One of the most successful incubators in the country is the Agriculture and Land-Based Training Association (ALBA), located in California. ALBA's website carefully lays out the long and patient evolution of the organization, and it now has a "Farm Incubator Toolkit," featuring several downloadable documents for groups considering starting an incubator (available at the USDA's Start2Farm website).[7] When I asked Gary Peterson, ALBA's director of communications, what he sees as the most critical considerations to take into account when conceiving of an incubator, he quickly prioritized the key issues:

- What are the market opportunities? Do the sales possibilities make sense? Will they be at the appropriate scale for long-term success? How much time will it take to develop these markets?
- Establish clear expectations of how to develop farmers and the program. The organization must be explicit about the skills and financial development necessary for the farmers themselves to be successful.[8]

ALBA has created an incubator model that fits the cultural diversity of the area. It maintains two farms in Monterey County where approximately seventy aspiring farmers from farmworker and low-income backgrounds currently engage in a six-month training program, after which they can lease land and work their way into agricultural markets. ALBA's base farm, the Rural Development Center, serves as the training center for PEPA (Programa Educativo para Pequeños Agricultores), which prepares graduates who can opt to enroll in the First-Year Farmer Apprenticeship program. The Farm Training and Research Center (Triple M Ranch) serves as the base for Latino farmers who would like to lease land for their first three years of farming on their own. In addition to sharing infrastructure for processing, storage, and distribution, these farmers receive training for launching their own farm operations in the near future.

Like the Intervale, ALBA saw a need to provide distribution support for farmers looking to gain a foothold in local markets. Thus far, its nonprofit distribution enterprise, ALBA Organics, has experienced 40 to 60 percent annual sales growth, with an impressive achievement of surpassing $3 million in sales in 2011. Its distribution radius is about fifty miles, and it services a variety of customers. When asked about how the enterprise has achieved such remarkable growth, ALBA Organics general manager Tony Serrano replied, "Organic is the trend, and we're growing with the trend. Chain stores are taking organic products out of segregated areas and putting [them] in the mainstream sections of the stores."[9]

Such incubator farm projects not only support immigrant and low-income populations, but they also take into account culturally appropriate diets and associated niche markets for new farm operations.[10] One example is Seattle Tilth Farm Works, which is similar to ALBA in that it serves a variety of immigrant populations, including Somali, Bantu, Ethiopian, and Burundian farmers. These growers begin with quarter-acre plots and participate in an intensive training program that includes visits to other farms in the region. The market destinations for these traditional foods tend to be ethnic neighborhoods, a strategy that increases access to culturally appropriate foods for those immigrant populations. Although this kind of initiative on behalf of immigrants may be somewhat hidden from the view of many Americans, the USDA recognized this project with a grant of nearly $500,000 in 2011 to expand its capacity. This project and others like it are a good reminder that the history of agriculture in the United States is ultimately a fusion of cultures and generations.[11]

Of course, communities can support new farm enterprises without necessarily providing them with a land base. Numerous organizations across the country are finding creative ways to offer start-up farm operations low-cost lease options, market-entry opportunities, and access to capital. Access to capital is one of the most complicated aspects of establishing a new farm, and the options range from traditional low-interest federal loans to new community-supported financing models. On the federal front, Start2Farm is a new federal program and website (http://start2farm.gov) designed to connect farmers with less than ten years of experience with an array of useful local, state, and federal resources.

Models for community-based financing are on the rise, thanks in large part to Woody Tasch's push for "slow money," a concept he lays out in his 2008 book *Inquiries into the Nature of Slow Money: Investing as if Food, Farms, and Fertility Mattered*. Tasch has proposed that one million people invest 1 percent of their assets into finance programs that support local and organic farming and food enterprises. This novel financing approach essentially incorporates soil fertility and the production of healthy food into return on investment (ROI) calculations. Tasch also proposed what he dubs "slow munis," tax-exempt bonds directed to enterprises dedicated to local food systems. Supporters of the idea have formed Slow Money chapters in different parts of the United States, and some proponents have created investment clubs that pool their investments for particular food and farm enterprises in their communities, including the conservation of organic farmland.[12] (For more detailed information on this topic, see the first book in this Community Resilience Guide series, Michael Shuman's *Local Dollars, Local Sense.*)

Other similar approaches are laid out in a new document put out by the University of Vermont Extension and Center for Sustainable Agriculture. *The Guide to Financing the Community Supported Farm: Ways for Farms to Acquire Capital within Communities* is geared toward Vermont, but many of the ideas can be replicated or already exist in other parts of the country.[13] While the mechanisms can be somewhat complex at times, the vocabulary in the document is clear. We're talking about seed money, greenbacks, and dollar values.

Fishing for Complements . . . to Local Farms

A fish story is the fisherman's version of a tall tale, but here are two fish stories worth telling. These two fish stories can either make the bad news appear overwhelmingly large or the good news big enough to make the stories worth telling over and over again. The fact that these two fish stories are separated by 3,285 terrestrial miles should assure most listeners of their veracity.

Port Clyde, Maine, and San Francisco, California, may not have a lot in common in terms of climate or population size, but fishers in both places have felt the claws of market concentration and consolidation squeeze the vitality out of small-scale fishing operations. The politics are just as complicated as the ecology in both situations, but the economic realities are fairly—or unfairly—simple. The intense concentration and consolidation of industrial fishing fleets created markets in which smaller fishers couldn't compete. Additionally, the industrial fleets' use of intensive, undiscriminating harvest methods generated unsustainable levels of bycatch and seafloor degradation, leaving behind fish populations in threat of collapse and coastal communities in decline. (For an excellent illustrated overview of different fishing methods and their impacts, see the *Sustainable Fisheries* guide from the Monterey Fish Market).[14]

There is hope, however. Port Clyde, Maine, has been a fishing village for more than two centuries. You wouldn't think such a small community, with a population of just 1,574, might confront global challenges with any hope of success. However, fishers in the area banded together to form a community-supported fishery (CSF) based on the twin values of community and ecological sustainability. Fishers in this CSF use environmentally conscious fishing methods, and they also process, package, and market their products together. By cutting out the middlemen and capitalizing on their ability to guarantee 100 percent supply-chain traceability, they quickly built a viable business model that brings them two to five times the price that they received for their fish through the previous channels.[15]

On the other side of the continent, a group of fishers decided that Fisherman's Wharf in San Francisco might benefit from having some actual fishers present once again. Small-scale fishers in the Bay Area were feeling the same squeeze as the Port Clyde fishers, so they worked for over six years until, in 2011, they formed a CSF called the San Francisco Community Fishing Association, with their facility based right at Fisherman's Wharf. This cooperative provides the fishers with more autonomy, from having their own hoist to controlling their markets and prices.

Remember EcoTrust, the nonprofit group discussed in chapter 10 that helped found FoodHub in Portland, Oregon? EcoTrust has helped advance CSFs, the nautical parallel. Not only has EcoTrust supported the development of individual CSFs, but in 2012 EcoTrust and the Island Institute in Maine announced the official launch of the National Community Fisheries Network, a coalition of fourteen community fishing associations, including the groups from Port Clyde and San Francisco. One of the network's first initiatives is to build a campaign for "community-caught" fish in local and regional markets.[16]

One of the most interesting aspects of this new coalition is that "community-based fishing organization" is an official classification in the national fisheries law, but this designation has yet to be utilized to any great degree in order to advance the interests of local fishing communities. Having this official status allows these groups a seat at the table as regional fisheries councils develop policy and management issues for specific sectors. Fishing communities are finding their voice, and consumers now have increased opportunities to vote for a fish story worth believing.

Farm to Institution

Since the birth of our third child, my wife and I have finally come to grips with the truth about choosing a hospital: it's all about the food. I know it's an outrageous notion, but I'm only half joking. Food is the foremost link in our healing process as we transition from passively being cared for to caring actively for ourselves—for the long term. Nutritious, tasty food is what patients in our health care system deserve. (I wish that we also more fully embraced the idea that institutions such as schools and senior centers are also part of health care system.)

A few years ago, we spent a harrowing summer night with our then four-year-old son, Ethan, at our local hospital. The care we received was superb: kind, caring, and competent. But when breakfast and lunch came the next day, Ethan took one look at the factory foods on his tray and left virtually all of it untouched. Not ones to waste food, my wife and I tried to finish Ethan's meal, but we couldn't do it either. Not only did we find the food visually unappealing (despite its packaging), but as farmers ourselves we understood the grim realities on the bland tray.

In stark contrast, consider the menu we found at another much more distant and smaller Vermont hospital a few months later during the birth of our third child. It was a community hospital that we chose for its midwives and chefs, not just its doctors and dieticians. Appetizers at Gifford Medical Center included scallion pancakes with thin-sliced crispy chicken breast served with a ginger-soy dipping sauce, smashed potato napoleons with smoked bacon and chive sour cream, and seared Maryland crab cakes with lemon aioli. Entrees included crisp tangerine chicken served with jasmine rice and Asian-style vegetables, a portabella mushroom tower served with polenta, and New York strip steak with charred tomatoes, green onions, and a balsamic reduction.

Hospital chef Ed Striebe, a former New England Culinary Institute instructor, feels that part of the healing and educational mission of the hospital has to be the quality of the food it served. He has struck up alliances with local farmers to procure local foods, and he emphasizes presentation and pride on the plate. Vegetables are carved, color is sprinkled on the plate as liberally as the fresh spices, and a signed card from the chef comes with each meal. The wording on the menu says it all: "This menu reflects our commitment to wholesome, locally grown goods to better serve you and our community."

In trading his white-tablecloth restaurant career for hospital work, Chef Ed seems to have found a new sense of purpose, not to mention regular working hours. I was particularly grateful for his guest tray policy. Chef Ed has found a way to purchase nearly a third of his food from local farmers, all within the tight budget constraints of a small community hospital. He even provided "his farmers" with extra money for seeds one year when they were hit hard by terrible growing conditions. And locals actually duck into the hospital cafeteria to grab a meal. It all works, not because of its efficiency of scale, but because of carefully cultivated relationships.

The lessons are obvious, but the fact that most of us view Chef Ed's experiment as radical while considering the usual substandard hospital fare as normal is indicative of the need for change. Perhaps we shouldn't expect much from food served on an airplane at thirty thousand feet. But hospital food is no joke. It is the first step in any promising recovery from illness or accident. Patients in our health care system deserve high-quality food.

When it comes to increasing demand for local foods, institutions such as schools, hospitals, colleges and universities, and senior centers can make a huge difference. An increase in supply generally follows an increase in demand, so involving institutions in local food purchasing can ultimately enhance a community's long-term capacity to feed itself. Remember, Americans are spending approximately half of their food dollars on "away-from-home" food, so farm-to-institution sales can have a big impact on a region's agricultural economy.[17] And community members often have more leverage than farmers in getting local foods into institutions because many of these institutions are supported by public dollars and have service to the broader community in their mission.

Farmers tend to appreciate steady accounts that can be trusted to order large and consistent volumes, since solid relationships and clear expectations help minimize marketing expenses and maximize farming time. Institutions, meanwhile, can demonstrate their commitment to their communities through local purchasing programs, while also maximizing fresh produce and minimizing highly processed foods. There are multiple benefits for farmers and institutions in these big "buy local" initiatives, as I have depicted in table 13-1.

TABLE 13-1. Benefits of Farm-to-Institution Programs

POTENTIAL BENEFITS FOR FARMERS	POTENTIAL BENEFITS FOR INSTITUTIONS
Reliability of established accounts	Increased freshness and quality
Predictable volume of sales	Increased nutritional value
Clear criteria for products	Enhanced traceability and food safety
Ability to focus on production	Less food and packaging waste
Decreased marketing efforts	Food with a face and a story

Perhaps the key to any successful farm-to-institution program is determining price points that provide fair prices to farmers and affordable prices for the institutions. Most institutions are working with the constraints of a specified food budget and don't have the luxury of tailoring costs charged to the consumer based on prices paid for the ingredients. Farmers selling within local markets are trying to make a reasonable profit instead of being passive price takers. Ideally, a farmer and a food service manager can recognize the constraints on both sides and find a fair middle ground.

Two additional elements critical to the success of a farm-to-institution program are the consistency of the products and the reliability of deliveries. Chefs have to have a product delivered as promised—clean, consistent, in the proper quantity, and on time. Frankly, the consistency and reliability have been a challenge in some of these programs, simply because there are lessons and hurdles for farmers who want to scale up to this level. As the farmers work to meet the demands, chefs have to shift from a mentality of assumed confidence in the industrialized system to earned trust in the local farmer. At the same time, communities need to consider ways in which they can support farmers in this process of scaling up. This support may come in the form of investing in aggregation opportunities, commercial processing facilities, access to land and capital, or other means.

Community investment in farm-to-institution programs has multiple payoffs that go beyond local economic vitality. These programs also serve a critical educational component. Serving local foods in school cafeterias is arguably the most important among the farm-to-institution programs, simply because farm-to-school programs have the greatest potential for fostering change. And that change doesn't come solely by virtue of what appears on a child's plate. I've long admired the work of Vermont FEED (Food Education Every Day), a national model for farm-to-school programs. Vermont FEED's success is based on the three Cs: cafeteria, classroom, and community.[18] Educators across the country have begun to recognize that all three elements need to be connected to maximize the impact for the students.

It's not just what is placed on the plate—it's also about serving up new expectations. Simply stated, if we want local food to be the new normal, then we have to push hard to make it the norm within the institutions where we spend so much of our lives—not just reserved for special dishes for occasional, isolated events. If we want hospitals to heal, then they should begin the process by serving the best food our communities have to offer. If we want our colleges and universities to educate consumers and health care professionals, then why create contradictions between what is taught in the classroom and what appears in the dining hall?

One of the most exciting developments that I've seen in my career as a college educator is the galvanizing and increasing momentum behind these issues from students across the country. The real drive behind dissolving the contradiction between what we are teaching and what we offer in campus dining facilities is fueled by a group called Real Food Challenge (RFC). One of the leaders of the RFC, Anim Steel, summarizes it this way: "We're not going back to the land; we're going forward to the land." And RFC is not unambitious in its goals. Anim reminds me that there are seventeen million college students and $5 billion spent on food at U.S. colleges and universities. That creates a substantial opportunity for academic institutions to invest in local, sustainable, and fair-trade agriculture. Anim and his colleagues want only a modest one-fifth of the pie by 2020; their national target is 20 percent "real food" purchasing by that date.

In speaking with one of the other leaders of the RFC, David Schwartz, I ask whether he was seeing a significant impact on reshaping local food systems in their work. He immediately comes up with an example of one student who was a model for such change. Bonnie May grew up in the area near Southeastern Louisiana University, located not far from Baton Rouge. It's an institution of thirteen thousand students, most of whom are commuter and working-class students. When Bonnie May tried to get Aramark, the university's contracted food service provider, to increase its purchasing of "real food," Aramark officials expressed little interest. So she decided to set up an on-campus farmers' market and began working with a local black farmers' cooperative. The produce sold out the first two times they ran the market, so the farmers' market is now held several times each semester. Oh, and the Aramark management team did make its way downstairs to see what all of the commotion was about. The result, David observes, was that "the student-farmer relationships began to chip away at the disinterest initially expressed by the 'business as usual' crowd."

The impact in dollars of the RFC thus far? David estimates that the group has already shifted $48 million in food purchases toward "real food." The impact on education? A graduate of Brown University, David gives his own perspective: "I honestly spent most of my undergraduate career working on the Real Food Challenge—and it was actually where I learned the most. I wouldn't trade that decision for anything." When I ask about his hopes for the near future, he doesn't hesitate: "This has been a student-powered movement from the beginning—so much of it is about empowerment and leadership development, not just food. It's about youth developing self-respect. In the end, I hope that this work is no longer called 'activism' but that it is viewed as education and doing action-oriented change."

Despite the dry and staid images conjured up by the word "institution," it should be obvious by now that institutions amplify the scale of our locally focused expectations. While these new scenarios may create some complications, they open wide the range of possibilities. Breaking down these institutional walls can provide some of the bricks that pave the way to community self-reliance.

Selling Local . . . at the Federal Level

In thinking about approaches to reconstructing local food systems, no institution seems more daunting than the federal government, and "local" and "federal" certainly seem like odd bedfellows. Nonetheless, there are points of intersection, and it is unwise to ignore that reality. Policy advances at the federal level depend not only upon concerted advocacy stemming from the local and regional levels, but also upon *successes* from these levels. Federal attention and dollars follow replicable successes at the local level.

There are ample valid arguments describing how federal policies and programs thwart the very existence of local food systems, much less the rebuilding of these systems. Federal subsidies have created enormous challenges to enhancing the growth of sustainable agriculture practices, and they have propped up a national food system hell-bent on big, efficient, and cheap. The tide may be shifting somewhat, however, as both sides of the political aisle have philosophical reasons to protest the negative consequences of these taxpayer giveaways. Whether you're for fair markets or free markets, agricultural subsidies don't add up—especially in the current economic downturn.

Even as subsidies seem to be waning in political popularity, another federally inspired push is perhaps an even greater threat to local food systems: food safety regulations that cater to the scale and capacity of the industrialized food system, at the expense of smaller producers and processors. Some practices and technologies that make sense in an industrialized food system not only are far beyond the financial means of smaller farmers and processors, but they are also sometimes irrelevant to these smaller operations that rely on local markets.

The Food Safety Modernization Act of 2010 brought this issue into the public spotlight during the legislative process that led to its ultimate passage. Fortunately, several U.S. senators, backed by the grassroots efforts of a number of food-related organizations, added amendments to the final bill that helped protect small farmers and food entrepreneurs from regulations that threatened their markets, if not their very existence. These amendments took into account the limited resources, relative transparency, and farmer-to-consumer relationships associated with many smaller farm operations. One of the most important amendments, cosponsored by Senators Tester (D-MT) and Hagan (D-NC), provided more accommodating regulatory alternatives to farmers who direct-market more than half of their products, have gross sales under $500,000, sell their products either in-state or within a 275-mile radius, and provide customers with their contact information.[19]

Nonetheless, both federal and state regulatory hurdles abound for small farmers and local food system entrepreneurs. The ability for community groups and issue-oriented advocacy organizations to communicate concerns and alerts by way of the Internet is critical; Internet updates and alerts from advocacy groups across the United States helped to prevent the Food Safety and Modernization Act from becoming a death knell for local food systems. The speed and pervasiveness of this kind of communication through a variety of social media means that individuals and organizations can work proactively and reactively on these fast-breaking regulatory issues.

Not all such issues are fast-breaking. In fact, the Farm Bill might be referred to as "slow Policy" if it had a better reputation for encouraging what Slow Food calls food that is "good, clean, and fair." The Farm Bill is the enormous omnibus legislation focusing on food and agriculture that Congress passes approximately every five years. In essence, it funds the comprehensive efforts of the USDA, so it dictates programs related to agricultural production, food security, environmental protection, nutrition, international trade, and many other critical areas. The Farm Bill may never do as much to support local food systems as it does to pose serious obstacles to them. Nonetheless, it is not to be ignored, as it does offer some resources for local food systems, while other provisions in this massive legislation need to be challenged for the damage they do. Numerous organizations are alert watchdogs and tireless advocates for change in the ever-evolving Farm Bill. Keeping an eye on their recommendations for change and action alerts is not only an excellent way to hone our understanding of the complex issues at play, but this practice can also help us identify potential funding opportunities for our local community. In the midst of all the chatter lie patterns, keywords, and funding trends. The following organizations provide sophisticated analyses (in straightforward language) and alerts of hot-button Farm Bill issues related to local and regional food systems:

- Center for Rural Affairs: http://www.cfra.org
- Environmental Working Group: http://www.ewg.org
- Institute for Agriculture and Trade Policy: http://www.iatp.org
- Leopold Center for Sustainable Agriculture: http://www.leopold.iastate.edu
- National Farmers Union: http://www.nfu.org

- National Sustainable Agriculture Coalition: http://www.sustainable agriculture.net
- Watershed Media (in particular its book on the Farm Bill, titled *Food Fight*): http://www.watershedmedia.org
- WhyHunger: http://www.whyhunger.org

Can the federal government really "go local"? In a sense, it can't afford not to do so, but there may be more gestures than dramatic turns. Utter cynicism, however, tends to lead to neglect, and there are too many possibilities—good and bad—to warrant turning our backs on local food systems initiatives at the federal level. There are also some good people in those federal offices trying to do good things. (For example, the USDA recently launched its innovative—at least for a governmental agency—Know Your Farmer, Know Your Food website, which provides an extensive resource center for ordinary citizens interested in these issues.)[20] Not many of them are farmers anymore, as Jefferson and others had hoped. However, they all come from communities that they care about, and some of them buy local as a concept worth fighting for—even at the federal level.

Bringing It to the Streets

If the government isn't making the best food choices, then perhaps it makes more sense to encourage those choices in the places where the average consumer encounters the most junk food—the local corner store. There might even be a way to get some state and federal dollars to support your good idea! (Just check out ChangeLab Solutions' excellent report *Green for Greens*, available online.)[21] Across the country, various agencies and organizations have been trying to find ways to address the food desert issue. One of the newest national collaborative efforts in this regard is the Healthy Corner Store Initiative, a project developed and coordinated by the Philadelphia-based nonprofit Food Trust.[22] Capitalizing upon the idea that bringing fresh and healthy foods into communities with little access to these items is a viable form of economic development, communities across the United States are developing strategies for linking producers to store owners. As with food hubs and incubators, the iterations of this concept vary widely, but at the core, the desire is to link nutrition, economic development, and more resilient local food systems.

One such initiative is in development in Washington State. Warren Neth, founder of Urban Abundance, is helping to create a marketing co-op for urban farmers, called the Backyard Bounty Co-op, in the Vancouver area. Thus far twelve farmers have been recruited, and they will sell at two farmers' markets, three retail markets, and one healthy neighborhood grocery, Oscar's Marketplace. According to Warren, "We need to find a way to make healthy food affordable for low-income residents—one way of doing it is to get Farm Bill dollars in the form of SNAP dollars to go to healthy local food. We also need to target these little convenience stores." Pioneering an effort like this requires both the farmers and the vendors to be flexible, and Warren sees the Backyard Bounty Co-op as "nimble" and committed in ways that other distributors would not be.[23]

Another approach is under way in Birmingham, Alabama. Taylor Clark is the Urban Food Project market coordinator for Main Street Birmingham, an organization dedicated to supporting local entrepreneurial activity, including new possibilities in the regional food system. Taylor's recent experiences in trying to promote economic development through local food markets in food-desert areas in Birmingham provide some interesting perspectives.[24] I called Taylor on a hot Alabama summer day, and we began our conversation by comparing thoughts on the particular challenges that culture and foodways can present in the South. "We need a toolkit for the South," Taylor observes. "What is developed in Seattle ain't gonna fly here." Both being of southern heritage, we swap ideas on the challenges. Taylor summarizes them quite well:

> Culturally we are very tied to our food in the South—and our identity is, unfortunately, tied to unhealthy food. If you strip away our identity, what will we have left? There's a general lack of education about these issues. It's a cultural shift we have to make. Also, the political mind-set here is completely different, and none of us want to be told what to do. And there's also the direct correlation between poverty and its influence on education. It's a tough one to figure out here.

Nonetheless, Taylor is someone who has an important skill set for tackling the issues. Before joining Main Street Birmingham, she oversaw a $13 million federal grant through the United Way of Central Alabama to combat obesity and tobacco addiction. She also has over a decade of marketing experience, so she is bringing her experience to bear on these challenging issues.

She began her work with the intent of building farmers' markets in food-desert areas after polling local residents and getting feedback as to the most effective approach for success. Despite the preemptive surveying and planning, the first farmers' market didn't work. "People just bypassed the fresh fruit and vegetables and went for the soda and chips," she says. "We thought, 'If you build it, they will come.' We looked at the models elsewhere. We did everything right in the planning. But they didn't come."

As any good marketer would do, Taylor reassessed and ran the numbers. As it turns out, putting a farmers' market in the community was going to have relatively minimal economic impact in contrast to establishing the market potential in a farm-to-corner-store initiative. Furthermore, it is more challenging to get farmers to develop all of the necessary skills to successfully market their products at a farmers' market, and it's a challenge for farmers to stick to the commitments required of an emerging farmers' market when it is not thriving from the outset. In contrast, the farm-to-corner-store approach allows farmers to tap into a market that has fewer time commitments. In fact, Taylor began the initiative thinking that it would be best to have farmers do direct sales and deliver the products themselves. However, the farmers were not consistent enough in their delivery schedules or in quality control of the perishable products, so she shifted to using a distributor. "Trying to create markets in challenged neighborhoods while prepping farmers, well, it's slow. It just takes time."

Bringing It All Home

It is slow sometimes. While technological developments are changing the food and agriculture sectors with astounding speed, creating viable and resilient *community* food systems takes time. Consider for a moment some of the most successful models reviewed in the last few chapters. Most of them have been around for at least one or two decades. The successes often span the length of a typical career, and organizations and initiatives are not static entities—they are a continuum of successes and failures built one upon another.

But they are well worth doing. We need the guidance of stories, we need the nourishment of successes (no matter how small), and we need the honesty and humility of occasional failures (let's keep them small, too).

Seldom are we going nowhere. Usually we're just staying close to home.

Chapter 14

Bridging the Divides

When I began researching and writing this book a little over a year ago, I had an image in mind of "local food" in the United States. Not that it was a correct image or even one that I liked, but the same visual cue kept coming to me, whether I was mulling over the issues in sequestered solace or discussing them with others. It was an image of a dot on a map with an almost mechanical radius that could be expanded or shrunk on a whim.

Over time, however, that image has shifted—for the better. The image that comes to mind these days is of dynamic, interlocking systems—a vast network of differently sized pulsing centerpoints connected to one other by means of surging flows that create exchanges of resources, ideas, and of course foods.

There are two problems with that initial image of a dot and a radius around it. First of all, it's static. Second, it's too tightly defined and isolated. Which leads me to another unfolding thought brewing over the course of this year of inquiry and writing. Thinking of our own local food systems as dots on a map is shortsighted, and it stymies the real potential of this critical work. We should be concentrating much more on the flows than the dots. Just as a good ecologist understands that organisms in and of themselves aren't really the point of study—the interactions between all the different organisms are the point—anyone interested in food systems should understand that we're really thinking about flows, not dots. The dots are ultimately part of the flows.

This dot-on-a-map notion means that we risk segregating ourselves from one another when, in fact, the ultimate driver of all this hard work is the desire to build healthier and more just communities by reframing our food systems. Community doesn't end with a line any more than it begins with a dot.

Recognizing the flows that we call community, commerce, and ecology ensures that we avoid the illusory island effect of a completely independent local food system—an illusion that ultimately leaves us stranded and less enriched. We can recognize the flows while celebrating our home communities and our connections with other communities, no matter how different. In the end, how we recast and redirect the flow of food and associated resources determines whether we will succeed in collectively finding healthy food, real food, fair food, food access, food security, food justice, food sovereignty, and all of those other critical phrases that comprise the stepping-stones for the path toward more just and healthy food systems.

Ultimately, the recent surge of interest in rebuilding community-based food systems is not just about resilience—it is about *just resilience*, planning and preparing for a fast-changing world in ways that leave no one behind. That is a much different vision from the modern militia mentality of hunkering down to ward off anything threatening or unknown. Adopting a just-resilience approach involves embracing the uncertain, the unknown, and the stranger who is actually a neighbor.[1] It also requires us to step out, beyond the realms of comfort and certainty, and into more distant communities—not simply to find answers but also to create them, together.

In sum, much of the conversation about local food over the past few years has created four false divides that I think need to be addressed, and I hope that these parting thoughts are in no way a conclusion but rather a foundation for thinking anew about this important work.

Urban/Rural

In our emphasis on local, I believe that we have obscured the reality of the deep interrelationship between cities and rural communities. In other words, we got hung up on the dots and pushed the flows aside in our zeal to relocalize our food systems in both settings, as well as within America's vast in-between, the suburbs. Some rural communities have approached rebuilding local food systems as a means of reclaiming their agrarian heritage and wedging it tightly into the cracks of the modern industrial food supply, eyeing food self-sufficiency as a means of self-preservation in the gales of globalization. Meanwhile, urban agriculture proponents have justifiably wanted to feed their own, reconnect people with their food supply, and create vibrant pockets of ecological entrepreneurism, sometimes substituting the concept of resilience with a more utopian idea of self-reliance.

Carolyn Steel, a Cambridge architect, has written a provocative and intensely researched book on the relationship between cities and food, titled *Hungry City: How Food Shapes Our Lives*. Although she began by examining this relationship in London, her research quickly took her far afield and deep into the annals of human history.

> Born of a new kind of food, the first cities represented man's emancipation from brute survival—the start of what we call civilization. Yet the freedom they brought was never complete. Beneath the surface of urbanity, the contract between man and land never changed. As our chosen habitat, cities express the inner contradictions of the human condition. They provide us with shelter, but not sustenance. They give us space to dream, yet obscure our place in the natural order. Neither good nor evil, cities represent the messy imperfectability of human life. For better or worse, they are our common home, and it is up to us to make them work. Going back to live in forests is not an option.[2]

Nathan McClintock, a geographer at Portland State University with extensive farming experience, has thought long and hard about the questions surrounding a city's ability to feed itself and the ultimate purpose of urban agriculture. McClintock began his research into urban agriculture in Oakland, California, where he utilized his agricultural savvy and his GIS mapping skills to examine the potential for food production in that city. He prefers to err on the more conservative side of the numbers in order to keep expectations in check and prevent disillusionment. When I ask him about his most recent research, he tells me that he estimates that Oakland has one hundred to five hundred acres of public land available for sound agricultural production, with one hundred acres being the more realistic number.

The next question Nathan faced was which farming techniques were most likely to be used and their productive capacity. He scaled the farming techniques as low, medium, and high production, with the highest being the use of the biointensive method (à la John Jeavons). Such calculations led him to production figures of 10 tons/acre on the low side and 25 tons/acre with biointensive production, assuming that 75 percent of each acre was devoted to production and 25 percent to infrastructure. Nathan makes a critical point, though: "To focus on urban agriculture solely in regard to quantitative methods, it's going down the wrong road. At best, we might get 5 percent of Oakland's vegetable production, maybe 10 percent—*if* you have it as a government priority and *if* you have professional farmers. But if you leave it to nonprofits, educators, etcetera, then it will be a fraction of that."

That's not to say that Nathan sees urban agriculture as unimportant—far from it. After all, he has built his academic career and his personal interests around it. He sees tremendous benefits in urban agriculture: "Education is a huge part of it, making food systems visible and allowing community members to interact with the food system. Urban agriculture is a connective tissue to the real bread baskets, as well as serving critical education and health benefits. . . . If we go down the road of production alone, we'll never compete."

Similar to Ramsey Cronk, the school garden designer mentioned in chapter 12, Nathan wonders whether we should always be asking urban agriculture to fund itself. "It may be like public transportation," he muses. "If it is to flourish, then we need to bolster it." He went on to provide one more caution: "The solution to hunger is to produce jobs and provide income." Job creation, in his view, is an important component of urban agriculture, and he is not alone in his thinking.[3]

Erika Allen, the Chicago projects director and national outreach manager for Milwaukee-based Growing Power, expresses a similar way of thinking about the importance of urban agriculture for getting people employed and on a path out of poverty. She describes Growing Power's work with Altgeld Gardens Urban Farm in the far South Side of Chicago as a means of providing a community that has suffered the consequences of environmental racism with job opportunities. The two-and-a-half-acre urban farm now has one acre in production, including a hoop greenhouse and large-scale composting and vermiculture programs. The Altgeld Gardens community is composed of just over three thousand people, many of whom descended from black veterans returning stateside from World War II. Their American dreams were soon tainted, however, when the community became a toxic waste dump by way of asbestos in construction materials and high levels of industrial toxins. The community remains nearly 99 percent black, cancer rates and other serious diseases are at tragic levels, and the nearest grocery store is six and a half miles away.[4]

The project was so daunting that it took Erika a little time to decide that she could take it on as part of Growing Power's Regional Outreach Training Center program. She got to the point, though, that she decided, "I'm ready for this one." Now, three years later, she is working with the Chicago Housing Authority to employ 150 adults and 40 youth during the summertime through workforce reinvestment dollars.[5]

The Altgeld Gardens project reflects the broad vision of Growing Power's ROTCs. Erika describes the development of the ROTC concept as a response to Growing Power's need to stretch out beyond its Milwaukee base; cultivating local citizens in the different ROTC project locations ensures that they maintain the necessary investment in and longevity of the programs. The ROTC programs fulfull the need for long-term commitment. As Erika's father, Will Allen, the founder of Growing Power, quickly learned, it takes years for a person to become a "farmer." People have the passion, but they need the skills, and the projects require daily maintenance by professional farmers. "It's like the heart," she notes. "The beating is monotonous, but it's important."

Growing Power took note of the necessary flows from rural to urban landscapes early on in the development of the Rainbow Farmers Cooperative, a marketing co-op Will Allen helped create when he observed that farmers with strong ecological and social values were having trouble breaking into the tightly controlled wholesale markets. According to Erika, "The co-op couldn't break into the wholesale market, so they tried to break in with cabbages and peppers through aggregating the products so that they could meet a weekly contract. The small farmers came together to meet the demands of the contract, but the wholesalers were using fluctuating prices, so at the end of the year, the small farmers were losing money." At that point, Will and these farmers formed the cooperative so that it—not the farmers—would take the hit if the prices fluctuated. The co-op now markets through Growing Power to diversified markets. "Dad found it hard to get a consistent product, so they have expanded their own program," says

Erika. "It's been so successful that now we're kicking off an urban agriculture aggregation project that will begin selling to Walgreen's, allowing us to reach a broader inner-city customer base that would otherwise not have access to such healthy food."

When I ask her about the realities of a rural/urban divide, she takes the question head-on: "Growing Power actually began with the name 'Farm to City Link.' It was changed to Growing Power, but Dad's modus operandi is that there is a vital link between the two—urban and rural. Take, for example, the Great Migration, the movement of predominately rural blacks to the northern cities."

Erika continues, "This cultural divide is ridiculous. Urban agriculture is critical for healthy ecosystems, job creation, cultural education, emergency preparedness for food security—but we need what the rural communities can offer us in food. But what many rural communities around us in the Midwest are just growing is 'commodities'—not real food! So the rural food crisis is real, too—there's just not the food out there. That story needs to be better told and understood. And the Midwest needs to create a better connection between these two worlds. If I lived in California, I'd see things very differently, I'm sure. Yet Illinois has some of the richest farmland in the world—and it has a food crisis in both the rural and the urban environments!"

Once again, flows. Not dots. Rather, flows of food, information, people, and inspiration.

Small Scale/Large Scale

The issue of scale arises time and again in local food conversations and in sustainable agriculture circles. "Small" usually wins, "medium" is sometimes mentioned, and "large" is typically castigated. However, I think some caution is in order. There are different kinds of medium- and large-scale agriculture, and even small is not always ideal. Oversimplifications can create dead ends in our local logic. We do have to feed everyone, and it's not so much the size of a farming operation or processor that matters as much as how it operates—whether it embodies sustainable agriculture or triple-bottom-line management.

Assessing the relationship between scale and sustainability can be a tricky business, but it is important, partly because we probably do need a diverse portfolio, even in terms of scale, in order to adequately feed our population. Perhaps that will change at some point, but it seems to be a given reality at this time in history. That idea flies directly in the face of the local aspirations of some food activists. However, I would like to offer a compromise solution that I think works, at least some of the time: cooperatives.

In a recent discussion with Ken Meter, one of the most experienced consultants in rebuilding community food systems in the country (his consulting and research organization, Crossroads Resource Center, can be found at http://www.crcworks.org), he noted that proponents of local food economies too often neglect cooperatives. As has been the case for Growing Power's Rainbow Farmers Cooperative, co-ops can help solve the scale issue through informal collaborations and formal agreements. These cooperatives can support local economies while stretching their geographical bounds in order to capture an increased market share—dollars that can then stream back into local economies, if managed well. That increased geographical spread can also help build resilience on farms and within communities by creating a diverse portfolio of economic activity for the individual entrepreneurs and the broader community.

I think the following story will convey my point. It's always a good idea to make sure college students have an opportunity to meet local farmers and have their reigning assumptions about what sustainable agriculture looks like "ground-truthed." (In fact, it's probably even more important to provide college professors with that kind of experience!) One of my favorite guest speakers to bring to our college community is John Malcolm, a retired dairy farmer and current state legislator. John is extremely intelligent and well spoken, and he is such a critical thinker and good listener that you never know on which side he may come down when there is an informal debate or a "call to question" in the Vermont statehouse. He is also one of the few remaining farmers left in our legislature—a dramatic shift from a generation ago, when farmers filled the legislative seats and voiced their grassroots opinions with flair and verve.

Before he retired from his life as a dairy farmer, John was on the board of Cabot, the dairy cooperative that makes Cabot cheese. In fact, his farm was depicted on the Cabot trucks making their way through the Northeast for several years. Established in 1919 for farmers in the area surrounding Cabot, Vermont, the Cabot farmers' cooperative merged with Agri-Mark in 1992 in order to increase its production and marketing opportunities. Like Cabot, Agri-Mark was a dairy cooperative, dating back to 1916. In 2003, Agri-Mark and Cabot merged with the Chateaugay Cooperative, another upstate New York cooperative, known for its quality McCadam cheeses. This merger opened up more markets for the cooperatives because of the diversified offerings made possible through McCadam's different styles of cheeses. Is this kind of growth reflective of market consolidation? Yes, although these cooperatives have been facing the enormous behemoths of the dairy industry, a fearful position in which to dwell as a relatively small player.[6] In the best of circumstances, the cooperative model is also a fairer business model than many of the privately owned dairy processors that are repeatedly charged with monopolistic practices.

On one speaking engagement a decade ago, John stated that Cabot's entry into Walmart was one of the best things that happened to the cooperative because it expanded market opportunities and the Cabot name, meaning that farmers like him would benefit, not only in income but in quality of life. An increase in income could mean an occasional day off for some farmers, a less stressful life in trying to make ends meet, and a reduced temptation to sell what for many farmers is their only retirement account—the farm itself, perhaps for high-dollar development opportunities. His point was clear: making dairy farming financially viable is central to keeping farmers like him in business. And John preferred the cooperative method not only for keeping dairy farms solvent but also for encouraging opportunities for new farmers. This approach is not so far afield from Growing Power's Rainbow Farmer Cooperative, although dairy farms are generally larger in scale and heavier in expensive infrastructure. Dairy farms are also more vulnerable to external shocks than farms in many other sectors, with the most challenging shocks usually coming in the form of the economy or the weather—two things dairy farms can't really control.

In August 2011, Tropical Storm Irene struck the Northeast with a vengeance, essentially following the climate change models predicted for the region. Irene caused one of the most serious flooding events in Vermont's recorded history—perhaps the worst, depending upon how you calculate it. More than a dozen Vermont towns were isolated due to the damage and even disappearance of highway infrastructure. Dairy farmers throughout the Northeast were left without power and, just as importantly, without the ability to transport their milk to the processors. It's important to realize that cows *have* to be milked, usually twice a day—otherwise the cows will suffer, and ultimately die a painful death. Of course, dairy farmers almost always have backup generators for use when the electricity fails, but an utter absence of roads that allow milk truck drivers to get to the farm's bulk tank (which holds the milk at the proper temperature prior to transport) created a crisis. Farmers struggled to milk their animals, and then many of them had to toss away thousands of gallons of perfectly good milk. It was a grim scene, made even more uncertain with telecommunications outages.

Several weeks after Irene, the Vermont Sustainable Agriculture Council met to discuss Irene and preparations for increasing the state's resilience in future scenarios such as this one. The group listened to several reports from different agencies, and members compared notes from their respective regions. Lunch came around, and I found myself sitting beside Jed Davis, Cabot's current director of sustainability. I asked him how the cooperative's farmers had fared. He recounted the washed-away infrastructure and the stranded farmers who had to dump their milk for several days or even longer, suffering both severe property and enormous financial losses. "So we made a decision to pay the farmers who couldn't ship their milk to our processing facilities as if they had shipped it, based on the average of their recent shipments." I'm sure that I looked astounded. He smiled and said, "Well, it was a good thing to do." Then he paused in reflection for a brief moment and looked back up: "Actually, it was the *right* thing to do. We take care of our own."

As I was driving home, I pondered Jed's words and found myself remembering the conversations with John Malcolm several years prior. The cooperative had a cushion that it could use to do the right thing for its members in a crisis, a cushion that might not have been there had the cooperative not made its way to *shared* profitability by increasing and diversifying its markets, thereby finding a way to compete against much less benevolent privately held companies. In 2011, Agri-Mark made $900 million in sales, and it markets over three hundred million gallons of farm-fresh milk annually for more than 1,250 dairy farm families throughout New England and New York.[7]

In the end, size and geographic scale still matter. But ownership and management are sometimes more important, if the intentions and commitments are appropriate and carefully monitored, from inside and out.

Local/International

A significant part of the human experience is the quest for food, first for nourishment and then for pleasure. That quest for familiar and exotic foods has defined much of the human experience, and that's where the interpretation of "locavore" as a purist doctrine runs into trouble. It is highly unlikely that most of us will cease our distant culinary wanderings and eat only local foods. However, the pursuit of local foods is also an act of discovery, and this pursuit is important for so many reasons beyond taste, as we've seen through the previous chapters.

Nonetheless, a shopper faces some perplexing ethical challenges when maintaining the importance of local foods but hitting the co-op or grocery store to pick up bananas and coffee. There is perhaps some middle ground here, however, and it involves thinking about supporting local food *systems*, no matter where they are. To tease it all out for myself, I had to call Mexico City. (Actually, I Skyped for the first time ever, and it is that kind of technology that will play a key role in this story.)

Pablo Muñoz Ledo, the founder and CEO of Greenexus, answered. Pablo has a diverse career—once a researcher in creative education and an amateur organic farmer, he ultimately created the best-known organic label and the largest organic distribution company in Mexico. Now he is embarking on what may be his most interesting enterprise yet: an effort to create an Internet-based platform to link appropriately scaled and sustainably oriented farmers in Mexico with American and eventually European consumers. Before we get into the details of what Greenexus is, however, it's worthwhile to consider the problem statement, as recounted breathlessly by Pablo (you'll need to recall some of the terms that were covered in previous chapters): "Products are still highly intermediated. The value chain is still hidden. There is a twisted and opaque farm-to-table connection. Fair Trade has made progress, but there is still no exact origin for the products. And Fair Trade has failed as an economic strategy in selling to the First World and giving value to these distant producers for bettering their communities."[8]

"What does any of this have to do with 'local'?" you might ask. "Offshore local," Pablo would respond. "We all agree—local is good. The problems are seasonality and diversity. But it can't be so simple as being geographically close-by; it should be about being close to the producers—creating connections between the livelihoods of producers and the livelihoods of consumers."

Pablo's experience in establishing his own experimental organic farm in the 1990s is seminal in his current thinking and innovative Greenexus model. Operating his own farm helped him to understand the realities of small family farms in Mexico. "What I learned was the way my neighbors lived, what they charged, the reality that they couldn't go to the markets. The system in Mexico is old and expensive. As I learned from my own farming efforts, I came to appreciate the price of food. We need to get back to the farmers—that's the real sense of local: reconnecting. We can bring health and balance back to the food system."

As a participant in the Slow Food movement, Pablo brought traditional Mexican farmers to Terra Madre, the biannual international gathering of farmers, conservationists, chefs, and educators from across the world in Turin, Italy. What he realized was that "even if the farmers had not succeeded economically, the value of what they are doing is now recognized." The question then becomes, "How can we make it economically viable?" Pablo's experience with his farmer compatriots at Terra Madre inspired him to look for new market solutions for farmers in situations such as theirs.

One of the "aha" moments came when he toured in Europe with these farmers to see the organic markets in which their goods were being sold. The big moment came when the farmers saw their organic bananas being sold for twenty times what they had sold them for to the distributors. "The farmers didn't know the prices that their goods were being sold for, and they thought—justifiably—that something was screwed up. They realized organic bananas were really appreciated by consumers."

On seeing how the long chain of intermediaries—buyers, warehousers, distributors, retailers—that separates producers from consumers undermines the farmers, Pablo began creating his model for Greenexus. He sums up the basic idea as follows: "These intermediaries try to hide the fact that they typically choose the most expensive logistics to make as much money as possible. They control their profits by keeping things secret. We want to bring transparency to the process by providing producers with all of the services they need to export. All of the things that the intermediaries do, we will provide as a service to the farmers. It will be a 'cloud-based' online transactional business."

The model also involves using trusted and prestigious organic certifiers, with two of Pedro's top choices being Oregon Tilth and Quality Assurance International. This kind of a marketplace creates what Pablo describes as "green direct exchange," where the producer and consumer are directly connected, with no intermediary. Consumers will enjoy virtually complete transparency—or perhaps that should be "completely virtual transparency." They can scan the QR code for the product they are buying with a smartphone, and it will link them directly to a profile of the farmer who raised the product. Consumers will also be able to contact the farmer through various social media, and Greenexus will publish the online *Green Village Journal,* telling the stories of both producers and consumers. The producers, on the other hand, will theoretically be able to track their product to the consumers, collecting valuable sales data.

Greenexus is composed of a team of savvy and talented individuals, and they are planning to launch the company's first operation in 2012, beginning with one or two products and communities and perhaps one of the big players in the U.S. retail sector. Pablo notes that this company, as a pioneer in the field, has many avenues of aggregation and marketing to explore, so having one operation go really well on a large scale at the beginning is critical to show that success is possible. But he's confident in the possibility of success because of Greenexus's differentiation in the marketplace. The company's principle of *exchange*—differentiated from *trade*—fulfills a need in the marketplace for export/import systems that connect consumers to producers at a local level and that treat *all* the players fairly.

Pablo points out that it's the Internet that makes this all possible. "We can now link local communities, and we need to stress the Internet. And so far, the food industry has not benefited and grown more efficient as a result." Indeed, a quick look at the numbers shows the enormous market potential for direct international exchange: Organic Trade Association data from 2011 estimates organic products and markets at $60 billion worldwide. The United States is at about one-half of that total, around $28 billion, with the European Union comprising much of the other half.

The Internet provides "disintermediation" power, according to Pablo, which allows Greenexus to attempt to create a model that goes beyond-fair-trade. Pablo is clear that he has great respect for the intentions of the fair-trade concept, as it brought the issues to the table, but ultimately fair trade is just a seal—somebody else's judgment. The transparency potentially afforded by Greenexus allows for truly direct exchange, in which producers determine the price that they need for a product while consumers decide what they're willing to pay—taking into consideration the need to sustain and support the farmers and their sustainability agendas. It allows for a coming to terms on a fair price.

The Greenexus concept brings us once again to the question of scale. It's important to be clear about scale, Pablo notes, because in Mexico and Latin America, there is a serious contrast in scale. In terms of "small" and "medium-size" producers, the Greenexus approach refers not to family farms but rather cooperatives:

> The cooperatives that face corporations need help—and we would like to make up for those who haven't been able to reap the rewards of the marketplace. We can give them some leverage and balance. These small and medium cooperatives comprise about 80 percent of the producers, so if we can give them the way to transact and go as far as possible into destination markets, this could trigger the producers in tropical countries and even reduce illegal immigration to the United States and elsewhere. They will stay home. Currently, it's heartbreaking when "country boys" don't go to work on their home farms but go elsewhere. We need to make it worth it for them to be on their own farms.

The model that Pablo and his colleagues are putting together may end up allowing those of us committed to rebuilding local food systems to assuage our guilt at buying the tropical products so many of us love. In the end, it seems to be distance that creates the largest contradiction for us—it carries both ecological and economic questions with it. The dilemma for us is largely one of values: Are we more concerned about the ramifications of distance or the ability to connect and ensure fairness on both sides of the "direct exchange"? Can this kind of direct exchange truly link local communities around the world while also helping us think more complexly about the foodsheds (the flows) of specific foods that cross our borders and our palates with considerable frequency? I leave these as important questions for us all to chew on.

All/Nothing

The final problematic divide is relatively easy to articulate but more complex to balance in a way that feels satisfactory and not overly contradictory. It is the fairly simplistic notion that it's "all or nothing" when it comes to local. While certain individuals and families are able to procure virtually all of their food from their land and local markets, that simply is not a possibility for the vast majority in our modern industrial society. The impediments are too great, whether they be access to land, markets, or money. But we can all make choices that lead to a balance that feels appropriate.

As geographer Nathan McClintock puts it, "We need to recognize the limits about what local can accomplish on its own. Ultimately, it has to be a 'both/and' resolution. The concept of food miles is breaking down. Fetishizing local for some spatial consideration is problematic. It's a naive way of understanding the food system. . . . Let's reframe it in terms of supporting local economies, creating jobs, [and] educating about food, health, and nutrition. There is something important to the resilience concept in this regard—we diversify our financial portfolios, so why not diversify the food system?"[9]

It should be clear by now that I believe local food systems are not only viable but also vital—and *community-based* food systems are even more important. Our national and global food systems are *not* diversified without strong community-based food systems—but they are also only part of the answer. When we dabble in hyperbole or make unrealistic claims, we are only providing the critics with strong counterarguments. More importantly, perhaps, such naïveté sets us up for disillusionment when we can't make it all happen.

Remapping Our Expectations

I have a final confession to make. When I began researching and writing this book, I was initially thinking primarily in terms of producers, consumers, and markets. Of course, energy, environmental impacts, food security, food justice, and biodiversity were important issues for me from the outset. However, it became clearer over time that the rebuilding of local food systems by any descriptor—resilient, sustainable, or community based—will succeed only if we begin this hard work as *citizens*, not as consumers or producers or entrepreneurs.

As a patient reader of this book, you have essentially walked with me on a long journey in circumambulating the outer edges of this thing we have been calling "local food," peering in to try and getter a better perspective from a variety of angles. If you have read the chapters in sequential order, you have essentially retraced the chronology of my thinking—influenced by people, places, writing of all sorts, and more than a few meals. In some ways, you have witnessed the evolution in my understanding of what really matters here, and you have perhaps noticed a progression from "local" to "local food systems" to "community-based food systems." Of course, I have retraced that circle many times now, with the guidance and critique of editors and several invited readers, and many people saw these realities well before I ever tried to articulate them.

My goal has been to present you with a means to help all of us get to the point that I referenced early in the book: "the Local Food Systems 2.0 campfire conversations" that require us to think a little longer, a little harder, and increasingly in the company of others who are strangers only because the campfire has not been bright enough or the circle around it wide enough. I think we are getting there, and I hope that this book and the associated information and networking opportunities at its online home, www.resilience.org, will play a role in helping to envision and create the food systems that we want.

There is such a surge of interest in food, farming, resilience, and all things local that I cannot help but believe that these campfire conversations will increase not only in size and number but also in effectiveness. It is time to get things done—and it's happening all across the country already in a way that demonstrates like no other recent social phenomena the extraordinary diversity that defines the United States. And rebuilding community-based food systems all across the country has the opportunity to change the way we view democracy and justice. Why? Because, as I mentioned in the preface, food is the one inalienable right that no one of any political persuasion dares to dismiss.

If we are wise, then I think that we will try to make sure that these campfire conversations reflect on the intersection—real and potential—between food and democracy. While I myself am skeptical about the possibility of creating a "democratic food system," I do believe that we must continually work to *democratize* food systems to the greatest extent possible. And the best chance for success in those efforts is at the local level, the scale at which it is easiest to translate *passion* for local foods into *compassion* (*com* meaning "with") for those living in our communities. Compassion and citizenship are at the root of community-based food systems.

The work ahead of us is not easy. It requires us to move from a sense of individual resignation to a spirit of collective resolve. The values that drive us to begin this community-based food systems work must also be values that can sustain us over the long haul. In the end, building resilient local food systems is a remapping of our expectations. It is a cartography of hope.

Notes

Preface

1. Jeff Nelson, "How My Family Started the Armour Meat Company," video, Vegsource.com, October 31, 2010, http://www.vegsource.com/news/2010/10/how-my-family-started-the-armour-meat-company.html. In this talk a figure of $6,000 is given, whereas the PBS documentary *Chicago: City of the Century* puts the sum at $8,000. Regardless, it was an enormous sum of money at the time.
2. Thomas Petraitis, "East St. Louis, Illinois: 'Hog Capital of the Nation,'" *Ecology of Action* (blog), Preservation Research Office, May 9, 2005, http://preservationresearch.com/2005/05/east-st-louis-illinois-hog-capital-of-the-nation/.
3. "People & Events: Philip Danforth Armour (1832–1901)," on the website for PBS's *American Experience*, accessed October 2, 2012, http://www.pbs.org/wgbh/amex/chicago/peopleevents/p_armour.html.
4. Ibid.
5. "Cattle Pens, Union Stock Yard, c.1920s," entry in the online *Electronic Encyclopedia of Chicago* (Chicago Historical Society, 2005), http://www.encyclopedia.chicagohistory.org/pages/3402.html.
6. Louis Carroll Wade, "Meatpacking," entry in the online *Electronic Encyclopedia of Chicago* (Chicago Historical Society, 2005), http://www.encyclopedia.chicagohistory.org/pages/804.html.
7. Petraitis, "East St. Louis, Illinois."
8. Rick Halpern, "Packinghouse Unions," entry in the online *Electronic Encyclopedia of Chicago* (Chicago Historical Society, 2005), http://www.encyclopedia.chicagohistory.org/pages/943.html.
9. Wade, "Meatpacking."
10. Petraitis, "East St. Louis, Illinois."
11. Barbara Krasner-Khait, "The Impact of Refrigeration," *History Magazine* (February/March 2000), http://www.history-magazine.com/refrig.html.

Introduction

1. "Shift Local," web page for an initiative of Support Local North Carolina, accessed June 14, 2012, http://www.supportlocalnc.com/ShiftLocal.html.
2. "What Is 2.0? A New Era Defined," *Duane's Dartboard* (blog), February 17, 2010, http://duanehallock.com/2010/02/17/what-is-2-0-a-new-era-defined/.

Chapter 2: The Geography of Local

1. For more information on food chain clusters, see the various works by William D. Heffernan, available on the website of the Food Circles Networking Project at http://www.foodcircles.missouri.edu/consol.htm.
2. Edward S. Casey, "Between Geography and Philosophy: What Does It Mean to Be in the Place-World?" *Annals of the Association of American Geographers* 91, no. 4 (2001): 684.
3. Robert Feagan, "The Place of Food: Mapping Out the 'Local' in Local Food Systems," *Progress in Human Geography* 31, no. 1 (2007): 33.

4. Amy Trubek, *The Taste of Place: A Cultural Journey into Terroir* (Berkeley: University of California Press, 2008).
5. Gary Paul Nabhan, interview by the author, September 15, 2007.
6. C. Clare Hinrichs and Tom Lyson, eds., *Remaking the North American Food System: Strategies for Sustainability* (Board of Regents of the University of Nebraska, 2007), 19.
7. Laura B. DeLind, "Are Local Food and the Local Food Movement Taking Us Where We Want to Go? Or Are We Hitching Our Wagons to the Wrong Stars?" *Agriculture and Human Values* 28, no. 2 (June 2011): 274.
8. Feagan, "The Place of Food," 27–28.
9. Michael W. Hamm and Anne Bellows, "Community Food Security: Background and Future Directions," *Journal of Nutrition Education and Behavior* 35, no. 1 (2003).
10. For more discussion about the agriculture of the middle, visit the website of the national Agriculture of the Middle initiative at http://www.agofthemiddle.org/.

Chapter 3: How Far Should Local Go?

1. Branden Born and Mark Purcell, "Avoiding the Local Trap: Scale and Food Systems in Planning Research," *Journal of Planning Education and Research* 26, no. 2 (2006): 195.

Chapter 4: Energy

1. Martin C. Heller and Gregory A. Keoleian, *Life Cycle-Based Sustainability Indicators for Assessment of the U.S. Food System*, report no. CSS00-04 (Ann Arbor: University of Michigan Center for Sustainable Systems, December 6, 2000), 42.
2. David Pimentel et al., "Reducing Energy Inputs in the U.S. Food System," *Human Ecology* 36 (July 15, 2008): 459.
3. It is important to note here that I also think it imperative that we consider the plights of those persons well beyond our local and national borders. The point here is that it is often easier to begin the caring process when there are direct and proximate relationships. "Local," in my view, is a starting point for caring—not an endpoint of any sort.
4. "Food-insecure" populations include persons who have limited or uncertain access to nutritionally appropriate foods.
5. Alan Alda, *Things I Overheard while Talking to Myself* (New York: Random House, 2008), 47.
6. David Pimentel and Marcia H. Pimentel, *Food, Energy and Society* (New York: CRC Press, 2008).
7. For more information on the number of calories expended in producing a single calorie of food in the U.S. food system, see Heller and Keoleian, *Life Cycle-Based Sustainability Indicators*. More information on this topic can also be found in Richard Heinberg and Michael Bomford, *The Food & Farming Transition* (Sebastopol, Calif.: Post Carbon Institute, Spring 2009), http://www.postcarbon.org/report/41306-the-food-and-farming-transition-toward.
8. John C. Jeavons, "Biointensive Mini-Farming," *Journal of Sustainable Agriculture* 19, no. 2 (2001): 81–83.
9. "Characterizing Ag of the Middle and Values-Based Food Supply Chains," Agriculture of the Middle website, January 2012, http://www.agofthemiddle.org.
10. It is not commonly understood that organic vegetable production can require increased field cultivation to reduce weed pressures and that organic fruit production often utilizes more frequent sprayings of less potent materials than in conventional fruit production. These energy tradeoffs can be mitigated to some degree with other management practices that do not require fossil fuels, but it is important to bear in mind that virtually every agricultural practice has its tradeoffs. Having a clear end goal can help a farmer determine which practices to adopt and reject.

11. Patrick Canning, Ainsley Charles, Sonya Huang, Karen R. Polenske, and Arnold Waters, *Energy Use in the U.S. Food System*, Economic Research Report no. 94 (Washington, D.C.: USDA Economic Research Service, March 2010), 18.
12. Ibid.
13. For more information about the company and its sustainability efforts, visit the website of Veritable Vegetable at http://veritablevegetable.com/.
14. "Distribution Models for Local Food," Center for Integrated Agricultural Systems at the University of Wisconsin–Madison, January 2009, http://www.cias.wisc.edu/uncategorized/distribution-models-for-local-food/.
15. "About Us," a history of FreeAire Refrigeration; http://freeaire.com/about-us/ (accessed October 2, 2012).
16. Alex Wilson, "The Energy Smart Kitchen," *Fine Homebuilding*, Fall/Winter 2007.
17. David M. Cutler, Edward L. Glaeser, and J. M. Shapiro. "Why Have Americans Become More Obese?" *Journal of Economic Perspectives* 17, no. 3 (2003): 103.
18. Jennifer Poti and Barry Popkin, "Trends in Energy Intake among US Children by Eating Location and Food Source, 1977–2006," *Journal of the Academy of Nutrition and Dietetics* 111, no. 8 (2011): 1156–64. The researchers indicate in their conclusions that these figures are, in fact, quite conservative, and the numbers are likely higher.
19. Pimentel et al., "Reducing Energy Inputs in the U.S. Food System," 459.
20. Ibid., 460.
21. The Food Research and Action Center website provides extensive data and publications documenting the disturbing hunger levels of residents living in agriculturally rich regions such as some counties in Florida. For more information, visit the center's "Data and Publications" page at http://frac.org/reports-and-resources/.
22. For information on apples and other U.S. agricultural commodities, see the Agricultural Marketing Resource Center "Commodity Apple Profile" pulled together by Malinda Geisler of Iowa State University, at http://www.agmrc.org/commodities__products/fruits/apples/commodity_apple_profile.cfm.
23. "WTO and U.S. Chicken Exports: China Puts Its Case," the Poultry Site, September 22, 2011, http://www.thepoultrysite.com/poultrynews/23599/wto-and-us-chicken-exports-china-puts-its-case.
24. For a superb overview of the impacts of our food system on energy consumption and climate change, see the Food Climate Research Network's report *Cooking Up a Storm*, by Tara Garnett (September 2008), available at http://www.fcrn.org.uk/fcrn/publications/cooking-up-a-storm. Tara Garnett also coauthored, with Tim Jackson, one of the best available histories of refrigeration in our food system in her paper "Frost Bitten: An Exploration of Refrigeration Dependence in the U.K. Food Chain and Its Implications for Climate Policy," presented to the 11th European Round Table on Sustainable Consumption and Production, Basel, Switzerland, June 2007, and available at http://www.fcrn.org.uk/fcrn/publications/frost-bitten.
25. "Food and Alcoholic Beverages: Total Expenditures," table 1 in the USDA Economic Research Service Food Expenditure Series, accessed November 26, 2011, http://www.ers.usda.gov/briefing/cpifoodandexpenditures/Data/Expenditures_tables/table1.htm.
26. Hayden Stewart, Noel Blisard, and Dean Jollife, *Let's Eat Out: Americans Weigh Taste, Convenience, and Nutrition*, Economic Research Service Economic Information Bulletin no. 19 (Washington, D.C.: USDA Economic Research Service, October 2006), 1, http://www.ers.usda.gov/publications/eib19/eib19.pdf.
27. For an overview of the terms "food loss" and "food waste," see Jenny Gustavsson et al., *Global Food Losses & Food Waste: Extent, Causes, & Prevention* (Rome: Food and Agriculture Organization of the United Nations, 2011), http://www.fao.org/fileadmin/user_upload/ags/publications/GFL_web.pdf.
28. Ibid., 2.

29. Ibid., 5.
30. "Basic Information about Food Waste," U.S. Environmental Protection Agency, accessed November 26, 2011, http://www.epa.gov/osw/conserve/materials/organics/food/fd-basic.htm.
31. William Yardley, "Cities Get So Close to Recycling Ideal, They Can Smell It," *New York Times*, June 27, 2012, http://www.nytimes.com/2012/06/28/us/a-recycling-ideal-so-close-cities-can-smell-it.html?_r=1&pagewanted=all.
32. "First Municipal Food Waste-to-Renewable Energy Facility to Connect to Power Grid in Urban Setting in U.S.," press release, August 23, 2012, http://www.prweb.com/releases/prweb2012/8/prweb9831566.htm.
33. Stephanie Pruegel, "Pioneering Partnership Optimizes Power Production," *BioCycle*, July 2010, 51.
34. Anya Kamanetz, "The Starbucks Cup Dilemma," *Fast Company* online, October 20, 2010, http://www.fastcompany.com/magazine/150/a-story-of-starbucks-and-the-limits-of-corporate-sustainability.html.
35. *2008 Fast Food Industry Packaging Report* (Asheville, N.C.: Dogwood Alliance, n.d.), http://www.nofreerefills.org/download-report/.

Chapter 5: Environment

1. "Pragmatic Visionary," *BioCycle*, June 2012, 4.
2. "Passion, Vision, and Grit: Jerome Goldstein, Ecopioneer, 1931–2012," *BioCycle*, June 2012, 39.
3. Ibid., 40.
4. Nerlita M. Manalili, Moises A. Dorado, and Robert van Otterdijk, *Appropriate Food Packaging Solutions for Developing Countries* (Rome: Food and Agriculture Organization of the United Nations, 2011), 4.
5. Morgan Erickson-Davis, "Fast-Food Industry Destroying Forests in the Southern U.S.," Mongabay.com, April 28, 2008, http://news.mongabay.com/2008/0428-davis_nofreerefills.html.
6. Center for Sustainable Systems, University of Michigan, 2011, *U.S. Food System Factsheet*, pub. no. CSS01-06.
7. "More Than 1 in 5 American Children Live at Risk of Hunger," Feeding America press release, September 29, 2011, http://feedingamerica.org/press-room/press-releases/fa-children-at-risk-of-hunger.aspx.
8. Rattan Lal, "Carbon Management in Agricultural Soils," *Mitigation and Adaptation Strategies for Global Change* 12 (2007): 304.
9. James Walcott, Sarah Bruce, and John Sims, *Soil Carbon for Carbon Sequestration and Trading: A Review of Issues for Agriculture and Forestry* (Australia Bureau of Rural Sciences, Dept. of Agriculture, Fisheries & Forestry, 2009), 3.
10. It is important to note here that ecosystems are never really "stable" or "undisturbed"—they are always in flux, although the chronology of those dynamics may be stretched well beyond what we as humans would consider to be frequent.
11. Lal, "Carbon Management in Agricultural Soils," 304.
12. Walcott et al., *Soil Carbon for Carbon Sequestration and Trading*, 15–16.
13. Lal, "Carbon Management in Agricultural Soils," 304.
14. "Soil Organic Matter," *Soil Biology Primer* (online), on the website of the USDA Natural Resources Conservation Service, accessed March 26, 2012, http://soils.usda.gov/sqi/concepts/soil_organic_matter/som.html.
15. Charles Henry, D. Sullivan, R. Rynk, K. Dorsey, and C. Cogger, "The Nitrogen Cycle," chapter 1 in *Managing Nitrogen from Biosolids*, Ecology Publication no. 99-508 (Washington State Department of Ecology, April 1999), 2.1.
16. E. Favoino and D. Hogg, "The Potential Role of Compost in Reducing Greenhouse Gases," *Waste Management & Research* 26 (February 1, 2008): 62.

17. *A Rock and a Hard Place: Peak Phosphorus and the Threat to Our Food Security* (Bristol, U.K.: Soil Association, 2010) , 4, http://www.soilassociation.org/innovativefarming/policyresearch/resourcedepletion.
18. "Phosphorus: A Vital Source of Animal Nutrition," Inorganic Feed Phosphates, accessed December 10, 2011, http://www.feedphosphates.org/guide/phosphorus.html.
19. "Phosphate Rock," entry in *Mineral Commodity Summaries 2010* (U.S. Geological Survey, 2010), 118, http://minerals.usgs.gov/minerals/pubs/mcs/2010/mcs2010.pdf.
20. *A Rock and a Hard Place*, 8.
21. David Tilman, Christian Balzer, Jason Hill, and Belinda Befort, "Global Food Demand and the Sustainable Intensification of Agriculture," *Proceedings of the National Academy of Sciences*, November 21, 2011, http://www.pnas.org/cgi/doi/10.1073/pnas.1116437108.
22. *A Rock and a Hard Place*, 14. Also see Mara Grunbaum, "Gee Whiz: Human Urine Is Shown to Be an Effective Agricultural Fertilizer," *Scientific American*, July 23, 2010.
23. *Municipal Solid Waste Generation, Recycling, and Disposal in the United States: Facts and Figures for 2010* (Washington, D.C.: U.S. Environmental Protection Agency, December 2011), http://www.epa.gov/osw/nonhaz/municipal/index.htm.
24. Ibid.
25. J. Mata-Alvarez, S. Macé, and P. Llabrés, "Anaerobic Digestion of Organic Solid Wastes: An Overview of Research Achievements and Perspectives," *Bioresource Technology* 74, no. 1 (2000): 3–16; *Solid Waste Management and Greenhouse Gases: A Life-Cycle Assessment of Emissions and Sinks*, 3rd ed. (Washington, D.C.: U.S. Environmental Protection Agency, September 2006), http://www.epa.gov/climatechange/wycd/waste/downloads/fullreport.pdf; Donald Gray, *Anaerobic Digestion of Food Waste*, Funding Opportunity no. EPA-R9-WST-06-004 (U.S. Environmental Protection Agency, East Bay Municipal Utility District, March 2008), http://www.epa.gov/region9/organics/ad/EBMUDFinalReport.pdf; Dominic Hogg, Adrian Gibbs, Enzo Favoino, and Marco Ricci, *Managing Biowastes from Households in the U.K.: Applying Life-Cycle Thinking in the Framework of Cost-Benefit Analysis* (WRAP, May 2007), www.eunomia.co.uk.
26. Mata-Alvarez, Macé, and Llabrés, "Anaerobic Digestion of Organic Solid Wastes," 11.
27. "National Program 206: Manure and Byproduct Utilization: FY 2005 Annual Report," USDA Agricultural Research Service, last updated October 28, 2008, http://www.ars.usda.gov/research/programs/programs.htm?np_code=206&docid=13337.
28. Victor Hugo, *Les Misérables*, transl. M. Jules Gray (London: H. S. Nichols, 1895).
29. For more information, see the website of the Ecological Sanitation Research Program at http://www.ecosanres.org and the website of the Sulabh International Social Service Organisation at http://www.sulabhinternational.org/.
30. *The Benefits of Anaerobic Digestion of Food Waste at Wastewater Treatment Facilities* (U.S. Environmental Protection Agency, n.d.), accessed December 10, 2011, http://www.epa.gov/region9/organics/ad/Why-Anaerobic-Digestion.pdf.
31. Ronald Leblanc, Peter Matthews, and Richard Roland, eds., *Global Atlas of Excreta, Wastewater Sludge, and Biosolids Management: Moving Forward the Sustainable and Welcome Uses of a Global Resource* (United Nations Human Settlements Program, 2008), 81.

Chapter 6: Food Security

1. The national Master Gardening program was started as part of Washington State University's Cooperative Extension Service in 1972 and has now spread to all fifty states and Canada. For more information on programming in your state, see the "Master Gardeners" page of the American Horticultural Society at http://www.ahs.org/master_gardeners/index.htm.
2. Michael J. Weiss, "Going from White House to Jail, William Smith Devises a New Chef D'oeuvre—Making Cooks of Crooks," *People* 23, no. 18 (October 28, 1985), http://www.people.com/people/archive/article/0,,20092045,00.html; Karen Haywood, "Armed with

Knives, Inmates Learn to Dish It Out," *Los Angeles Times*, October 22, 1995, http://articles.latimes.com/1995-10-22/news/mn-59735_1_cooking-class.

3. Francis Koster, "Culinary Arts Program for Ex-offenders Places 85% of Graduates; During Training They Feed 5000 Meals a Week to Disadvantaged," Optimistic Futurist, December 6, 2012, http://theoptimisticfuturist.org/index.php/americas-correctional-system/culinary-arts-program-for-ex-offenders-places-85-of-graduates-during-training-they-feed-5000-meals-a-week-to-disadvantaged.html.
4. Shannon Mullen, "Prison Meal Deal: Where the Staff Serves Lunch . . . and Time," National Public Radio broadcast, February 3, 2012, as featured on the website of WBUR (Boston), http://www.wbur.org/npr/146110728/prison-meal-deal-where-the-staff-serves-lunch-and-time.
5. Chris Camire, "Mass. Inmates Cook Up a Future after Prison," *Lowell Sun*, October 27, 2010, as featured on the website CorrectionsOne.com, http://www.correctionsone.com/re-entry-and-recidivism/articles/2853164-Mass-inmates-cook-up-a-future-after-prison/.
6. Alisha Coleman-Jensen, Mark Nord, Margaret Andrews, and Steven Carlson, *Household Food Security in the United States in 2010,* Economic Research Report no. 125 (Washington, D.C.: USDA Economic Research Service, September 2011), 11.
7. Ibid.
8. This is not intended as a reference to the Lexicon of Sustainability project, but readers are encouraged to look into this project, as it is quite relevant to many of the discussions in this book. See the project's website at http://www.lexiconofsustainability.com for more information.
9. "Food Security: Concepts and Measurement," chapter 2 in *Trade Reforms and Food Security: Conceptualizing the Linkages* (Rome: Food and Agriculture Organization of the United Nations, 2003), available at http://www.fao.org/docrep/005/y4671e/y4671e06.htm). The term "food security" first appeared in the mid-1970s at a time of a global food crisis, so early definitions focused on food supply issues as they impacted food availability and costs. In the early 1980s, the notion of access was added to the prevailing definitions; in other words, it didn't matter if there was a supply of food if the people who needed it could not access it. About a decade later, food safety and nutritional balance were added as important components of food security. Then the new millennium approached and the importance of cultural background and personal choices emerged, bringing food preferences into the picture. The moving target of such definitions can be challenging, but bear in mind that it's a result of our understanding and more inclusive conversations. The FAO document cited here provides an excellent overview of the evolution of the understanding of the food security concept internationally.
10. "Food Security in the United States: Measurement," USDA Economic Research Service, accessed March 17, 2012, http://www.ers.usda.gov/Briefing/FoodSecurity/measurement.htm#what.
11. Ibid.
12. Michael W. Hamm and Anne Bellows, "Community Food Security: Background and Future Directions," *Journal of Nutrition Education and Behavior* 35, no. 1 (2003). As with the term "food security," it is interesting to note the aspects of the definition of "community food security" that go beyond the goals of full bellies and sufficient calories. Community food security initiatives work to address systemic concerns regarding the availability, affordability, and accessibility of healthy food for everyone in the community.
13. "Food Security in the U.S.: Community Food Security," USDA Economic Research Service, accessed September 3, 2012, http://www.ers.usda.gov/topics/food-nutrition-assistance/food-security-in-the-us/community-food-security.aspx.
14. "Glossary," Growing Food & Justice for All Initiative, accessed March 17, 2012, http://www.growingfoodandjustice.org/Glossary.html.
15. "Food Justice," Just Food, accessed March 17, 2012, http://www.justfood.org/food-justice.
16. WhyHunger home page, accessed August 26, 2012, http://www.whyhunger.org/.
17. For more details on the Community Food Security Coalition's transition, see the group's website at http://www.foodsecurity.org.

18. Theresa Snow's blog posts during her trip across the country depict in words and images the stunning creativity of food security advocates and social entrepreneurs across the United States, including a peek into Canada. Visit her blog, called *Salvationfarms*, at http://salvationfarms.wordpress.com/2011-research-trip/.
19. Theresa Snow, "More Progress through Partnerships," *Salvationfarms* (blog), July 28, 2012, http://salvationfarms.wordpress.com/2012/07/28/more-progress-through-partnerships/.
20. These quotations come from a presentation by Josh Slotnick at Green Mountain College on November 10, 2011. For more information on Garden City Harvest, see the group's website at http://www.gardencityharvest.org/. *Growing a Garden City* by Jeremy Smith (New York: Skyhorse Publishing, 2010) is a magnificent telling of the story through the words of Missoula citizens and the stunning photographs of Chad Harder and Sepp Jannotta.
21. "Bisphenol A (BPA): Questions and Answers about BPA," National Institute of Environmental Health Sciences, accessed September 5, 2012, http://www.niehs.nih.gov/news/sya/sya-bpa/.
22. "Bisphenol A (BPA) Action Plan Summary," U.S. Environmental Protection Agency, last updated July 31, 2012, http://www.epa.gov/oppt/existingchemicals/pubs/actionplans/bpa.html#concern.
23. "Bisphenol A (BPA) Information for Parents," U.S. Department of Health and Human Services, accessed September 3, 2012, http://www.hhs.gov/safety/bpa/.
24. "Bisphenol-A," Environmental Working Group, accessed September 5, 2012, http://www.ewg.org/chemindex/chemicals/bisphenolA.
25. "General Mills to Pull BPA from Organic Tomato Cans," GreenBiz.com, April 19, 2010, http://www.greenbiz.com/news/2010/04/19/general-mills-pull-bpa-organic-tomato-cans.
26. Elisa Crouch, "Agent of Change Retiring from Maplewood-Richmond Heights Schools," *St. Louis Post-Dispatch*, May 21, 2012, http://www.stltoday.com/news/local/education/agent-of-change-retiring-from-maplewood-richmond-heights-schools/article_b57146b9-fc75-56bc-8d40-0daedaec1a0c.html.
27. "Our History: Sappington Farmers' Market," accessed September 5, 2012, http://www.sappingtonfarmersmkt.com/about.htm.
28. "Farm to School Workshop," *Healthy Eating with Local Produce (H.E.L.P)* (blog), February 18, 2011, http://healthyeatingwithlocalproduce.blogspot.com/2011/02/farm-to-school-workshop.html.
29. Cathy Farnworth and Jessica Hutchings, *Organic Agriculture and Womens' Empowerment*, (Bonn, Germany: International Federation of Organic Agriculture Movements, April 2009). Typographic errors in the quoted text were corrected per its cited data source in Lohr and Park, "Gender Effects on Adoption of Organic Weed Management Techniques," 16th IFOAM Organic World Congress, Modena, Italy, June 16-20, 2008, archived at http://orgprints.org/12109.
30. "Our Story," FoodCorps, accessed March 20, 2012, http://foodcorps.org/about/our-story.
31. "About Montana's FoodCorps: Frequently Asked Questions," Grow Montana, accessed March 20, 2012, http://www.growmontana.ncat.org/foodcorps_faq08.php.
32. "About DBCFSN," Detroit Black Community Food Security Network, accessed March 20, 2012, http://detroitblackfoodsecurity.org/.
33. Aba Ifeoma, interview by the author, July 2, 2012.
34. "About DBCFSN," Detroit Black Community Food Security Network, accessed March 20, 2012, http://detroitblackfoodsecurity.org/ (March 20, 2012); Monica M. White, "D-Town: African American Farmers, Food Security, and Detroit," *Black Agenda Report* online, June 22, 2010, http://www.blackagendareport.com/?q=content/d-town-african-american-farmers-food-security-and-detroit; "Detroit Urban Agriculture Movement Looks to Reclaim Motor City," an interview with Malik Yakini on *Democracy Now!* June 24, 2010, http://www.democracynow.org/2010/6/24/detroit_urban_agriculture_movement_looks_to. Growing Power home page, accessed March 20, 2012, http://www.growingpower.org.

35. For more information on the role of Beulah Land Farms as well as another large-scale operation, Muhammad Farms in Georgia, see the websites for both organizations (http://theshrineonline.org/beulah-land and http://www.muhammadfarms.com/) and read Priscilla McCutcheon's essay on these initiatives, "Community Food Security 'For Us, By Us': The Nation of Islam and the Pan African Orthodox Christian Church," chapter 8 in *Cultivating Food Justice: Race, Class, and Sustainability*, ed. Alison Hope Alkon and Julian Agyeman (Boston: MIT Press, 2011).
36. Monica White, "Sisters of the Soil," *Soil2Soul* (blog), May 25, 2011, http://soil2soul.blogspot.com/2011/05/queens-council-of-detroits-urban-ag.html.
37. Monica White, *Soil2Soul* (blog), accessed March 20, 2012, http://soil2soul.blogspot.com/.
38. "Garden Resource Program", Greening of Detroit, accessed August 23, 2012, http://detroitagriculture.net/urban-garden-programs/garden-resource-program/.
39. Kathryn Colasanti, Charlotte Litjens, and Michael Hamm, *Growing Food in the City: The Production Potential of Detroit's Vacant Land* (C.S. Mott Group for Sustainable Food Systems at Michigan State University, June 2010), 3.
40. "Malik Yakini, Detroit Food Justice Activist," Institute for Agriculture and Trade Policy Food and Community Fellows, April 2, 2012, http://foodandcommunityfellows.org/blog/2012/malik-yakini-detroit-food-justice-activist.

Chapter 7: Food Justice

1. Erika Allen, conversation with the author, August 1, 2012.
2. Even though I find any term that categorizes people predominately by skin color problematic in some ways, I have opted to use the term "communities of color" in lieu of "minorities" for two reasons: 1) the racial groups described in some of these regions are not minorities in those regions, and 2) "whites" will soon be a minority within the United States.
3. For more information on these programs, begin with these resources: National Hunger Clearinghouse, http://www.whyhunger.org/findfood; National Commodity Supplemental Food Program Association, http://www.csfpcentral.org; Feeding America, http://feedingamerica.org/how-we-fight-hunger/programs-and-services/public-assistance-programs.aspx; American Commodity Distribution Association, http://www.commodityfoods.org/about_acda.php.
4. Patricia Allen, "Reweaving the Food Security Safety Net: Mediating Entitlement and Entrepreneurship," *Agriculture and Human Values* 16, no. 2 (1999): 117.
5. "Most Children Younger than Age 1 Are Minorities, Census Bureau Reports," press release from the U.S. Census Bureau, May 17, 2012, http://www.census.gov/newsroom/releases/archives/population/cb12-90.html.
6. For more information about the National Farm to School Network, its mission, and its programs, visit its website at http://www.farmtoschool.org/.
7. Alison Alkon and Julian Agyeman, eds., *Cultivating Food Justice: Race, Class, and Sustainability* (Boston: MIT Press, 2011).
8. Ibid., 7.
9. Ibid., 12.
10. Michael Windfuhr and Jennie Jonsén, *Food Sovereignty: Towards Democracy in Localized Food Systems* (FIAN/ITDG Publishing, 2005), 45, http://www.ukabc.org/foodsovpaper.htm. This paper provides an invaluable overview of the evolution of the early origins of the concept of food sovereignty.
11. "Definition of Food Sovereignty," from the 2007 Food Sovereignty Forum Declaration of Nyéléni, on the website of the International Planning Committee for Food Sovereignty, http://www.foodsovereignty.org/FOOTER/Highlights.aspx.
12. Ibid.

13. "Local Food and Community Self-Governance Ordinance," Town of Sedgwick, Maine, 2011, http://www.sedgwickmaine.org/content/view/634/41/.
14. "NAAO Water Rights Settlement Program," U.S. Department of the Interior Bureau of Reclamation, accessed March 28, 2012, http://www.usbr.gov/native/naao/water/index.html; Felicity Barringer, "Indians Join Fight for an Oklahoma Lake's Flow," *New York Times*, April 11, 2011, http://www.nytimes.com/2011/04/12/science/earth/12water.html?pagewanted=all; Karl Puckett, "Indian Tribes Exercising Water Rights," *USA Today*, February 25, 2008, http://www.usatoday.com/news/nation/2008-02-25-water-rights_N.htm.
15. Hispanos are natives or residents of the southwestern United States descended from Spaniards settled there before annexation.
16. *Hispano and Native American Farmers in New Mexico*, submitted by the Center of Southwest Culture, Inc., to the Bioneers Dreaming New Mexico Project (March 15, 2009), 4–5.
17. Ibid., 9–10.
18. Nathan McClintock and Jenny Cooper, *Cultivating the Commons: An Assessment of the Agricultural Potential of Oakland's Public Land* (Department of Geography at the University of California–Berkeley, 2009), http://www.oaklandfood.org/media/AA/AD/oaklandfood-org/downloads/27621/Cultivating_the_Commons_COMPLETE.pdf.
19. *Inventory of Farmworker Issues and Protections in the United States*, United Farm Workers and Bon Appétit Management Company Foundation (March 2011), 5.
20. Ibid., 3–4.
21. Ibid., 6.
22. Ibid.
23. Victor J. Oliveira, J. Runyan Effland, and S. Hamm, *Hired Farm Labor Use on Fruit, Vegetable, and Horticultural Specialty Farms* (Washington, D.C.: USDA Agricultural Research Service, 1993).
24. "America's Agricultural Labor Crisis: Enacting a Practical Solution," official hearing notice and witness list of September 27, 2011, for a hearing before the Senate Judiciary Committee Subcommittee on Immigration, Refugees, and Border Security on October 4, 2011, http://www.judiciary.senate.gov/hearings/hearing.cfm?id=0bd5589287f5bbb3d229c1850f7b44e2.
25. Sarita Chourey, "Illegal Immigration Law Worries South Carolina Farmers, *Augusta* [Georgia] *Chronicle*, October 7, 2011, http://chronicle.augusta.com/news/business/2011-10-07/illegal-immigration-law-worries-south-carolina-farmers#comment-1017043.
26. For a fuller understanding of the challenges in finding data on farmworkers in the United States, see Vera Chang's blog posting on this topic: "Hungry for Data: My Thoughts on the Inventory of Farmworker Issues and Protections," Bon Appétit Management Company, May 3, 2011, http://www.bamco.com/blog/archives/thoughts_on_the_inventory. Chang was one of the lead researchers for the *Inventory of Farmworker Issues and Protections in the United States* (cited above).
27. "About America's Farmworkers: Population Demographics," National Center for Farmworker Health, accessed March 27, 2012, http://www.ncfh.org/?pid=4&page=3.
28. *Inventory of Farmworker Issues and Protections in the United States*, v.
29. "Enumeration and Population Estimates," National Center for Farmworker Health, September 8, 2012, http://www.ncfh.org/?pid=23. This web page helps us see some of the methodological challenges associated with research related to farmworker populations.
30. "The Agricultural Economy," National Center for Farmworker Health, September 8, 2012, http://www.ncfh.org/?pid=4&page=4; Doris P. Slesinger and Steven Deller, *Economic Impact of Migrant Workers on Wisconsin's Economy,* CDE Working Paper no. 2002-06 (Center for Demography and Ecology, University of Wisconsin–Madison, 2003), http://www.ssc.wisc.edu/cde/cdewp/2002-06.pdf.
31. *Inventory of Farmworker Issues and Protections in the United States*, 6.

32. Mary Bauer and Mónica Ramírez, *Injustice on Our Plates: Immigrant Women in the U.S. Food Industry* (Montgomery, Ala.: Southern Poverty Law Center, 2010), 30, http://cdna.splcenter.org/sites/default/files/downloads/publication/Injustice_on_Our_Plates.pdf; Denise VanDeCruze and Melinda Wiggins, *Poverty and Injustice in the Food System: Report for Oxfam America* (Durham, N.C.: Student Action with Farmworkers, June 2008), http://ducis.jhfc.duke.edu/wp-content/uploads/2010/06/Poverty-and-Injustice-in-the-Food-System.pdf.
33. Mary Bauer, *Close to Slavery: Guestworker Programs in the U.S.* (Montgomery, Ala.: Southern Poverty Law Center, 2007), http://www.gpn.org/splcenter.org.SPLCguestworker.pdf; Bauer and Ramírez, *Injustice on Our Plates; Picked Apart: The Hidden Struggles of Migrant Worker Women in the Maryland Crab Industry* (International Human Rights Law Clinic of the American University Washington College of Law and Centro de los Derechos del Migrante, Inc., [2010]); VanDeCruze and Wiggins, Poverty and Injustice in the Food System.
34. VanDeCruze and Wiggins, *Poverty and Injustice in the Food System*, appendix A, table 6.
35. Bauer and Ramírez, *Injustice on Our Plates.*
36. VanDeCruze and Wiggins, *Poverty and Injustice in the Food System*, appendix A, table 6.

Chapter 8: Biodiversity

1. "Pigs," American Livestock Breeds Conservancy, accessed April 13, 2012, http://albc-usa.org/cpl/pigs.html; the ALBC's description is an excerpt (pages 71–73) from the organization's own book, *A Rare Breeds Album of American Livestock*, by Carolyn J. Christman, D. Philip Sponenberg, and Donald E. Bixby (1998).
2. It is important to note that the idea of crossbreeding rare breeds with other breeds is considered problematic for some conservation breeders who prefer to keep the breeds separate and distinct, lest those specific rare-breed genetics get lost in the mix. Others argue that crossbreeding is precisely how we got our current breeds and offers new possibilities for livestock producers.
3. Elizabeth Meister, "Battle Over Michigan's New Swine Rules Goes Hog Wild," National Public Radio broadcast, August 31, 2012, as featured on the website of WBUR (Boston), http://www.wbur.org/npr/160394513/battle-over-michigans-new-swine-rules-goes-hog-wild.
4. Darwin L. Booher, "DNR Order Threatens Heritage Swine Farmers Because of How Their Pigs Look," editorial in the *Manistee News Advocate*, February 27, 2012, http://www.misenategop.com/readarticle_printable.asp?id=4863&District=35.
5. For more information on Renewing America's Food Traditions (RAFT), see http://www.albc-usa.org/RAFT/. That website also offers inventories of threatened foods and foodways of each region.
6. Sarah McClellan-Wech, "Original Flour Corn Seeds Returned to the Tribe," *Cherokee One Feather*, June 1, 2011, http://theonefeather.com/2011/06/original-flour-corn-seeds-returned-to-the-tribe/.
7. "Putting Culture Back in to Agriculture," Cherokee Preservation Foundation, accessed September 8, 2012, http://www.cpfdn.org/cultural-preservation-connect/success-stories/99-putting-culture-back-in-to-agriculture.
8. "Treasure of the Sierra Foothills: Heritage Fruit and Nut Trees Discovered and Preserved," an interview with Amigo Cantisano, Bioneers website, accessed April 13, 2012, http://www.bioneers.org/programs/food-farming-1/articles-interviews/treasure-of-the-sierra-foothills-a-cornucopia-of-heritage-fruit-and-nut-trees-discovered-and-preserved.
9. "Design," Beacon Food Forest, accessed April 13, 2012, http://beaconfoodforest.weebly.com/design.html.
10. "The Project," Boston Tree Party, accessed April 13, 2012, http://www.bostontreeparty.org/about/party/; "Design."
11. "SF Environment: Urban Orchards," SFEnvironment.org, accessed September 12, 2012, http://sfenvironment.org/article/types-of-urban-ag/urban-orchards.

12. "Food Forest," Community Food Bank of Southern Arizona blog post, reprinted from *Nourishing News*, Summer 2009, http://communityfoodbank.com/2009/07/14/food-forest/.
13. Portland Fruit Tree Project home page, accessed September 10, 2012, http://portlandfruit.org/.
14. For more information about Prairie Heritage Farm, visit the farm's website at http://www.prairieheritagefarm.com/.
15. For more information about The Land Institute, visit their website at http://www.landinstitute.org.
16. "From I'itoi's Garden: Traditional Tohono O'odham Foods," Desert Rain Café, accessed April 13, 2012, http://www.desertraincafe.com/www.desertraincafe.com/Traditional_Foods.html.
17. Tara Pascual with Jessica Powers, *Cooking Up Community: Nutrition Education in Emergency Food Programs* (New York: WhyHunger, 2012), 47, http://www.whyhunger.org/uploads/fileAssets/CUCFINAL1.pdf.
18. "An Interview with Captain Plyler," UNC (University of North Carolina) TV, http://www.unctv.org/ncrising/projects/ocracoke/overview.html#top.
19. Sundae Horn, "Ocracoke Fish House—Catch of the Day!" *OcracokeCurrent*, April 5, 2012, http://www.ocracokecurrent.com/28203.
20. Ocracoke Working Watermen's Association home page, accessed September 11, 2012, http://www.ocracokewatermen.org/.
21. "North Carolina Rising: Feature Project: Preserving Coastal Traditions," UNC (University of North Carolina) TV, http://www.unctv.org/ncrising/projects/ocracoke/overview.html.

Chapter 9: Market Value

1. Special thanks to Dr. Siegfried de Rachewiltz, my previous employer and longtime mentor in all things Alpine and Mediterranean.
2. Bruce Blythe, "Wal-Mart's U.S. Grocery Sales Rise 2.1%" *The Packer*, March 31, 2011, http://www.thepacker.com/fruit-vegetable-news/fresh-produce-retail/wal-marts_us_grocery_sales_rise_21_122014269.html.
3. Meredith Lepore, "Here's How Walmart Became the #1 Grocery Store in the Country," *Business Insider*, February 11, 2011, http://www.businessinsider.com/walmart-biggest-supermarket-2011-2?op=1.
4. Jennifer Chait, "6 Largest Organic Retailers in North America 2011," About.com, accessed September 12, 2012, http://organic.about.com/od/marketingpromotion/tp/6-Largest-Organic-Retailers-In-North-America-2011.htm.
5. Patrick Canning, *A Revised and Expanded Food Dollar Series: A Better Understanding of Our Food Costs*, Economic Research Report no. 114 (Washington, D.C.: USDA Economic Research Service, 2011). For this report, "local foods" were defined as foods that were sold either direct to consumer or direct to grocer/restaurant.
6. Some of these programs include Farmigo, www.farmigo.com; Foodzie, www.foodzie.com; Local Dirt, www.localdirt.com; Local Harvest, www.localharvest.org; Locally Grown, www.locallygrown.net; Small Farm Central, www.smallfarmcentral.com; and Your Farm Stand, www.yourfarmstand.com.

Chapter 10: Marketplace Values

1. Sarah Low and Stephen Vogel, *Direct and Intermediated Marketing of Local Foods in the United States*, Economic Research Report no. 128 (Washington, D.C.: USDA Economic Research Service, November 2011), 10.
2. Value Chain Partnerships home page, accessed September 12, 2012, http://www.valuechains.org/.
3. "Introduction to ACEnet in Athens, Ohio," video, NEO Food Web, accessed August 21, 2012, http://www.neofoodweb.org/video/introduction-acenet-athens-ohio.

4. "Walmart Commits to America's Farmers as Produce Aisles Go Local," Walmart press release, July 1, 2008, http://www.walmartstores.com/pressroom/news/8414.aspx.
5. Ibid.
6. Julie Davis, David Merriman, Lucia Samayoa, Brian Flanagan, Ron Baiman, and Joe Persky, *The Impact of an Urban Wal-Mart Store on Area Businesses: An Evaluation of One Chicago Neighborhood's Experience* (Chicago: Loyola University, December 2009).
7. *Food for Thought: A Case Study of Walmart's Impact on Harlem's Healthy Food Retail Landscape* (New York: Office of Scott Stringer, Manhattan Borough President, November 2011).
8. Tom Philpott, "Is Walmart the Answer to 'Food Deserts'?" *Mother Jones*, January 12, 2012, http://www.motherjones.com/tom-philpott/2012/01/walmart-answer-food-deserts. The author of this article, Tom Philpott, has been following the Walmart story as it relates to food issues for some time, and any reader interested in this topic should keep an eye out for his articles on Walmart's evolving initiatives.
9. Local Harvest Grocery home page, accessed August 21, 2012, http://www.localharvestgrocery.com/.
10. Good Food Store home page, accessed August 21, 2012, http://www.goodfoodstore.com/.
11. People's Grocery home page, accessed August 21, 2012, http://www.peoplesgrocery.org/.
12. The Good Table home page, accessed August 21, 2012, http://thegoodtable.com/.
13. Denison Farm home page, accessed August 21, 2012, http://www.denisonfarm.com/.
14. Farmers Web home page, accessed August 21, 2012, https://www.farmersweb.com/.
15. Jennifer Goggin, cofounder of Farmers Web, interview by the author, March 29, 2012.
16. Bon Appétit Management Company home page, accessed August 20, 2012, http://www.bamco.com/.
17. "Helene York," short biography on the website of *The Atlantic*, accessed August 20, 2012, http://www.theatlantic.com/helene-york.
18. "Fedele Bauccio—Fair Treatment of Workers," video, YouTube, uploaded March 1, 2011, http://www.youtube.com/watch?v=BFAmAdZYnwg&feature=related.

Chapter 11: Bringing It All Back Home

1. Wild Gourmet Food home page, accessed July 21, 2012, http://wildgourmetfood.com/.
2. "Fruit Harvesting in the City," *Food Mapping* (blog), August 26, 2008, http://foodmapper.wordpress.com/2008/08/26/fruit-harvesting-in-the-city/.
3. Fallen Fruit home page, accessed August 20, 2012, http://www.fallenfruit.org/.
4. "Fruit Gleaning," Salt Lake City government website, accessed August 20, 2012, http://www.slcclassic.com/slcgreen/food/fruitgleaning.htm.
5. "Zero Waste FAQ," SFEnvironment, accessed October 11, 2012,http://sfenvironment.org/zero-waste/overview/zero-waste-faq .
6. Mireya Navarro, "Bring On the Bees," *City Room* (blog) on the *New York Times* website, March 16, 2010, http://cityroom.blogs.nytimes.com/2010/03/16/bring-on-the-bees/.
7. http://sierraclub.typepad.com/scrapbook/2012/05/daily-green-acts-ripple-out-from-350-home-and-garden-challenge.html (accessed October 11, 2012).
8. "2012 Transition Challenge," Transition United States, accessed September 16, 2012, http://transitionus.org/actions/transition-challenge.
9. "Healthy Zoning Regulations," Healthy Eating Active Living Cities Campaign, accessed September 12, 2012, http://www.healcitiescampaign.org/healthy_zone.html.
10. Grow It Eat It home page, accessed July 21, 2012, http://www.growit.umd.edu/.
11. Patrick Canning, Ainsley Charles, Sonya Huang, Karen R. Polenske, and Arnold Waters, U.S. Department of Agriculture Economic Research Service (ERR-94), March 2010.
12. "County of Los Angeles Department of Public Works," http://ladpw.org/epd/drp/beveragecontainers.cfm.
13. Bill Sheehan and Helen Spiegelman, "Climate Change, Peak Oil, and the End of Waste," in Richard Heinberg and Daniel Lerch, eds., *The Post Carbon Reader: Managing the 21st Century's Sustainability Crises* (Healdsburg, Calif.: Watershed Media, 2010), 377–78.

14. *Six-Month Report of Beverage Container Recycling and the Significance of Carbon Reductions* (California Department of Conservation, 2008), 4. This report also includes a helpful life-cycle analysis of aluminum, plastic, and glass beverage containers.
15. "Environmental Benefits," Reduce, Reuse, Refill!, a website maintained by the Institute for Local Self-Reliance, accessed July 20, 2012, http://refillables.grrn.org/content/environmental-benefits; "Reviving Refilling in the United States," Reduce, Reuse, Refill!, a website maintained by the Institute for Local Self-Reliance, accessed July 20, 2012, http://refillables.grrn.org/content/reviving-refilling-united-states.
16. "Facts and Figures," Montana Brewers Association, accessed July 19, 2012, http://montanabrewers.org/about-us/facts-and-figures/.
17. "Montana Brewery Trail," Montana Official State Travel Site, accessed April 20, 2012, http://visitmt.com/experiences/food_and_beverage/breweries/montana_brewery_trail/.
18. "Apple Project," Glynwood Foundation, accessed September 12, 2012, http://www.glynwood.org/the-apple-project/.
19. Ibid.
20. Hotlips Soda home page, accessed April 20, 2012, http://hotlipssoda.com/.
21. Longmont Dairy Farm home page, accessed April 20, 2012, http://www.longmontdairy.com/.
22. "About Us," Crescent Ridge Dairy, April 20, 2012, http://www.crescentridge.com/about_us.cfm; for information on Ward's Berry Farm CSA, see the farm's web page at http://www.wardsberryfarm.com/BOXES.html.
23. "Save Money When You Cook for Good," Cook for Good, accessed September 12, 2012, http://www.cookforgood.com/save-money/.

Chapter 12: Collaborative Possibilities

1. W. P. Hedden, *How Great Cities Are Fed* (Boston: D.C. Heath and Company, 1921).
2. Toby Hemenway, "Understanding Our Foodshed and Securing it for the Future," *Connections: The Journal of Coalition for a Livable Future*, Winter 2006, vol. 8 no. 2.
3. "Local Foodshed Mapping Tool for New York State," Cornell University Department of Crop and Soil Sciences, accessed April 20, 2012, http://www.cals.cornell.edu/cals/css/extension/foodshed-mapping.cfm.
4. Dreaming New Mexico home page, accessed September 12, 2012, http://www.dreamingnewmexico.org/.
5. Alethea Harper, Annie Shattuck, Eric Holt-Giménez, Alison Alkon, and Frances Lambrick, *Food Policy Councils: Lessons Learned*, Development Report no. 21 (Food First/Institute for Food and Development Policy, 2009), http://www.foodfirst.org/en/foodpolicycouncils-lessons.
6. "City of Hartford Food Commission," accessed October 12, 2012, hartfordfood.org/publications/FoodPolicy2011.pdf.
7. Oakland Food Policy Council home page, accessed August 8, 2012, http://www.oaklandfood.org/home.
8. "Statement on Urban Agriculture," Oakland Food Policy Council, April 2011, http://www.oaklandfood.org/home/ua_statement.
9. Cleveland–Cuyahoga County Food Policy Coalition home page, accessed July 27, 2012, http://cccfoodpolicy.org/.
10. Northeast Ohio (NEO) Food Web home page, accessed July 27, 2012, http://www.neofoodweb.org/.
11. Brad Masi, Leslie Schaller, and Michael Shuman, *The 25% Shift: The Benefits of Food Localization for Northeast Ohio and How to Realize Them* (December 2010), http://www.neofoodweb.org/sites/default/files/resources/the25shift-foodlocalizationintheNEOregion.pdf. For a video describing some of the highlights of this work, see "Michael Shuman at City Club of Cleveland," video, November 16, 2010, http://www.neofoodweb.org/video/michael-shuman-city-club-cleveland.

12. Masi, Schaller, and Shuman, *The 25% Shift*, 117–19.
13. "History of Agriculture in Hawaii," Hawaii Department of Agriculture, accessed April 12, 2012, http://hawaii.gov/hdoa/ag-resources/history.
14. Heather Wooten and Amy Ackerman, *Seeding the City: Land Use Policies to Promote Urban Agriculture* (Public Health Law & Policy, October 2011), http://www.phlpnet.org/childhood-obesity/products/urban-ag-toolkit.
15. Mindy Goldstein, Jennifer Bellis, Sarah Morse, Amelia Myers, and Elizabeth Ura, *Urban Agriculture: A Sixteen City Survey of Urban Agriculture Practices across the Country* (Turner Environmental Law Clinic at Emory University, 2011), http://www.georgiaorganics.org/Advocacy/urbanagreport.pdf; Wooten and Ackerman, Seeding *the City.*
16. Kimberly Hodgson, Marcia Campbell, and Martin Bailkey, *Urban Agriculture: Growing Healthy, Sustainable Places* (American Planning Association Planning Advisory Service, 2011).
17. "Design Competition Guidelines," Trinity Avenue Farm Design Competition, Mayor's Office of Sustainability (Atlanta, Georgia), accessed September 12, 2012, http://www.trinityavenuefarm.org/?page_id=30.
18. "About the Mayor's Office of Sustainability," Trinity Avenue Farm Design Competition, Mayor's Office of Sustainability (Atlanta, Georgia), accessed September 12, 2012, http://www.trinityavenuefarm.org/?page_id=32.
19. "Community Gardens," Richmond [Va.] Grows Gardens, accessed June 20, 2012, http://www.richmondgov.com/CommunityGarden/.
20. Kubi Ackerman, *The Potential for Urban Agriculture in New York City: Growing Capacity, Food Security, and Green Infrastructure* (New York: Urban Design Lab at the Earth Institute [Columbia University], 2012), http://www.urbandesignlab.columbia.edu/?pid=nyc-urban-agriculture.
21. Ibid.
22. Ibid.; Wooten and Ackerman, *Seeding the City*, 11.
23. "The City of Cleveland Land Bank Property Search: Available Lots," City of Cleveland, accessed July 27, 2012, http://cd.city.cleveland.oh.us/scripts/cityport.php.
24. Lila Zautner, ed., *Re-imagining Cleveland: Ideas to Action Resource Book* (Cleveland, Ohio: Kent State University's Cleveland Urban Design Collaborative and Neighborhood Progress, Inc., January 2011), http://reimaginingcleveland.org/about/links-and-resources/.
25. "About NGA," Neighborhood Gardens Association / A Philadelphia Land Trust, accessed April 23, 2012, http://www.philadelphialandtrust.org/history.html.
26. "History of the Texas Agricultural Land Trust," Texas Agricultural Land Trust, accessed April 23, 2012, http://www.txaglandtrust.org/index.php?option=com_content&view=article&id=48&Itemid=67.
27. "Vermont Land Trust Helps Young Farmers Buy First Farm," Vermont Land Trust press release, July 9, 2010, http://www.vlt.org/news-publications/publications-archive/archived-press-releases/elworthy-pr.
28. Farm-Based Education Association home page, accessed September 16, 2012, http://www.farmbasededucation.org/.
29. "Adopt a School Garden Program," National Gardening Association, accessed September 16, 2012, http://assoc.garden.org/ag/asg/.
30. Life Lab home page, accessed September 16, 2012, http://www.lifelab.org/.

Chapter 13: Farmland Security

1. Richard Heinberg, "Fifty Million Farmers," speech given as part of the Twenty-Sixth Annual E. F. Schumacher Lectures, October 2006, Mount Holyoke College, South Hadley, Mass., http://neweconomicsinstitute.org/publications/lectures/heinberg/richard/fifty-million-farmers.

2. Joanna Gangi, "You Don't Have to Move Out of Your Neighborhood to Live in a Better One," *Yes!* magazine online, May 11, 2011, http://www.yesmagazine.org/happiness/majora-carter-how-to-bring-environmental-justice-to-your-neighborhood.
3. For national listings of farm apprenticeships, see the "Sustainable Farming Internships and Apprenticeships" page of the National Sustainable Agriculture Information Service, https://attra.ncat.org/attra-pub/internships/; see also the World Wide Opportunities on Organic Farms home page, http://www.wwoof.org/.
4. Growing Growers home page, accessed June 26, 2012, http://www.growinggrowers.com/.
5. "Impacts of Intervale Food Hub," Intervale Center, accessed June 21, 2012, http://www.intervale.org/what-we-do/intervale-food-hub/food-hub-impacts/.
6. "Lowcountry Local First Incubator Farm and Apprentice Program," GrowFood, accessed June 21, 2012, http://www.growfood.org/es/farm/24254.
7. For more information on ALBA's incubator start-up information, see http://www.start2farm.gov/resources/alba-agriculture-and-land-based-training-association-farm-incubator-toolkit.
8. Gary Peterson, interview by the author, July 5, 2012.
9. Tony Serrano, interview by the author, July 2, 2012.
10. "Our Farms/Nuestras Fincas/Ranchos," Agriculture and Land-Based Training Association, accessed July 1, 2012, http://www.albafarmers.org/farms.html.
11. "Seattle Tilth Farm Works," Seattle Tilth, accessed July 1, 2012, http://seattletilth.org/about/seattletilthfarmworks; "Seattle Tilth Receives USDA Funding for Farm Business Training Program for Underserved Communities," Seattle Tilth press release, October 25, 2011, http://seattletilth.org/press/press-releases/seattle-tilth-receives-usda-funding-for-farm-business-training-program-for-underserved-communities.
12. Slow Money home page, accessed July 1, 2012, http://www.slowmoney.org/.
13. *Guide to Financing the Community Supported Farm: Ways for Farms to Acquire Capital within Communities* (University of Vermont Center for Sustainable Agriculture, University of Vermont Extension, 2012), http://www.uvm.edu/newfarmer/?Page=business/community-supported-farm-guide.html&SM=business/sub-menu.html.
14. Monterey Fish Market, *Sustainable Fisheries Guide* (Monterey, Calif.: Monterey Fish Market, n.d.).
15. "From Sea to Plate: Port Clyde, Maine," Nature Conservancy, accessed July 5, 2012, http://www.nature.org/ourinitiatives/regions/northamerica/unitedstates/maine/explore/from-sea-to-plate-port-clyde-maine.xml.
16. Community Fisheries Network home page, accessed July 5, 2012, http://www.communityfisheriesnetwork.org/.
17. "Food Away from Home as a Share of Food Expenditures," table 10 in the USDA Economic Research Service Food Expenditure Series, accessed July 1, 2012, http://www.ers.usda.gov/briefing/cpifoodandexpenditures/Data/Expenditures_tables/table10.htm.
18. Vermont FEED home page, accessed July 20, 2012, http://www.vtfeed.org/.
19. "Update: Food Safety Modernization Act," Rodale Institute, accessed July 20, 2012, http://www.rodaleinstitute.org/20110110_update-food-safety-modernization-act.
20. Know Your Farmer, Know Your Food home page, on the USDA website, accessed July 20, 2012, http://www.usda.gov/wps/portal/usda/knowyourfarmer?navid=KNOWYOURFARMER.
21. Serena Unger, Hannah Laurison, and Christine Fry, *Green for Greens: Finding Public Funding for Healthy Food Retail* (ChangeLab Solutions, 2012), http://changelabsolutions.org/publications/green-for-greens.
22. "Healthy Corner Store Initiative," Food Trust, accessed July 2, 2012, http://www.thefoodtrust.org/php/programs/corner.store.campaign.php.
23. Warren Neth, personal interview by the author, July 1, 2012.
24. Main Street Birmingham home page, accessed July 3, 2012, http://mainstreetbham.org/wp/.

Chapter 14: Bridging the Divides

1. Julian Agyeman provides a similar perspective through his thinking on the term "sustainability": "Just sustainability is more accurately described as just sustainabilities because the singular form suggests there is one prescription for sustainability that can be universalized. The plural, however, acknowledges the relative place and culturally bound nature of the concept." See his autobiographical page (accessed September 16, 2012) on the Tufts University website at http://sites.tufts.edu/julianagyeman/.
2. Carolyn Steel, *Hungry City: How Food Shapes Our Lives* (London: Vintage Books, 2009), 291.
3. Nathan McClintock, interview by the author, July 31, 2012.
4. Erika Allen, interview by the author, August 1, 2012.
5. "Chicago Farms and Projects," Growing Power, accessed September 12, 2012, http://www.growingpower.org/chicago_projects.htm.
6. Agri-Mark, Inc., home page, accessed September 12, 2012, https://www.agrimark.net/.
7. "Agri-Mark Honored as 2012 'Business of Distinction' by New York Dairy of Distinction Organization," Cabot Creamery Cooperative press release, August 20, 2012, http://www.justmeans.com/press-releases/Agri-Mark-Honored-as-2012-Business-of-Distinction-by-New-York-Dairy-of-Distinction-Organization/9889.html.
8. Pablo Muñoz Ledo, interview by the author, August 3, 2012. Pedro is the founder and CEO of Greenexus, formerly known as Greenimbus (http://www.greenimbus.com). For more information on Greenexus, see the YouTube video "Greenexus: The Time for Green Direct Exchange," http://www.youtube.com/watch?v=OKp4K7JzgM4&feature=channel&list=UL.
9. Nathan McClintock, interview by the author, July 31, 2012.

Index

Note: page numbers in *italics* refer to photos and figures; page numbers followed by *t* refer to tables; and *map* pages refer to the color insert.

Get excerpts, resources, and more from the

Community Resilience Guides series at

resilience.org

post carbon institute

About the Author

Philip Ackerman-Leist, author of *Rebuilding the Foodshed* and *Up Tunket Road,* is a professor at Green Mountain College, where he established the college's farm and sustainable agriculture curriculum and is director of the Green Mountain College Farm & Food Project. He also founded and directs the college's Masters in Sustainable Food Systems (MSFS), the nation's first online graduate program in food systems, featuring applied comparative research of students' home bioregions. He and his wife, Erin, farmed in the South Tirol region of the Alps and North Carolina before beginning their sixteen-year homesteading and farming venture in Pawlet, Vermont. With more than two decades of experience working on farms, in the classroom, and with regional food systems collaborators, his work is focused on examining and reshaping local and regional food systems from the ground up.

About Post Carbon Institute

Post Carbon Institute provides individuals, communities, businesses, and governments with the resources needed to understand and respond to the interrelated economic, energy, environmental, and equity crises that define the twenty-first century.

About the Foreword Author

Chef, author, and local food advocate Deborah Madison, founding chef of Greens Restaurant in San Francisco, has authored nine cookbooks, including *The Greens Cookbook, Vegetarian Cooking for Everyone,* and *Local Flavors.* She has received the M.F.K. Fisher Award, the IACP Julia Child Cookbook of the Year Award, and three James Beard Foundation Awards. She writes on food and farming for such magazines as *Gourmet, Saveur,* and *Orion*, and has long been active in Slow Food and other groups involved in local food issues.

the politics and practice of sustainable living

CHELSEA GREEN PUBLISHING

Chelsea Green Publishing sees books as tools for effecting cultural change and seeks to empower citizens to participate in reclaiming our global commons and become its impassioned stewards. If you enjoyed reading *Rebuilding the Foodshed*, please consider these other great books related to food and agriculture.

UP TUNKET ROAD
The Education of a Modern Homesteader
PHILIP ACKERMAN-LEIST
9781603580335
Paperback • $17.95

INQUIRIES INTO THE
NATURE OF SLOW MONEY
Investing as if Food, Farms,
and Fertility Mattered
WOODY TASCH
9781603582544
Paperback • $15.95

FARMS WITH A FUTURE
Creating and Growing
a Sustainable Farm Business
REBECCA THISTLETHWAITE
9781603584388
Paperback • $29.95

RAISING DOUGH
The Complete Guide to Financing
a Socially Responsible Food Business
ELIZABETH Ü
9781603584289
Paperback • $19.95

For more information or to request a catalog,
visit **www.chelseagreen.com** or
call toll-free **(802) 295-6300**.